T0220890

Introduction to Numerical Electrostatics Using MATLAB®

Introduction to Numerical Electrostatics Using MATLAB®

Lawrence N. Dworsky

 WILEY

Published by John Wiley & Sons, Inc., Hoboken, New Jersey

Published simultaneously in Canada

For general information on our other products and services or for technical support, please contact our Customer Care Department within the United States at (800) 762-2974, outside the United States at (317) 572-3993 or fax (317) 572-4002.

Wiley also publishes its books in a variety of electronic formats. Some content that appears in print may not be available in electronic formats. For more information about Wiley products, visit our web site at www.wiley.com.

Library of Congress Cataloging-in-Publication Data:

Dworsky, Lawrence N., 1943– author.
 Introduction to numerical electrostatics using MATLAB® / Lawrence N. Dworsky
   pages   cm
 Includes index.
 ISBN 978-1-118-44974-5 (hardback)
1. Electromagnetism–Data processing. 2. Electrostatics–Data processing. I. Title.
   QC760.54.D86 2014
   537′.20151–dc23
                                                        2013018019

Printed in the United States of America

10 9 8 7 6 5 4 3 2 1

Contents

Preface

My graduate work was in the area of microwave oscillation mechanisms in semiconductor devices. My contribution was the prediction, analysis, and verification of yet another mode of semiconductor microwave oscillation. After graduation, I taught for a brief period (2 years) and worked part-time for a local electronics firm designing positive–intrinsic–negative (PIN) diode attenuators for their line of microwave signal generators. Then, in 1974, I went to work for Motorola, Inc., in southern Florida.

The part of Motorola that I joined designed and manufactured two-way portable radios (for police, fire department, etc.) and radio pagers, along with the supporting infrastructure systems. The developmental push at the time was to extend the product base up to the (then new) 900-MHz bands.

This organization had no interest whatsoever in my semiconductor physics background. They were, however, keenly interested in the skills that I had acquired during my graduate and part-time work designing stripline and microstrip (transmission line) circuitry. Motorola needed capability with stripline and microstrip filters, interconnects, materials, and so on, and my job was to help develop this capability. I never returned to semiconductor physics, and 40 years after finishing graduate school, I think it's safe to say that I never will.

Transmission line circuit design consists of two parts: (1) the actual circuit design based on the transmission line parameters and (2) the relationships between these transmission line parameters and the physical structure and materials.

As will be explained in Chapter 2, transmission line parameters can be described in terms of their DC (direct current; i.e., electrostatic) capacitance. A significant part of my effort therefore was devoted to performing electrostatic analyses of stripline and microstrip structures to predict the electric fields and capacitances in these structures. I didn't realize it at the time, but I was developing a skill that I would continue using and improving for the rest of my career.

While working on stripline and microstrip circuits, I also worked on piezoelectric (quartz) resonator and filter technology. The piezoelectric device models involved coupling of mechanical motion to electric fields, but the very high ratio of acoustic to electric wave velocities in the materials of interest (approximately 10^5) allows the electrical part of the analysis to consist of electrostatic analysis.

The 1980s saw the introduction of analog cellular telephone technology in the United States. These cellular telephones and base stations required a complex two-filter system called a *duplexer* that would enable a cellular telephone to transmit and receive simultaneously on two nearby frequencies using the same antenna. (The challenge was to keep your own transmitter's signal out of your receiver.) These duplexers were realized using blocks of high-dielectric-constant ceramic, with partially metalized surfaces, acting as interconnected resonators. Once again, electrostatic modeling was required, and new modeling programs and approaches had to be perfected.

In the late 1980s we did exploratory work in the newly emerging field of micromachined electromechanical devices. Using semiconductor industry processing technologies, it was becoming possible to build extremely small accelerometers, switches, and resonators whose operation is based on electrostatic forces. This was a new area of electrostatic modeling for me. Everything I had done before had involved electrodes that stayed in place, and we never cared about the physical forces involved. Now, we had to calculate the forces and keep track of the fields and forces as the electrodes moved. Again, I was extending my experience base in electrostatic modeling.

In the early 1990s we became interested in vacuum microelectronics, particularly in a structure called the *field emission display*, an electronic display whose operation is based on electron emission from millions of very small, sharp, metal tips due to high local electric fields. Once again, I was extending my electrostatic modeling experience to include structures with vastly different scales (submicromillimeter resolution near the tips to millimeter resolution near the screen). Structural capacitances were of interest in that they could limit circuit switching speeds, but the principal issues were the magnitude and uniformity of the fields at the emitter tips and then the electron trajectory control (both desired and undesired) due to these fields as the electrons traveled to the screen, striking the light emitting phosphors when they arrived. In these models the electrodes remained immobile, so the fields didn't change; electron trajectories were the principal subject of interest.

Putting my history together, although I didn't realize it at the time, I have spent more time creating and working with electrostatic analyses than with any single other electrical engineering discipline. These analyses were never a goal unto themselves. They were an engineering tool. The simplest approach that could do the job was always the chosen approach.

The philosophy of this book follows from my personal experience. There is an incredibly long list of mathematical approaches to numerical electrostatic modeling, but in terms of learning the electrostatics and choosing a modeling approach to study a given situation, I try to avoid using more exotic schemes simply "because they're there." This doesn't mean that all of the approaches in the literature aren't interesting, important, and valuable, but in any given circumstance the simplest tool that can do a job is probably the best tool for that job.

LAWRENCE N. DWORSKY

Introduction

An introductory treatment of electrostatics usually begins with Coulomb's law, the concepts of charge, the electric field and energy stored in the field, potential, capacitance, and so on. Poisson's and Laplace's equations soon appear.

Unfortunately, almost no real world problem can be solved in closed form using the latter equations. The most basic of electrostatic device analysis, the electric fields surrounding and the capacitance of a simple parallel plate capacitor, cannot be found.

Typically, a few interesting solution techniques, such as the separation of variables, are presented. Then the author has to choose a path. Other solution techniques such as conformal mapping can be shown; if the book is to be more than an introductory text, more formal materials such as Greene's functions can be introduced. In any case, the practitioner with real world geometries to be analyzed has been abandoned.

A book about numerical analysis techniques typically presents just that – numerical analysis techniques. The few examples presented are usually based as much on the ease of their presentation as on their ultimate usefulness.

My goal in writing this book is to present enough basic electrostatic theory as necessary to get into real world problems, then to present several of the available numerical techniques that are applicable to these problems, and finally to present numerous, detailed, examples showing how these techniques are applied. In other words, I am presenting the basics of electrostatics and several relevant numerical analysis techniques, with the emphasis on practical geometries.

The numerical analysis of problems in fields such as electrostatics typically have three distinct phases:

1 *Pre-processing.* The conversion of the physical description of the problem to a data set that is meaningfully digestible to the numerical analysis program chosen for the job.

2 *Numerical analysis.* The calculations, based upon the data set describing the geometry and the chosen boundary conditions, resulting (typically) in an approximate solution for the voltage distribution over the chosen space.

3 *Post-processing.* Calculations and programming necessary to provide summary data such as capacitance and computer visualization of the results.

This book will concentrate on item 2, the numerical analysis. Three fundamental techniques – method of moments, finite difference, and finite elements – are introduced. Sample problems are presented and computer code that solves the problems is developed.

In order to accomplish the above, some pre-processing capability is necessary. Rather than develop this capability or simply request that the reader "get it done," several freely available packages are introduced and basic tutorials on their usage are presented. These tutorials are not exhaustive; they introduce enough of the capabilities of these packages to allow the student to follow, replicate, and extensively modify the examples to the student's own needs. No claims are being made that the packages chosen are the best possible choices for the job. They are, however, choices that work well.

Post-processing of numerical analysis results, in this book, is done on an ad-hoc basis. Calculating the capacitance of a structure is often very useful because, if the example structure has been analyzed by other techniques and the results published, accuracy of results can be compared. This comparison allows for a convenient, one number figure of merit for choices of resolution, approximate boundary conditions, and so on. Often, graphical interpretations of field, voltage, and/or charge distributions are presented. These are useful as a quick visual check on the boundary conditions and on gaining insight into the electrostatic properties (high field points, etc.) of the structure being studied.

With only one exception, all the numerical analysis, post-processing and graphics were created using MATLAB®. This type of work is what MATLAB is designed to do. The analysis techniques presented convert partial differential equations into sets of linear equations with coefficients and variables represented by matrices and vectors. MATLAB is a scientific programming language designed with the matrix as the fundamental data type. The language is expressive, the available function list extensive and the easily used graphics superb.

MATLAB is fundamentally an interpreted computer language. It achieves impressive processing speeds by providing a liberal assortment of precompiled functions. Preparing user-written code for these functions involves a procedure that MATLAB calls "vectorizing." Vectorizing means not writing explicit loops to process array elements because the language itself allows processing of all the array elements simultaneously. From an authoring point of view, this raises a question: Should demonstration code be vectorized as much as possible in the name of program execution speed or should demonstration code be written to explicitly parallel the derivations of the formulas in the text in the name of pedagogical clarity?

There is no absolutely right answer to the above question. In every case in this book judgment calls were made – vectorized code is shown when the algorithm seems "clear enough." This compromise, like all compromises, won't please everybody every time; ideally, it will please enough readers enough of the time.

The problems at the end of the chapters were written, as much as possible, to be extensions of the chapters, rather than "verify XXX or put numbers into YYYY." Many of the problems involve modifying existing or writing new MATLAB code, leading to capabilities that were not presented in the chapters' materials. When the thrust of the problem isn't to extend the modeling capability, it is to make a point about the electrostatics issues involved in the example being treated. In all cases, solving the problems and reading the solution discussions will be an integral part of the learning process.

Solutions to end-of-chapter exercises may be found at the book companion site, www.wiley.com/go/numerelectrostatics. Additional resources may be found at www.lawrencedworsky.com.

Acknowledgments

I certainly could not have done this job alone. I thank Kari Capone and George Telecki (retired) at John Wiley & Sons: Kari for her very good natured help in teaching me how to prepare my manuscript then reviewing it, and George for his initial support and encouragement. The folks at MathWorks were very generous in providing me with MATLAB software and support. Without the love, patience, and support of my wife Suzanna, I might never have started this project and I certainly never would have finished it.

L. N. D.

1

A Review of Basic Electrostatics

Electric and magnetic phenomena, including electromagnetic wave propagation, are described by Maxwell's equations.[1] When nothing is changing with time, that is, when all derivatives with respect to time are zero, the electric and the magnetic phenomena decouple and become separate electric and magnetic phenomena. These are referred to respectively as *electrostatics*, which describes the properties of systems with separated static regions of positive and negative electric charge (although the entire system is charge-neutral), and *magnetostatics*, which describes the properties of systems with electric currents and/or magnetized materials.

In this book we shall consider only electrostatics. This subset of a subset of topics describes a vast number of real-world situations. Chapter 2 describes some practical needs and uses of electrostatic analyses, the remainder of the book will be dedicated to examining several techniques for performing these analyses.

The materials to follow are intended to be a quick review of the relationships that will be used throughout this book. The intent here is to provide a consistent set of notation using all the relationships that will be needed going forward. Many of these relationships are stated without derivation or proof. A more complete electrostatics theory text is recommended for newcomers to the subject. There are very many excellent texts available. The references list at the end of this chapter is certainly not exhaustive, but the texts cited are considered standards in the field.

1.1 CHARGE, FORCE, AND THE ELECTRIC FIELD

Electric charges exert forces on one another. This is the basis of electrostatics. The characteristics of these forces are summarized in Coulomb's law:

Introduction to Numerical Electrostatics Using MATLAB, First Edition. Lawrence N. Dworsky.
© 2014 John Wiley & Sons, Inc. Published 2014 by John Wiley & Sons, Inc.

1   Electric charge carries a polarity, or sign. The choice of sign was originally arbitrary, but now is established by tradition—the electron, the most common charged subatomic particle, carries a negative charge.

2   For point charges q_1 and q_2, measured in coulombs, the (coulomb) force, measured in newtons, in a uniform medium, is given by

$$\vec{F} = \frac{q_1 q_2}{4\pi\varepsilon r^2}\, \vec{a_r} \tag{1.1}$$

3   In equation (1.1) and elsewhere ε is the permittivity of the material in farads per meter (F/m). In free space, $\varepsilon = \varepsilon_0 = 8.854$ F/m. For other linear, isotropic, homogeneous materials, $\varepsilon = k\varepsilon_0$, where k is the relative permittivity, the relative dielectric constant, or sometimes simply the dielectric constant, of the material. *Farads* per se are defined as coulombs per volt (C/V). In this book we shall consider only linear, isotropic, homogeneous dielectric materials, and going forward this will be assumed.

4   In equation (1.1) r is the distance between q_1 and q_2.

5   Also, $\vec{a_r}$ is a unit vector along the line connecting q_1 and q_2. If q_1 and q_2 have the same sign, then $\vec{F}$ is pushing q_1 and q_2 apart. If q_1 and q_2 have opposite signs, $\vec{F}$ is pulling them together.

Equation (1.1) is expressed in the rationalized meter-kilogram-second (mks) system of units. The derivation of this set of units is an interesting discussion in itself.[2]

When a test charge is in the area of a collection of charges and the magnitude of these latter charges is sufficient, relative to the test charge, to render negligible any perturbation of the situation due to the test charge, then the force on the test charge divided by its charge is defined to be the electric field at that point (typically called the *field point*). The electric field at (the field point) p due to a charge q is therefore

$$\vec{E_p} \equiv \frac{\vec{F_p}}{q_p} = \frac{q}{4\pi\varepsilon r^2}\, \vec{a_r} \tag{1.2}$$

where $\vec{a_r}$ is the unit vector along the line from charge q to point p and r is the distance from charge q to point p. The values of $\vec{E}$ are expressed in volts per meter (V/m).

Since the test charge at p in the preceding example doesn't disturb the electric field, the electric field is considered to be a consequence of q; in other words, the test charge doesn't have to be present for the field to exist.

The term $\vec{E}$ is a vector with both magnitude and direction. The direction of $\vec{E}$ anywhere in space is identically the direction of the force that would be experienced by a (positive) test charge at that point. We can look at the *field lines* of $\vec{E}$ as a representative of the direction of the force on a test charge due to q. For a single-point charge, the field lines are simply radial lines pointing away from the charge. The lines point away because a positive test charge placed anywhere would feel a force pushing it away from the (source) charge. The magnitude of the field decreases with the square of the distance from the charge.

For a collection of charges, the electric field at any point is the sum of the contributions of all of the charges in the collection. Figure 1.1, for example, shows electric field lines in the

(a)

(b)

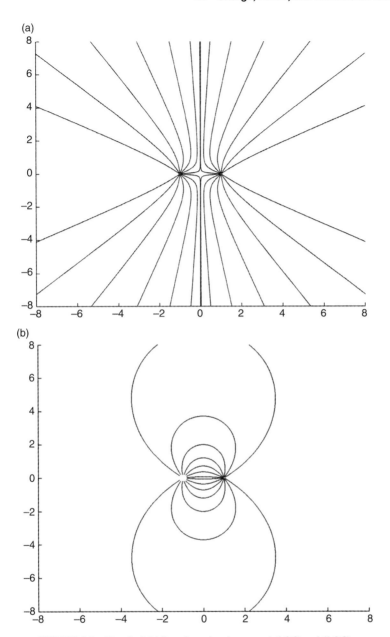

FIGURE 1.1   Electric field lines for point charges at (−1,0,0) and (1,0,0).

X–Y plane for two point charges placed at (−1,0,0) and (+1,0,0). In part (a) the two charges are identical; in part (b) they are the same in magnitude but opposite in sign.

In order to calculate $\overrightarrow{E}$ directly we must keep track of the vector components of every charge contributing to it. Continuing with the example of Figure 1.1, the simple MATLAB function charges.m shown here calculates the field anywhere in the X–Y plane. Calculation of the field components from the geometry is shown in the equations in this program. Setting q_2 to +1 or −1 produces the two cases discussed above.

```
function [ Ex, Ey, Emag ] = charges(x,y)
% This function calculates the electric field for the 2 charge
%   layout of Figure 1.1

q2 = -1;          % Set this to +1 or -1 as needed
eps = 8.854;      % pFd/m in free space

theta1 = atan2(y,x + 1);           theta2 = atan2(y,x-1);
rsq1 = (x + 1).^2 + y.^2;           rsq2 = (x-1).^2 + y.^2
Emag1 = 1./(4*pi*eps*rsq1);        Emag2 = 1./(4*pi*eps*rsq2);
Ex1 = Emag1.*cos(theta1);          Ex2 = q2*Emag2.*cos(theta2);
Ey1 = Emag1.*sin(theta1);          Ey2 = q2*Emag2.*sin(theta2);


Ex = Ex1 + Ex2;       Ey = Ey1 + Ey2;
Emag = sqrt(Ex.^2 + Ey.^2);

end
```

If both charges are equal to +1 in this example, then along the y axis E_x must always be zero. This can be deduced from the symmetry of the situation without consulting the equations. On the other hand E_y is zero only at $y = 0$ and must be an odd function of y. $E_y(0,y,0)$ is shown in Figure 1.2.

If the right-hand charge (in Figure 1.2) is changed to −1, then, along the y axis E_y must always be zero—again, from symmetry considerations. E_x in this case is an even positive function of y, as shown in Figure 1.3.

If a small charged mass such as an electron is placed near charge(s), as in part (a) or (b) of Figure 1.1, it would immediately start moving. Its trajectory would not be along a field line.

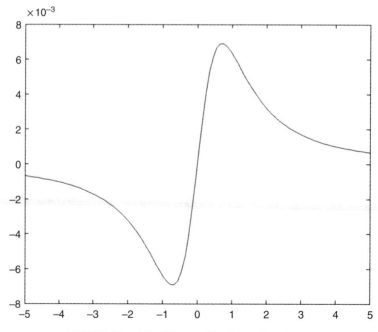

FIGURE 1.2 $E_y(0,y,0)$ for two identical positive charges.

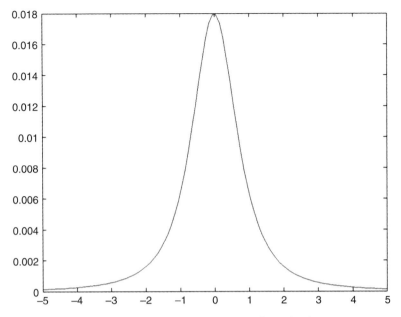

FIGURE 1.3 $E_x(0,y,0)$ for two charges of opposite sign.

Since the electron has mass, it gathers momentum as it moves and a proper description of its motion requires solving Newton's equation with the electric field as the driving force. Electron trajectories in various electric field profiles will be examined in Chapter 17.

Inspection of Figure 1.1a shows that the field lines emanating from both charges start out radially. They then bend rather than cross and leave the region, going instead to infinity. This characteristic is identical to the radial field lines from a single charge which also go to infinity. In Figure 1.1b, however, each field line travels from the positive (left-hand) charge to the negative (right-hand) charge and terminates. This is characteristic of an electrically neutral structure, and we can extract a general rule: Electric field lines originate at and terminate at charge; a neutral structure will have no field lines going to infinity. This will be expressed as a mathematical relationship in Section 1.2.

1.2   ELECTRIC FLUX DENSITY AND GAUSS'S LAW

Let us define a (vector) quantity $\vec{D}$ as follows:

$$\vec{D} = \varepsilon \vec{E} \tag{1.3}$$

Combining this definition with equation (1.2), we obtain

$$\vec{D} = \frac{q}{4\pi r^2} \vec{a_r} \tag{1.4}$$

which is independent of the dielectric constant.

The $\vec{D}$ in these equations is the *electric flux density*. The rationale for using this term will become clear shortly. Consider a point charge q surrounded by a virtual spherical shell of radius r_0. The surface area of this shell is $4\pi r_0^2$. Since $\vec{D}$ is a function only of r, it is a constant everywhere on this shell; also, since it is pointed radially outward everywhere, it is normal to the shell at intersection. The integral of (the magnitude of) $\vec{D}$ over the surface of the shell is

$$\iint_s |\vec{D}| \, ds = \iint_s \frac{q}{4\pi r^2} 4\pi r^2 = q \tag{1.5}$$

By examining the situation for an arbitrary collection of charges and an arbitrary surface surrounding them, we can generalize this result to Gauss's law[3]

$$q = \iint_s \vec{D} \cdot ds \tag{1.6}$$

where the integral is over the entire surface. $\vec{ds}$ is a differential area with vector direction normal to the plane of the area and q is the total charge enclosed.

Returning to equation (1.5), we have

$$\varepsilon \iint_s |\vec{E}| \, ds = q \tag{1.7}$$

For a spherical shell centered at q, we obtain $\vec{E} = \vec{E}(r)$ only, pointing radially outward, and therefore

$$\varepsilon \vec{E} \left(4\pi r^2\right) = q \, \vec{a_r} \tag{1.8}$$

which is essentially identical to equation (1.2). In other words, Gauss's and Coulomb's laws are equivalent.

Suppose that we have a sphere of charge of radius a, centered at the origin, of uniform charge density ρ [expressed in coulombs per cubic meter (C/m^3)] (see Figure 1.4).

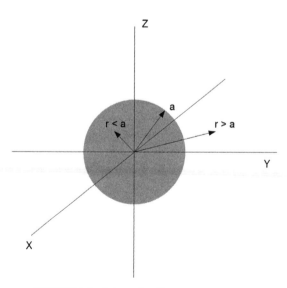

FIGURE 1.4   Sphere of uniform charge density ρ.

From the symmetry of the situation, we know again that $\vec{E} = \vec{E}(r)$ only, pointing radially outward. For any $r \leq a$, the charge enclosed is

$$Q_{enc} = \rho \int_0^r \int \int dv = \tfrac{4}{3}\pi r^3 \rho \qquad (1.9)$$

Putting this result into Gauss's law, we have

$$Q_{enc} = \tfrac{4}{3}\pi r^3 \rho = 4\pi r^2 \varepsilon E \qquad (1.10)$$

or

$$E = \frac{r\rho}{3\varepsilon} \qquad (1.11)$$

The field goes to 0 at $r = 0$, because there is no charge enclosed. It increases with increasing r. At $r = a$ all of the charge is enclosed and again using Gauss' law, for $r \geq a$, we obtain

$$Q_{enc} = \tfrac{4}{3}\pi a^3 \rho = 4\pi r^2 \varepsilon E \qquad (1.12)$$

and then

$$E = \frac{a^3}{3\varepsilon r^2} \qquad (1.13)$$

If ρ is not a constant but is instead a function of r (and only r), then it must be brought inside the integral of equation (1.9) and the integral properly evaluated. The electric field outside the sphere of charge ($r \geq a$) depends only on the total charge in the sphere, irrespective of the details of $\rho(r)$. This latter point is significant because it tells us that $E(r)$ (see Figure 1.5) will be the same (again, for $r \geq a$), if all the charge is concentrated at a point at the origin, is spread uniformly through the volume of the sphere, or is distributed in whatever other configuration that can be imagined. An important case we will consider (in Section 1.3) is the case where all of the charge resides in a thin shell at $r = a$.

1.3 CONDUCTORS

An ideal conductor of charge is a material in which the charge carriers are free to move about under the influence of electrostatic forces (Coulomb's law). Good examples of this are metals such as copper and silver—they are not ideal conductors but they are very good conductors. The very mobile charge in metals is the electrons in the outer shell of the metallic atoms; how charge mobility comes about is an important topic of solid-state physics.[4] How charge is arranged in conductors in different situations will be a central theme in discussion of the method of moments (MoM) in later chapters

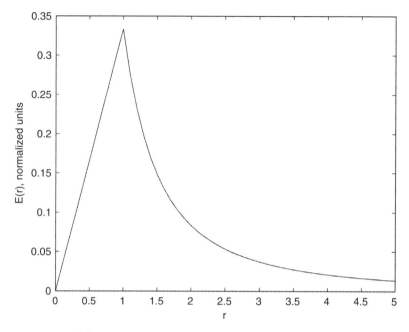

FIGURE 1.5 $E(r)$ for a sphere of radius a, charge density ρ.

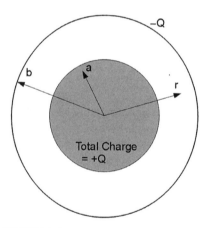

FIGURE 1.6   Two concentric spherical shells.

(Chapters 3, 4, etc.). Right now we will consider only situations with geometries whose symmetries require that charge distributions be uniform.

Consider Figure 1.6. A metal sphere has been placed at $r = a$, and a spherical metal shell has been placed at $r = b$. A charge $-Q$ equal to the total charge enclosed by the inner shell $(+Q)$ has been placed on the outer shell so that the entire system is now charge-neutral. The symmetry of the structure implies that charge must be uniform in terms of angle. The charges on the inner sphere repel each other and are attracted to the charges on the outer sphere. This means that the charges on the inner sphere will all move to the outer surface of the inner sphere, which, in turn, means that there is no electric field inside the inner sphere.

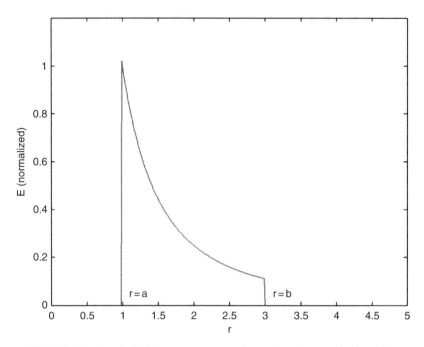

FIGURE 1.7 Electric field between two concentric opposite-charge conductive shells.

A further conclusion is that, in terms of the electric field between the outer shell and the inner sphere, the inner sphere can be either a solid conductor or simply a thin conductor shell at $r = a$.

This latter characteristic of electrostatic systems is put to very good use in ultra-high-voltage systems such as the Van de Graaff generator.[5] The safest place for people to be is inside one of the large metal spheres used in the device, as it is a field-free region.

Returning to Figure 1.6, in the region $a \leq r \leq b$, charge $+Q$ is enclosed, and

$$\overrightarrow{E} = E_r = \frac{Q}{4\pi\varepsilon r^2} \tag{1.14}$$

When $r \geq b$, the sum of the charge on both the inner shell and the outer shell is zero, so that there is no net charge enclosed and E abruptly drops to zero (Figure 1.7).

Next, consider the structure shown in Figure 1.8. Two large parallel conductor plates have surface charge densities $+\sigma$ and $-\sigma$ [expressed in coulombs per square meter (C/m^2)]. The plates are separated by a distance d.

Near the center of these plates, far from the edges, the charge density on both plates is uniform. The only possible electric field distribution in this region is uniform, directed from the positively charged plate toward the negatively charged plate. The figure shows a virtual right circular cylinder extending from the bottom plate up to some point between the plates. The actual shape of the virtual structure is insignificant as long as its walls are directed normal to the plates' surfaces (i.e., parallel to the electric field lines).

If the area of the top and bottom surfaces of the virtual structure is A, the charge enclosed by the structure, as long as the top surface is somewhere between the surfaces, is σA. Because the sidewalls of the structure are parallel to the electric field lines, no lines cross

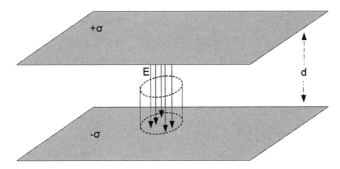

FIGURE 1.8 Electric field between two large parallel plates, near the center.

the surfaces, and therefore the only contribution to the right-handside of equation (1.6) is the top surface. Thus Gauss's law tells us that

$$\sigma A = \varepsilon E A \tag{1.15}$$

or

$$E = \frac{\sigma}{\varepsilon} \tag{1.16}$$

Gauss's law can also be expressed in differential, or point, form as[3]

$$\nabla \cdot \overrightarrow{D} = \rho \tag{1.17}$$

where $\nabla \cdot$ is the divergence operator. In rectangular coordinates this is

$$\frac{\partial D_x}{\partial x} + \frac{\partial D_y}{\partial y} + \frac{\partial D_z}{\partial z} = \rho \tag{1.18}$$

where $\rho = \rho(x,y,z)$ is the charge density, that is

$$q = \iiint_V \cdot \rho \, dV \tag{1.19}$$

where V is the total volume enclosed by s.

1.4 POTENTIAL, GRADIENT, AND CAPACITANCE

Since there is a force on a charged body in an electric field, moving that body through the field must require work. (If energy is transferred to the body, we'll consider it negative work done.) This is analogous to the work done lifting a mass in a gravitational field. As in the case of work done in a gravitational field, we can define a potential difference as the work done in moving the body, where $\overrightarrow{dl}$ is a differential length element along the path from p to q:

$$\phi_q - \phi_p = -\int_p^q \vec{E} \cdot \vec{dl} \tag{1.20}$$

The electrical potential φ is also called the *voltage V*, so equation (1.20) can equivalently be written

$$V_q - V_p = -\int_p^q \vec{E} \cdot \vec{dl} \tag{1.21}$$

As the preceding equations show, only a voltage difference between two points is defined. Strictly speaking, the voltage at a point has no meaning. It is common, however, to define the voltage at some point as zero, often called the *ground* or *reference* voltage or potential. It is then possible to refer to the voltage at any point using a single number—the implied meaning is that we are talking about the voltage difference between that point and the reference point.

Returning to the example of the concentric spheres (Figure 1.6), we can easily find the voltage difference (commonly called the *voltage*) between the two spheres by integrating equation (1.14):

$$V(r) = \frac{-Q}{4\pi\varepsilon} \int_a^r \frac{d\bar{r}}{\bar{r}^2} = \frac{Q}{4\pi\varepsilon} \left(\frac{1}{a} - \frac{1}{r} \right) \tag{1.22}$$

Here, we have chosen $V(a) = 0$ as the voltage reference.

The voltage between the two metal shells is then

$$V_b = \frac{Q}{4\pi\varepsilon} \left(\frac{1}{a} - \frac{1}{b} \right) \tag{1.23}$$

From a circuital perspective, we are often more interested in voltages (and fields) at different places in terms of the applied voltage. We obtain this result by dividing equation (1.22) by equation (1.23):

$$V(r) = V_b \frac{1/a - 1/r}{1/a - 1/b} \tag{1.24}$$

For the parallel plate structure (Figure 1.8), taking $z = 0$ as the bottom plate and $z = d$ as the top plate, with the bottom plate at ground and the top plate at V_0, integrating equation (1.16), and repeating the same procedure as above, we obtain

$$V(z) = \frac{\sigma}{\varepsilon} z = V_0 \frac{z}{d} \tag{1.25}$$

Again analogous to the mass in a gravitational field, the potential difference between two points is path independent, it is inconsequential which path the integral takes from point p to point q. This implies that the electrostatic field is conservative—any path leading from point p back to point p will yield a zero-voltage difference. In other words, electrostatic energy is neither gained nor lost going around a closed path. An important point to make here, even though it is beyond the purview of this book, is that this is not a general electromagnetic system property—it is valid only in the electrostatic case.

Restating equation (1.21) to yield the field in terms of the voltage difference, in rectangular coordinates, we have

$$\vec{E} = -\nabla V = -\left[\frac{\partial V}{\partial x}\vec{a_x} + \frac{\partial V}{\partial y}\vec{a_y} + \frac{\partial V}{\partial z}\vec{a_z}\right] \tag{1.26}$$

The operator ∇ is called the *gradient operator*. This equation shows clearly why an arbitrary reference voltage choice has no effect on the electric field.

Suppose that there is a charge q at the origin of our coordinate system. If q is the only charge present, then no work was required to bring q from anywhere else to the origin. Now, let us bring a test charge from infinity (where the field due to q is zero) to some radius a. Using equation (1.21), we obtain

$$V_a = -\int_{\infty}^{a} \vec{E}\cdot\vec{dl} \tag{1.27}$$

and using equation (1.2), the potential at a is

$$V_a = -\int_{\infty}^{a} \frac{q}{4\pi\varepsilon r^2}dr = \frac{q}{4\pi\varepsilon a} \tag{1.28}$$

Equation (1.28) is a scalar equation, which is almost always easier to work with than is a vector equation. Also, once the voltage is known, it is a straightforward job to calculate the field. Consequently, we will concentrate on finding voltages and then (if necessary) finding the field, not the other way around.

For the single-point charge of equation (1.28), we already know that the field lines point radially outward (from the charge), going to infinity. Figure 1.9. shows surfaces of constant potential, known as equipotential surfaces or more commonly equipotentials. These surfaces cross the field lines normally and in this situation are spheres.

If, instead of a single charge q, we have a collection of (discrete) charges, we must replace equation (1.28) by the sum of the contributions of all the charges, and a is replaced by the distances from each of the charges (x_i,y_i,z_i) to the measurement point $p = (x_p,y_p,z_p)$. In other words,

$$r_{ip} = \sqrt{\left(x_i-x_p\right)^2 + \left(y_i-y_p\right)^2 + \left(z_i-z_p\right)^2} \tag{1.29}$$

and then

$$V_p = \sum_i \frac{q_i}{4\pi\varepsilon r_{i,p}} \tag{1.30}$$

The gradient [equation (1.26)] operating on V produces en electric field vector whose direction is the same as that of the maximum change in V. Since the direction of maximum

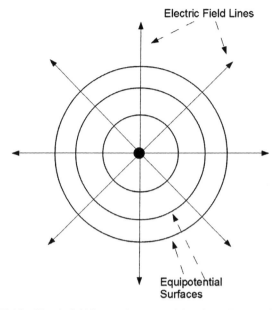

FIGURE 1.9 Electric field lines and equipotential surfaces about a point charge.

change from an equipotential contour is always normal to the contour, the electric field lines will always be normal to the equipotential contours. Figure 1.10 is essentially a repeat of Figure 1.1, which shows the electric field lines and the equipotential surfaces about a two-charge system.

The MATLAB program `linesofforce.m` generates the curves shown in Figure 1.10 (and Figure 1.1):

```
% lines of force.m

%   The lines of force between two equal charge
%   q2 may be switched between +1 and –1

close

[x,y] = meshgrid ( –10:.01:10, –10:.01:10 );

q1 = 1; q2 = –1;

theta1 = atan2 (y,x + 1);     theta2 = atan2 (y,x – 1);
r1sq = (x + 1).^2 + y.^2;     r2sq = (x – 1).^2 + y.^2;

E1x = q1*cos(theta1)./r1sq;   E1y = q1* sin(theta1)./r1sq;
E2x = q2*cos(theta2)./r2sq;   E2y = q2*sin(theta2)./r2sq;

Ex = E1x + E2x;     Ey = E1y + E2y;

startx = [];     starty = [];

for theta = .10: pi/8: .85*pi
```

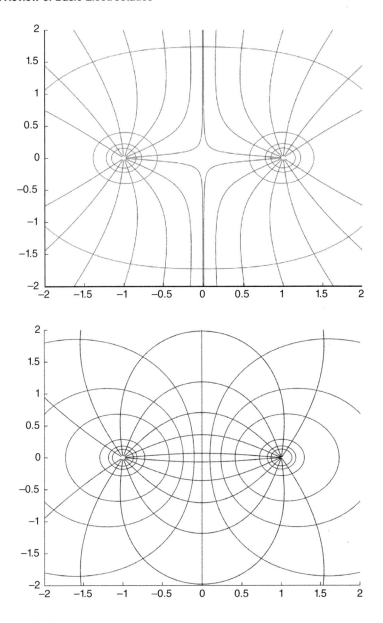

FIGURE 1.10   Repeat of Figure 1.1 with equipotential lines superimposed.

```
    startx = [startx, -1 + .03*cos(theta)];
    starty = [starty, .03*sin(theta)];
end

xy = stream2 ( x, y, Ex, Ey, startx, starty );
figure (1)
axis ([-2, 2, -2, 2])
streamline ( xy );
hold on
```

```
xy = stream2 ( x, y, Ex, Ey, startx, -starty );      % symmetry
streamline ( xy );

if q2 >0
   xy = stream2 ( x, y, Ex, Ey, -startx, starty );
   streamline ( xy );

   xy = stream2 ( x, y, Ex, Ey, -startx, -starty );
   streamline ( xy );
end

r1 = sqrt (r1sq); r2 = sqrt (r2sq);

V = (q1./r1 + q2./r2);

V_lines = [[-7:2:7],[-.5,.5, -.2, .2, 0]];
contour (x, y, V, V_lines)
```

For a distribution of charge with density $\rho(x_i,y_i,z_i)$, we have

$$V_p = \frac{1}{4\pi\varepsilon}\iiint\limits_{v_i} \frac{\rho(x_i,y_i,z_i)dv}{r_{i,p}} \tag{1.31}$$

where the integral is over the volume containing the charge distribution.

If we connect a battery between two electrodes, the electrodes assume the potential difference (i.e., the voltage) of the battery. In this process, some electrons leave the electrode that is connected to the positive terminal of the battery and flow into the battery; an equal number of electrons leave the negative terminal of the battery and flow into its connected electrode. If we were to then remove the battery, this charge imbalance and the voltage difference would remain at the electrodes. The electrode that lost electrons is now positively charged at some charge q, and the electrode that gained electrons is now negatively charged at $-q$. Both the electrode system (the two electrodes) and the battery remain electrically neutral. Some work has been done in *charging* the capacitor, and electrochemical changes in the battery that supplied the energy to do this have resulted in some *discharging* of the battery.

The ratio of the charge q to the voltage difference between the electrodes is called the *capacitance* of the structure, measured in farads:

$$C \equiv \frac{q}{\Delta V} \tag{1.32}$$

For the concentric spherical shells of Figure 1.6, directly from equation (1.23), we obtain

$$C = \frac{Q}{V_b} = 4\pi\varepsilon\frac{ab}{b-a} \quad \text{F} \tag{1.33}$$

For the central region of a large parallel plate structure (capacitor), we must consider capacitance per unit area. From equation (1.25), we obtain

$$C = \frac{\varepsilon}{d} \ \text{F/m} \tag{1.34}$$

Equation (1.34) is most often written as the total capacitance for an electrode area A:

$$C = \frac{\varepsilon A}{d} \ \text{F} \tag{1.35}$$

Equation (1.35) is known as the *ideal parallel plate capacitor relationship*. We will see how closely it approximates the capacitance of a real parallel plate capacitor in one of the first examples of calculations using the method of moments, presented later in the book.

1.5 ENERGY IN THE ELECTRIC FIELD

Using equation (1.2), the work expended moving a test charge q in the electric field of a capacitor is

$$U = \int \vec{F} \cdot \vec{dl} = q \int \vec{E} \cdot \vec{dl} = qV \tag{1.36}$$

If an incremental extra charge dq were then added, the incremental work would be

$$dU = V \, dq = \frac{q}{C} dq \tag{1.37}$$

The work expended in bringing up a total charge Q is

$$U = \int dU = \frac{1}{C} \int_0^Q q \, dq$$

$$= \frac{Q^2}{2C} = \frac{1}{2} CV^2 \tag{1.38}$$

where V is the voltage difference between the capacitor electrodes, that is, the voltage difference to which the capacitor is charged.

In the simple case of an ideal parallel plate capacitor, the electric field is a constant and

$$U = \frac{1}{2} \frac{\varepsilon A}{d} \left(E^2 d^2 \right) = \tfrac{1}{2} \varepsilon E^2 (\text{A}d) = \tfrac{1}{2} \varepsilon E^2 \ (\text{vol}) \tag{1.39}$$

The energy density is then

$$u = \tfrac{1}{2} \varepsilon E^2 \tag{1.40}$$

Although this relationship has been derived only for the special case of the ideal parallel plate capacitor, with the help of some vector relationships it can be shown to be true in general.[3] This general relationship, for $\vec{E} = \vec{E}(x, y, z)$ is then

$$U = \tfrac{1}{2}\iiint \varepsilon \,|\vec{E}|^2 dv \tag{1.41}$$

where the integral is over all space.

Setting equations (1.38) and (1.41), two expressions for the total energy stored in the electric field, equal to each other and bringing in equation (1.26) to express the electric field in terms of the voltage between two electrodes

$$U = \tfrac{1}{2}\iiint \varepsilon \,|\vec{E}|^2 dv = \tfrac{1}{2}\iiint \varepsilon \,|\nabla V(x,y,z)|^2 dv = \tfrac{1}{2}CV_0^2 \tag{1.42}$$

gives us a way to find the capacitance of a structure without ever explicitly worrying about the charge on the electrodes or the electric field:

$$C = \frac{\iiint \varepsilon \,|\nabla V(x,y,z)|^2 dv}{V_0^2} \tag{1.43}$$

The notation in this equation deserves some clarification: V_0 is the applied voltage at one electrode referred to another electrode of a structure. $V(x,y,z)$ is the voltage distribution throughout the volume (possibly all space) where the electric field is nonzero due to V_0.

Finding the capacitance directly from the voltage distribution will be shown to be a very useful technique when dealing with finite difference or finite element solutions for the voltage distribution.

For an ideal parallel plate capacitor, starting with equation (1.41), we obtain

$$V(z) = V_0 \frac{z}{d} \tag{1.44}$$

where $0 \le z \le d$, V_0 is the applied voltage, and the area of the plates is A. Continuing, we obtain

$$|\nabla|^2 = \left(\frac{\partial V}{\partial z}\right)^2 = \left(\frac{V_0}{d}\right)^2 \tag{1.45}$$

and finally

$$C = \frac{\varepsilon (\partial V/\partial d)^2 A d}{V_0^2} = \frac{\varepsilon A}{d} \tag{1.46}$$

which, of course, agrees with equation (1.35).

For the concentric spheres, from equation (1.24), we have

$$V(r) = V_b \frac{1/a - 1/r}{1/a - 1/b} \tag{1.47}$$

so that

$$|\nabla V|^2 = \left(\frac{\partial V}{\partial r}\right)^2 = \frac{V_0^2}{r^4}\left(\frac{ab}{b-a}\right)^2 \qquad (1.48)$$

and then, as expected, we obtain

$$C = \varepsilon \frac{ab^2}{b-a}\int\limits_0^{2\pi}\int\limits_0^{\pi}\int\limits_a^b \frac{1}{r^4}r^2\sin(\theta)\,dr\,d\theta\,d\varphi = 4\pi\varepsilon\frac{ab}{b-a} \qquad (1.49)$$

These examples of finding capacitance using the stored energy might seem pointless—the voltage distributions were found from solutions for the electric field in terms of the charge; these distributions were used to find the stored energy and then to find the capacitance "without referring to the field or the charge." Section 1.6 should clarify this issue. The voltage will be found directly from the structure description, without first looking at the field or the charge.

1.6 POISSON'S AND LAPLACE'S EQUATIONS

Combining equations (1.17) and (1.26), again in rectangular coordinates, we have

$$\nabla\cdot\vec{D} = \nabla\cdot\varepsilon\,\vec{E} = -\varepsilon\nabla\cdot\nabla V = \rho \qquad (1.50)$$

The operator $\nabla\cdot\nabla$, written ∇^2, is called the *Laplacian*, and in rectangular coordinates this equation becomes

$$\nabla^2 V = \frac{\partial^2 V}{\partial x^2} + \frac{\partial^2 V}{\partial y^2} + \frac{\partial^2 V}{\partial z^2} = \frac{-\rho}{\varepsilon} \qquad (1.51)$$

This is Poisson's equation.

In a charge-free region, Poisson's equation becomes *Laplace's* equation:

$$\nabla^2 V = \frac{\partial^2 V}{\partial x^2} + \frac{\partial^2 V}{\partial y^2} + \frac{\partial^2 V}{\partial z^2} = 0 \qquad (1.52)$$

Laplace's equation will prove to be a workhorse throughout much of this book, so it is useful to write out the Laplacian in cylindrical and spherical coordinates. In cylindrical coordinates, we obtain

$$\nabla^2 V = \frac{1}{r}\frac{\partial}{\partial r}r\frac{\partial V}{\partial r} + \frac{1}{r^2}\frac{\partial^2 V}{\partial\varphi^2} + \frac{\partial^2 V}{\partial z^2} \qquad (1.53)$$

and in spherical coordinates, looking only at systems with spherical (ϕ and θ) symmetry, we get

$$\nabla^2 V = \frac{1}{r^2}\frac{d}{dr}r^2\frac{dV}{dr} \qquad (1.54)$$

Laplace's equation, at first blush, seems irrelevant. If there is no charge, what is there to calculate? In many situations, however, we have a structure with two (or more) charged electrodes in an otherwise charge-free region. Consider, for example, two concentric metal shells. While we may never examine the charge explicitly, we can still solve Laplace's equation for the voltage distribution in the charge-free region between these shells using the shells as boundary conditions. From equation (1.54), we have

$$\frac{d}{dr} r^2 \frac{dV}{dr} = 0 \tag{1.55}$$

which integrates to

$$\frac{dV}{dr} = \frac{K_1}{r^2} \tag{1.56}$$

where K_1 is a constant of integration. This, in turn, integrates to

$$V = \frac{K_1}{r} + K_2 \tag{1.57}$$

where K_2 is a second constant of integration.

Fortunately, we have two boundary conditions to satisfy: $V(a) = 0$ and $V(b) = V_0$. Substituting these conditions into equation (1.57) gives us two equations in the two unknown constants. Solving for these constants and putting the results back into equation (1.57) gives us identically equation (1.47).

In Chapter 2 we will see that in many practical situations a structure is very long and uniform in one dimension and that therefore a cross section of the structure gives us excellent results for V, E, U, and C, the latter two on a per-unit-length basis.

An example of such a structure is the circular coaxial cable shown in Figure 1.11. This is an example of a transmission line; transmission lines will be discussed in Chapter 2. The salient point here is that these are two concentric circular conductors of radii $a < b$. As long as we are not near the ends of the cylinders, the electrostatic problem is essentially a two-dimensional problem.

Laplace's equation in cylindrical coordinates with both circular symmetry and no length (z) dependence is as follows, from equation (1.53):

$$\nabla^2 V = \frac{d}{dr} r \frac{dV}{dr} = 0 \tag{1.58}$$

Again, integrating twice and using the boundary conditions $V(a) = 0$ and $V(b) = V_0$, we get

$$V = V_0 \frac{\ln(r/a)}{\ln(b/a)} \tag{1.59}$$

and evaluating equation (1.43):

$$C = \frac{2\pi\varepsilon}{\ln(b/a)} \ \text{F/m} \tag{1.60}$$

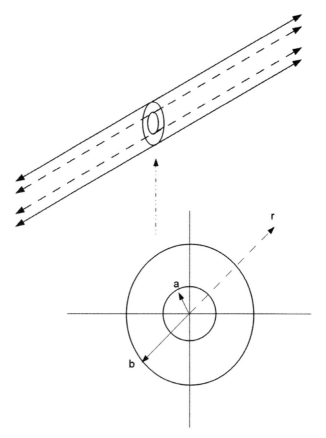

FIGURE 1.11   Circular coaxial transmission line and a cross section of the line.

1.7   DIELECTRIC INTERFACES

When a space is uniformly filled with a dielectric material, solution procedures are straight-forward. All the conditions and principles discussed and presented above still apply, except for the number actually used for ε. The circular coaxial cable, for example, is often manu-factured using a flexible plastic dielectric with a (relative) dielectric constant of approxi-mately 2 separating the inner and outer conductors. In equation (1.60), then, $\varepsilon = 2\varepsilon_0$, and the analysis is complete.

This is not the case, however, when there are multiple dielectrics, that is, there are dielectric interfaces, present in a structure. Figure 1.12 shows three examples of dielectric interfaces.

Figure 1.12a shows the cross section of a coaxial cable with two concentric different die-lectric materials between the two electrodes. Figure 1.12b shows a three-dimensional par-allel plate capacitor with a slab of dielectric material separating the electrodes. Figure 1.12c shows a block of dielectric material being used as a *standoff* insulator to separate two power carrying lines. In the latter two cases the dielectric interface is between the dielectric material and the surrounding air.

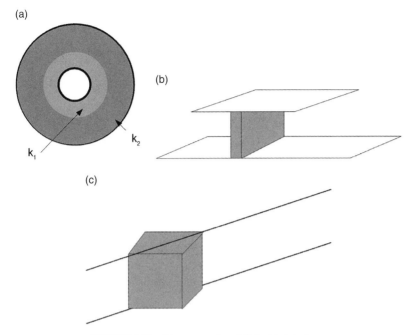

FIGURE 1.12   Three examples of dielectric interfaces.

In Figure 1.12a the direction of the electric field is known — symmetry dictates that it be radial. In Figure 1.12b,c the electric field direction is not known a priori; both the field strength and direction are unknown functions of position and must be found.

Dielectric material regions influence electrostatic solutions in two ways: their dielectric constant itself and the perturbation of the voltage and field distributions in a structure due to the dielectric interfaces. The dielectric interfaces are characterized by boundary conditions that arise at these interfaces.

Consider Figure 1.13a. Shown in cross section, a small virtual pillbox sits between two different dielectric materials. The top and bottom surfaces of the pillbox are of area A. The actual shape of these surfaces doesn't matter; it could be square, round, or in another configuration. The thickness of the pillbox is Δz. The normal D field components to the top and bottom surfaces is shown.

In the limit as Δz goes to zero (shrinking from the top and bottom toward the middle), contribution to the Gaussian surface whose shape is the pillbox from the sides vanishes, and since there is no charge inside the pillbox, we obtain

$$D_{n1}A + D_{n2}A = 0 \tag{1.61}$$

or

$$D_{n1} = D_{n2} \tag{1.62}$$

Now consider Figure 1.13b. In this case the dashed-line rectangle is a two-dimensional path. The integral of the tangential electric field around the path must be zero (the electric field is conservative) so that, as Δz goes to zero, either

(a)

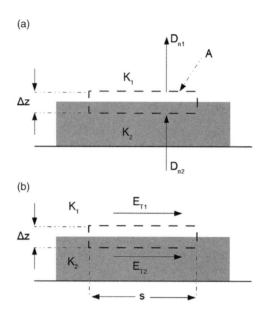

(b)

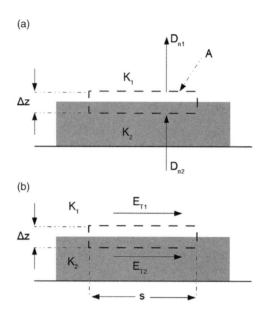

FIGURE 1.13 Boundary between two different dielectric materials.

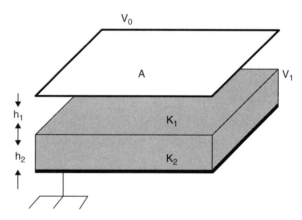

FIGURE 1.14 Two-dielectric-layer parallel plate capacitor.

$$E_{T2}s - E_{T1}s = 0 \qquad (1.63)$$

or

$$E_{T1} = E_{T2} \qquad (1.64)$$

Equations (1.62) and (1.64) represent the boundary conditions that we are looking for — across a dielectric interface, D normal and E tangential are continuous.

Figure 1.14 shows an example of a parallel plate capacitor containing two dielectric layers. We shall assume that the electrode separation is much less than either electrode dimension and use the ideal parallel plate capacitor approximation. Also, a new circuit symbol has been introduced in this figure. The small "rake" connected to the lower electrode near its left edge is the conventional circuit symbol, stating that the lower electrode is connected to ground, that is, is at zero, or reference, voltage. The upper electrode is at voltage V_0, which is an applied voltage (boundary condition). The dielectric interface is at voltage V_1, which at present we do not know.

For the ideal parallel plate capacitor, D is normal to the electrodes, and for a surface charge density σ on the bottom electrode, the total charge on the electrode is

$$Q = \sigma A = D_z A \tag{1.65}$$

from which

$$D_z = \sigma = \frac{Q}{A} \tag{1.66}$$

everywhere. The electric field is then

$$E_n = \begin{cases} \dfrac{D_n}{K_1} = \dfrac{Q}{AK_1}, & z \le h_1 \\[2ex] \dfrac{D_n}{K_2} = \dfrac{Q}{AK_2}, & h_1 < z \le h_2 \end{cases} \tag{1.67}$$

The voltages across the two layers are

$$V_1 = \frac{Qh_1}{K_1 A} \tag{1.68}$$

and

$$V_0 - V_1 = \frac{Qh_2}{K_2 A} \tag{1.69}$$

from which

$$V_0 = \frac{Q}{A} \frac{h_1}{K_1} + \frac{h_2}{K_2} \tag{1.70}$$

Since V_0 is the applied voltage to the top electrode, the capacitance is

$$C = \frac{Q}{V_0} = \frac{A}{(h_1/K_1) + (h_2/K_2)} \tag{1.71}$$

From equation (1.46), the capacitance of the two layers as individual capacitors is

$$C_1 = \frac{k_1 A}{h_1}$$
$$C_2 = \frac{k_2 A}{h_2} \tag{1.72}$$

Substituting equations (1.72) into (1.71), we have derived the relationship for two capacitors in series:

$$C_{ser} = \frac{1}{(1/C_1) + (1/C_2)} \qquad (1.73)$$

1.8 ELECTRIC DIPOLES

The dual-opposite-charge system shown in Figure 1.1b has something important missing. If two opposite-charge bodies are placed near each other, the Coulomb force will start pulling them together. What is missing is a mechanical support, or separator, of some sort that prevents the charges from moving. This separator can take many forms. It could be a dielectric slab through which the charges cannot move. It could exist at an atomic level — for instance, positive and negative charges may somehow be "pinned" in separate locations in a molecule of some sort. The molecule is electrically neutral, but it is made up of opposite charges somehow separated and held in place by the molecular structure itself.

However, when the charges are held apart, the electrical component of this structure is called an *electric dipole* or in context, simply a *dipole*.

Consider such a structure with the mechanical separating components having a (relative) dielectric constant of one. This lets us examine the electrical properties of dipoles without getting entangled in considerations of dielectric interfaces.

Figure 1.15 details an electric dipole with the supporting structure not shown. Charges $+q$ and $-q$ are located on the Y axis at $Y = +L/2$ and $Y = -L/2$, respectively.

The voltage at any point (x,y) is simply

$$V(x,y) = \frac{Q}{4\pi\varepsilon_0}\left[\frac{1}{\sqrt{x^2 + (y-L/2)^2}} - \frac{1}{\sqrt{x^2 + (y+L/2)^2}}\right] \qquad (1.74)$$

The same expression may, of course, be written in polar coordinates:

$$V(x,y) = \frac{Q}{4\pi\varepsilon_0}\left[\frac{1}{\sqrt{r^2 - Lr\sin(\theta) + L^2/4}} - \frac{1}{\sqrt{r^2 + Lr\sin(\theta) + L^2/4}}\right] \qquad (1.75)$$

If r is much larger than L, equation (1.75) may be approximated by

$$\frac{4\pi\varepsilon_0}{Q}V(x,y) \approx \left[\frac{1}{\sqrt{r^2 - Lr\sin(\theta)}} - \frac{1}{\sqrt{r^2 + Lr\sin(\theta)}}\right] \approx \left[\frac{1}{r} + \frac{Lr\sin(\theta)}{2r^3} - \frac{1}{r} + \frac{Lr\sin(\theta)}{2r^3}\right]$$

$$(1.76)$$

or

$$V \approx \frac{Q}{4\pi_0}\frac{L\sin(\theta)}{r^2} \qquad (1.77)$$

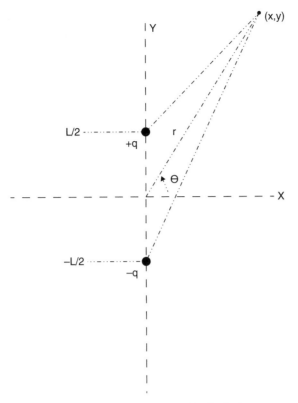

FIGURE 1.15 Structure of an electric dipole.

The voltage is a function of r and θ, as expected. At $\theta = 0$ or π (along the X axis), it is exactly zero, as symmetry demands. For any value of θ, the voltage falls off with the square of r. Far from an electric dipole [where distance ("far") is measured in units of L], the dipole does not have much electrostatic influence as compared to separated charge ("separation" is also measured in units of L).

If an electric dipole is placed in a uniform electric field, there will be no net translation (movement of the center of gravity of the system) force on the dipole. The force trying to move the dipole in the direction of the field will be exactly canceled by the force trying to move the dipole against the direction of the field.

A dipole will rotate in a uniform electric field so as to align itself with the field. If we define the vector L as the line from charge $-Q$ to charge $+Q$, then we can define the dipole moment as

$$\vec{P} = Q\,\vec{L} \tag{1.78}$$

and then the dipole torque in the uniform field is

$$\vec{\tau} = \vec{P} \times \vec{E} \tag{1.79}$$

As equation (1.79) predicts, the torque is maximum when the dipole is normal to the field (along the X axis in Figure 1.16) and zero when the dipole is aligned with the field (along the X axis in Figure 1.17).

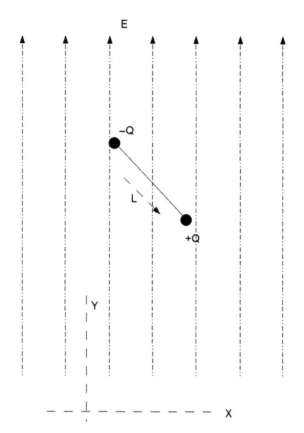

FIGURE 1.16   Electric dipole in a uniform field.

Figure 1.17 depicts a situation with a nonuniform electric field. Concentric circular (cross sections of long cylinders) electrodes were chosen for this example because the field lines can be drawn exactly without getting embroiled in calculations.

In this geometry all of the field lines are identical in that they are radial and all have the same magnitude–radius relationship. The nonuniform characteristic of these lines is that no two of them will point in the same direction; remember that two vectors have to agree in magnitude and direction for them to be identical. More complex geometries can be chosen in which the field lines differ in magnitude and direction, but this "extra" characteristic is not necessary for the point of this discussion.

In the situation depicted in the figure, two effects take place concurrently. First, as in the previous example, the dipole will rotate so that the $-Q$ side is pointing toward the higher potential electrode—in this case the inner circle.

The electric field between two circles exhibits a monotonically decreasing magnitude (with increasing r). Once the dipole has rotated so that its $-Q$ side is at a smaller radius than is its $+Q$ side, the $-Q$ side will feel a stronger force pulling it toward the inner circle than the $+Q$ side will feel pulling it to the outer circle. The dipole, while aligning itself with the field lines, will move towards the center electrode (assuming, of course, that it is not being held in place in some way). This effect is called *dielectrophoresis*.[6]

Uncharged particles in a fluid will move in a uniform electric field, due to a complex interaction of the surface of the particle collecting a charge from the fluid while the fluid

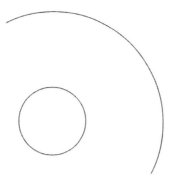

FIGURE 1.17 Section of concentric circles demonstrating a nonuniform field.

immediately surrounding the particle becomes charged opposite to the particle; the entire system, of course, remains neutral. This phenomenon is called *electrophoresis.*[7]

1.9 THE CASE FOR APPROXIMATE NUMERICAL ANALYSIS

All of the examples presented thus far have been for simple structures with helpful symmetries. The availability of such structures is good for exemplifying solving the equations that have been presented and examining some of these solutions. On the other hand, we must admit that these examples were not chosen only because they provide compact illustrative examples but because it is almost impossible to solve these equations formally for anything except very simple, highly symmetric, structures.

Classical texts are replete with solution techniques and problems that have been solved.[8] These are important in that they extend our knowledge of the mathematics and the physics of electrostatics. On the other hand, most situations that commonly arise in practice are either totally unsolvable by any of these techniques or would require months of analysis to produce a solution.

As an example of this situation, consider Figure 1.18. We begin with a strip of metal on the top of a slab of dielectric that is fully metalized on its bottom face. For reasons that will be discussed in Chapter 2, we are interested in the capacitance between these two electrodes and the peak electric field (as a function of the applied voltage). This fairly straightforward situation is already a very difficult problem to attack analytically.

Now let's add a narrow upper conductor strip branching off the original center conductor. What has happened to the capacitance and the peak electric field? Finally, we'll add some holes in the upper conductor metalization. At this point we are beyond the capability of an analytic solution.

This example, although somewhat contrived, is actually indicative of real-world issues. Modern electronic devices, such as cellphones, make use of multilayer circuitboard technology. Many layers of dielectric and electrode patterns are sandwiched together into a thin circuitboard structure. Scattered about the board are holes passing from some layer to another. The walls of these holes are metalized; the metal on intermediate electrode layers is cleared away from the holes to prevent accidental contact. Again, the capacitances and

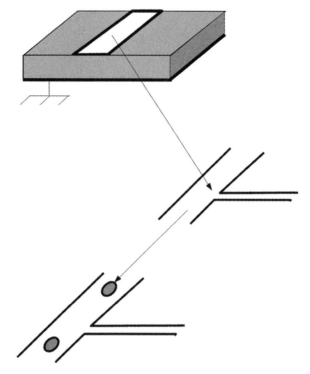

FIGURE 1.18   A common conductor–dielectric structure.

peak electric fields are of significant interest: not only the nominal values but the variations in these values with manufacturing tolerances of metal patterns, dielectric thickness, metal patterning alignment errors, and so on.

In general, we need techniques for studying these issues. Changes in geometry must be input parameters that can be varied easily.

The numerical techniques to be presented in the following chapters are not the only techniques available for solving electrostatic problems. They are, however, indicative of two basic approaches: approximating either (1) the charge on electrodes for specified voltages on the electrodes or (2) the voltages in space for specified voltages on the electrodes. In both cases we replace a partial differential equation with a set of linear algebraic equations, which we know very well how to solve. The results are, of course, approximate, but by solving problems that have been solved (or approximated) analytically, we know that excellent accuracy can be obtained.

The approach chosen to present these materials will be somewhat different from that used in many texts. Rather than moving quickly to mathematical sophistication and generalization, we will stay as much as possible with the basic approaches and exploit these approaches to demonstrate the problems that they can solve. The goal here is not to present an exhaustive treatment of the approaches but rather to set up and solve many useful problems with the least sophisticated method possible in each case. If we succeed, this book should complement rather than duplicate the other books available in the literature.

Many concepts in electrostatics, such as image charges, are introduced in the chapters where these concepts fit logically with the analyses being performed. The decision to do this rather than introduce them in this chapter was based on the desire for brevity of this initial chapter—only those topics that are needed to proceed were covered.

PROBLEMS

1.1 Two small spheres of mass m are suspended from weightless strings of length L. The suspension points are a distance S apart. The spheres are charged as a capacitor; that is, Q is removed from one sphere and placed on the other. The spheres will move toward each other as shown in Figure P1.1.

    (a) By establishing for equilibrium the force of gravity and the Coulomb force, find the suspension angle α as a function of Q.

    (b) What is the voltage between the spheres as a function of Q?

    (c) What is the smallest angle α can assume (for nonzero Q)?

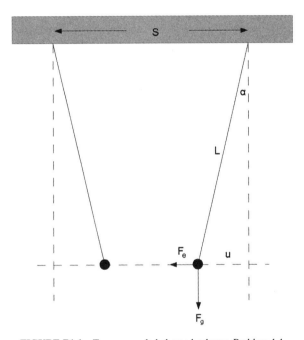

FIGURE P1.1   Two suspended charged spheres, Problem 1.1.

1.2 Repeat Problem 1.1a by finding the total potential energy of the structure and minimizing it.

1.3 Three small charged spheres are located as shown in Figure P1.3. One sphere is fixed at $X = 0$ with a charge of -1 (arbtirary units), one sphere is fixed at $X = 10$ with a charge of -3, and the third sphere is free to move on the X axis, with a charge q. What is the equilibrium position x for this system, and does it matter whether q is positive or negative?

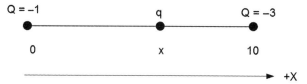

FIGURE P1.3   Layout for Problem 1.3.

1.4   Miniature chip capacitors are often constructed by building up successive layers of conductors and dielectric (typically a ceramic material) and then interconnecting the conductors, as shown in Figure P1.5. In these devices, the thickness : lateral dimensions ratio is low enough that we may use the ideal parallel plate capacitor calculation with good results. If the dielectric thickness between successive conductor layers is h, then the capacitance (per unit area) of the device is

$$\frac{C}{A} = \frac{2k\varepsilon_0 n}{h} \qquad (1.80)$$

where
k = relative dielectric constant
h = dielectric thickness between the metal layers
n = number of dielectric layers

Suppose that we are allowed a total capacitor thickness of 3 mm. Each metal layers is 0.1 mm thick. We have a choice of dielectrics with relative dielectric constant ranging from 1 to 100. Unfortunately, however, the breakdown field of the dielectrics varies with k as

$$E_p = 1 \times 10^6 - 1 \times 10^4 k \qquad (1.81)$$

and the capacitor must be able to withstand 100 V. Find the maximum attainable capacitance per unit area, along with the values of n and k to achieve this maximum capacitance.

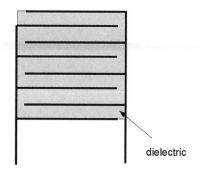

dielectric

FIGURE P1.5   A multilayer chip capacitors.

1.5 Figure P1.6 shows a quadrupole charge configuration (Figure P1.6). Note that is also possible to define a linear quadrupole configuration. For $r >> L$, derive an approximation for the voltage as a function of r and φ.

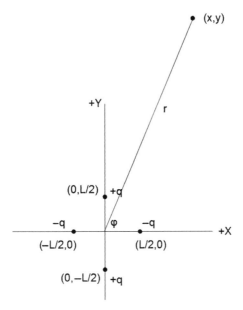

FIGURE P1.6 A quadrupole charge configuration.

1.6 An ideal parallel plate capacitor has a plate separation y. Calculate the force pulling the plates together under two conditions: **(a)** constant charge on the plates; **(b)** constant voltage on the plates.

REFERENCES

1. J. D. Jackson, *Classical Electrodynamics*, 3rd ed., Wiley, New York, 1999.

2. G. F. Nicholson, An introduction to the rationalized M.K.S. system of units, *Br. J. Appl. Phys.* **2**: 177 (1951).

3. S. Ramo, J. Whinnery, and T. Van Duzer, *Fields and Waves in Communication Electronics*, Wiley, New York, 1965.

4. J. McKelvey, *Solid-State and Semiconductor Physics*, Harper & Row, New York, 1966.

5. http://en.wikipedia.org/wiki/Van_de_graaf_generator.

6. http://en.wikipedia.org/wiki/Dielectrophoresis.

7. http://en.wikipedia.org/wiki/Electrophoresis.

8. W. Smythe, *Static and Dynamic Electricity*, McGraw-Hill, New York, 1950.

2

The Uses of Electrostatics

Each topic discussed below is a sophisticated discipline in itself. The purpose of this chapter is to show where actual use of electrostatic analysis is involved in actual structures. Section 2.1, on basic circuit theory, is included because this is a discipline based on electrostatic (and magnetostatic) approximations whose use is implied in the functioning of the various structures discussed in subsequent sections.

2.1 BASIC CIRCUIT THEORY

The speed of an electromagnetic wave in a vacuum (and essentially the same number in air) is 3×10^8 m/s. A radio wave with a frequency of 1 MHz (one million cycles per second) has a wavelength of

$$\lambda = \frac{3 \times 10^8}{1 \times 10^6} = 300 \, \text{m} \approx 1000 \, \text{ft} \tag{2.1}$$

If we are dealing with frequencies in this range or lower and with distances of, say, ≤1 ft, then we can ignore the propagation times along wires connecting various components and we can idealize components by pretending they're infinitely small.

These assumptions allow us to define a branch of electrical science known as *circuit theory*, in which wires are ideal, lossless equipotential surfaces, with infinitely high signal propagation speed. This lets us define a list of components which are characterized by voltage–current *terminal* conditions.[1,2] Although we're introducing simple circuit theory

Introduction to Numerical Electrostatics Using MATLAB, First Edition. Lawrence N. Dworsky.
© 2014 John Wiley & Sons, Inc. Published 2014 by John Wiley & Sons, Inc.

only because some of its tools will be useful going forward, in principle all of circuit theory is a topic in electrostatics (and magnetostatics).

Circuits are composed of interconnected components and obey two basic laws known as the *Kirchhoff voltage* and *current laws*: (1) Kirchhoff's *voltage* law states that the total voltage drop around a closed loop in a circuit is zero; (2) Kirchhoff's *current* law states that the total current flowing into (or out of) any point, or node, is zero.

For our purposes, we will be dealing only with simple circuits and four basic components (the standard symbols for these components are shown in Figure 2.1):

1 *The Ideal Voltage Source.* This is two-terminal device with a defined voltage drop, or potential difference (and polarity), between its terminals. The ideal voltage source can supply (source) or absorb (sink) energy depending on the direction of the current through it. The ideal voltage source can never be shorted-circuited (its terminals cannot be connected to each other) because this would cause a violation of Kirchhoff's voltage law. When the current through a voltage source is flowing into the positive (+) terminal [and out the negative (−) terminal], the voltage source is absorbing energy. When the current is flowing into the − terminal (and out the + terminal), the voltage source is delivering energy. The rate of energy absorption (or delivery) is the power

$$P = VI \tag{2.2}$$

2 *The Ideal Current Source.* This is two-terminal device with a defined current (and polarity) flowing through it. The ideal current source can supply or absorb energy depending on the polarity of the voltage across it. The ideal current source can never be open-circuited because this would cause a violation of Kirchhoff's

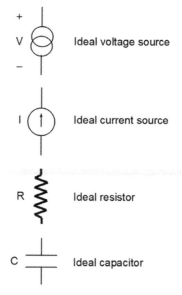

FIGURE 2.1 Four basic circuit elements.

current law. When the terminal with outflowing current is positive (and the terminal with current inflowing is negative), the current source is supplying power, and vice versa. Again.

$$P = VI \tag{2.3}$$

3   *The Ideal Resistor.* This is two-terminal device whose resistance R is defined by *Ohm's law*, where the ratio of the voltage across the resistor's terminals to the current flowing through it are expressed as follows:

$$R = \frac{V}{I} \tag{2.4}$$

When current is flowing into the positive terminal (and out of the negative terminal), the resistor is absorbing power. A resistor cannot deliver or store power; rather, it dissipates power as heat:

$$P = VI = \frac{V^2}{R} = I^2 R \tag{2.5}$$

4   *The Ideal Capacitor.* This is two-terminal device. From Chapter 1, the definition of capacitance is

$$Q = CV \tag{2.6}$$

where Q is the charge stored on an electrode of the capacitor when V is the voltage across the capacitor's terminals.

The current flowing into (and out of) a capacitor is the rate of change of charge on the capacitor's electrodes:

$$I = \frac{dQ}{dt} = C\frac{dV}{dt} \tag{2.7}$$

This equation is taken as the defining circuit equation of a capacitor. It may be inverted to express the voltage in terms of the current:

$$V = \frac{1}{C}\int I\, dt \tag{2.8}$$

The limits of integration (or boundary conditions in the case of the derivative equation) are discussed below.

A capacitor neither absorbs nor delivers power. It can, however, store energy (in the electric field), as described in Chapter 1:

$$U = \tfrac{1}{2}CV^2 \tag{2.9}$$

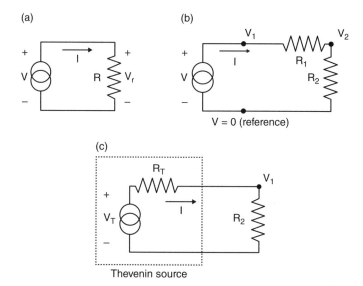

FIGURE 2.2   Some simple circuit examples using voltage sources.

Several examples (and further definitions) will complete our very abbreviated and truncated introduction to circuit theory.

Figure 2.2a shows a very simple circuit. A resistor is connected to an ideal voltage source. The circuit consists of one closed loop, so only one current (I) can be defined. The direction of the current is arbitrary, but it is conventional to define the direction of current as leaving the + terminal of a voltage source. The formal Kirchhoff voltage equation for this circuit is as follows, where V_r is the voltage across the resistor:

$$V + V_r = 0 \tag{2.10}$$

Substituting Ohm's law for the resistor voltage and noting the proper current direction conventions, we obtain

$$I = \frac{V}{R} \tag{2.11}$$

Figure 2.2b shows a simple circuit with two resistors, R_1 and R_2. These resistors are connected in *series* (a connection where the currents flowing through both resistors are identical). Again, there is only one current path, with current I. Writing Kirchhoff's voltage law, we have

$$V = V_{r1} + V_{r2} = IR_1 + IR_2 = I(R_1 + R_2) \tag{2.12}$$

$$I = \frac{V}{R_1 + R_2} \tag{2.13}$$

In the circuit shown in Figure 2.2b there are three possible places to label a voltage, or three *nodes*. Since only voltage differences are physically meaningful, however, it is convenient to arbitrarily label one of the nodes as the zero-, or *reference* or *ground*, voltage node, and then the other (two) nodes are understood to be voltages with respect to the reference node.

Voltage V_1 is, by the definition of a voltage source, equal to V. Voltage V_2 is known because the current I is known:

$$V_2 = IR_2 = \frac{VR_2}{R_1 + R_2} \tag{2.14}$$

Voltage V_2 is identically the voltage across R_2. The voltage across R_1 is found either as IR_1 or $V_1 - V_2$.

Although the ideal voltage source is an important abstraction for circuit models, it does not exist in the real world. All real voltage sources have some internal losses. Resistive losses can be modeled as resistors in series with the ideal voltage source. The resulting compound element, shown in the dashed lines in Figure 2.2c, is called a *Thevenin equivalent voltage source*. The internal connection between R_T and V_T is not accessible.

Since Figure 2.2c is identical, circuitwise, to Figure 2.2b, there is no need to re-solve the circuit equation for the current. There are, however, some important interpretive distinctions:

1 Since R_T cannot be changed, there is a maximum current that a Thevenin source can supply. This is the current in Figure 2.2c (or Figure 2.2b) when $R_2 = 0$.

2 There is no paradox created by shorting-circuiting (connecting together) the two terminals of a Thevenin source.

3 When a Thevenin source is supplying power to an outside resistor (or resistor network), it is dissipating power internally in R_T. A Thevenin source will become hot when it is being used.

4 An open-circuited (with nothing connected to it) Thevenin source is not dissipating power.

Figure 2.3 shows several simple examples using current sources.

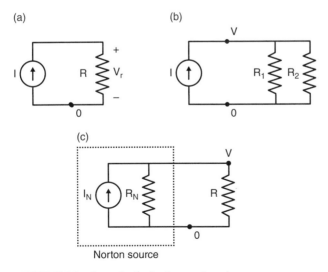

FIGURE 2.3 Some simple circuit examples using current sources.

Figure 2.3a shows an ideal current source connected to a resistor. There is only one (nonreference) voltage point (node) and

$$V = IR \tag{2.15}$$

Figure 2.3b shows two resistors connected in *parallel* (where the voltages across both resistors are identical). Writing Kirchhoff's current law at the voltage node V, we have

$$I = I_1 + I_2 = \frac{V}{R_1} + \frac{V}{R_2} = \left(\frac{1}{R_1} + \frac{1}{R_2} \right) \tag{2.16}$$

As in the case of voltage sources, ideal current sources also do not exist. A real current source, called a *Norton equivalent source*, is represented by an ideal current source in parallel with an internal resistance:

1   Since R_N cannot be changed, there is a maximum voltage that a Norton source can supply. This is the current in Figure 2.3c (or Figure 2.2c) when R goes to infinity (an open circuit).
2   There is no paradox created by opening (leaving unconnected) the two terminals of a Norton source.
3   When a Norton source is supplying power to an outside resistor (or resistor network), it is dissipating power internally in R_N. A Norton source will become hot when it is being used.
4   A short-circuited Norton source isn't dissipating power.

It can be shown that, insofar as the external circuit is concerned, it is impossible to tell whether the real source "inside the box" is a Norton or a Thevenin source, provided that

$$R_T = R_N$$
$$V_T = I_N R_T = I_N R_N \tag{2.17}$$

Figure 2.4 shows a Thevenin voltage source connected to a capacitor.
A new circuit element, the switch, has been introduced in this figure. The figure indicates that the switch is *closed* (i.e., a connection is made) at time $t = 0$. For $t < 0$ there is no connection. There is no current path, so $I = 0$. But what about V_C?

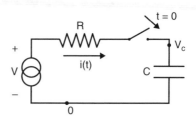

FIGURE 2.4   Simple *RC* circuit.

Since C is an ideal capacitor, it will retain charge on its electrodes (absent a discharge current path) forever. If there is charge Q on the electrodes, then our basic definition of a capacitor tells us that

$$V_C = \frac{Q}{C} \tag{2.18}$$

There is no way of deriving or calculating this value of V_C. It is an initial condition that we must be given.

At $t = 0$ the switch is closed and we can write Kirchhoff's voltage equation

$$v(t) = Ri + \frac{1}{C} \int i \, dt \tag{2.19}$$

We can convert equation (2.19) to a differential equation by taking the time derivative of both sides. Since the voltage across an ideal voltage source never changes, its time derivative must be zero:

$$\frac{dV}{dt} = 0 = R\frac{di}{dt} + \frac{1}{C}i \tag{2.20}$$

This equation may be solved by integration:

$$i(t) = i(0)e^{-t/RC} \tag{2.21}$$

We now know the form of the solution, but we aren't finished until we can specify the initial current $i(0)$.

We knew V_C before $t = 0$, when it was a given piece of information. Suppose that at $t = 0$, V_C suddenly changes value. This implies that the time derivative of the capacitor voltage dv_C/dt is, at that instant, infinite. In turn, this means that the current through the capacitor, using equation (2.7), is instantaneously infinite.

Since the current through the capacitor is identical to the current through the resistor and the voltage source, we conclude that the resistor is instantaneously dissipating an infinite amount of power that the voltage source is supplying. This conclusion is physically unacceptable. This leads us to a "law" of circuit theory: The voltage across a capacitor (in the circuit theory abstract world) cannot change instantaneously.

If, for example, the capacitor had zero voltage before the switch was thrown, at the instant that the switch is thrown the capacitor voltage would remain zero. At $t = 0$, if there is zero voltage drop across the capacitor, then the instantaneous current is defined by the voltage source–resistor circuit

$$i(0) = \frac{V_C}{R} \tag{2.22}$$

and therefore

$$i(t) = \frac{V_C}{R} e^{-t/RC} \tag{2.23}$$

The product RC in equation (2.23) has the units of time, and is usually defined as the *time constant* of the RC pair

$$\tau = RC \qquad (2.24)$$

so that

$$i(t) = \frac{V}{R} e^{-t/\tau} \qquad (2.25)$$

The voltage across the capacitor is simply the voltage source V minus the voltage across the resistor:

$$V_C(t) = V_C - Ri(t) = V_C\left(1 - e^{-t/\tau}\right) \qquad (2.26)$$

Figure 2.5 shows the voltage and current versus time for the capacitor. At $t = 0$, the uncharged capacitor briefly appears as a short circuit. For t large (as compared to the time constant) the capacitor appears as an open circuit. After several time constants the switch may be opened and the capacitor will remained *charged* at voltage V until or unless another opportunity for current flow is provided.

Before $t = 0$, if the capacitor had been charged to a voltage V_{C0}, then the preceding equations would be modified to

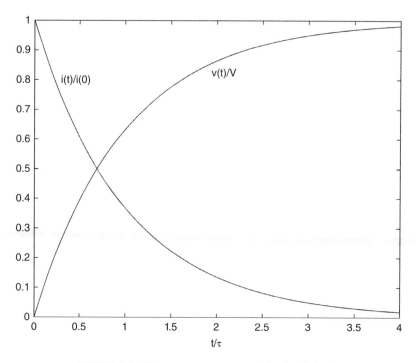

FIGURE 2.5   Voltage–current response of simple RC circuit.

$$i(t) = \frac{V_C - V_{C0}}{R} e^{-t/\tau} \tag{2.27}$$

and

$$v(t) = V - (V - V_{C0})e^{-t/\tau} \tag{2.28}$$

If the capacitor is *fully charged* (to V), the switch is opened and then a new resistor (R_2) is connected across the capacitor at time t_2, by following the same analysis as above:

$$i(t) = \frac{V}{R_2} e^{-t/R_2 C} \tag{2.29}$$

According to this analysis, a capacitor is never fully charged to a connected voltage. In later chapters we will be considering capacitors charged to some specified voltage. This may be construed to mean that a capacitor was connected to either (1) a charging voltage for so many time constants that the error is negligible or (2) a higher voltage and then disconnected at the right time as the voltage swept past the desired charging voltage.

This abbreviated introduction to circuit theory has ignored several basic circuit elements, including the *inductor* — an idealized coil of wire that stores energy in its magnetic field when current flows through it. This omission has, in turn, left us without a definition or explanation of resonance and resonance frequencies.

The most significant omission in this chapter is any discussion of sinusoidal alternating current (AC) analysis. This is a marvelously easy-to-present exercise in operational calculus that leads to many common applications of circuit theory (home electricity included) in the *sinusoidal steady state*, an important concept of which concerns impedance. *Impedance* is a property of circuit elements that is a generalization of resistance to include frequency-dependent effects of capacitance and/or inductance. An impedance can dissipate energy and/or cyclically store and release energy.

2.2   RADIO FREQUENCY TRANSMISSION LINES

This is a brief introduction to transmission line theory that is presented in handbook format, with only the necessary equations, with no supporting derivations. There are many transmission line texts available for learning more about the subject.[3,4]

Electrical transmission line theory is not an electrostatics topic. The transmission line equation is a one-dimensional wave equation and the study of transmission lines is the study of waves propagating on a transmission line. However, the properties of the typical propagating mode on a transmission line can all be calculated directly from the capacitance of the line (per unit length); this capacitance is the same as all other capacitances and is found from the stored electric field energy or the charge on the conductors. From the perspective of calculating transmission line properties, therefore, we can do our work in the electrostatic domain.

A *transmission line* is a length of two or more conductors with a net zero flow of current across a cross section of the line. Let us begin with a two-conductor line. Assume that the line is lossless and that there are no magnetic materials present. We will, however, allow for dielectric materials.

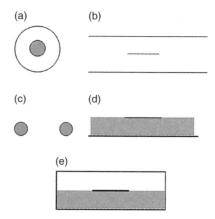

FIGURE 2.6   Some common transmission line cross sections: (a) coaxial cable; (b) stripline; (c) open-wire line; (d) open microstripline; (e) boxed microstripline. Note that there is nothing necessarily *micro* about the microstrip line — the name is historical.

Some typical transmission lines are shown in cross section in Figure 2.6.

The coaxial line shown has concentric cylindrical conductors (i.e., round conductor cross sections). This is not a necessary attribute of coaxial cable, although this is by far the most typical construction. What is special about the coaxial cable is that since the currents in the two conductors are equal in magnitude and opposite in direction, no electric or magnetic fields are present outside the coaxial cable.

While the open wire line also has equal and opposite conductors flowing in the two lines, the electric and magnetic fields extend to infinity. The stripline (and microstripline) might actually be a form of coaxial cable, or it might be an open system with fields extending to infinity.

Stripline is typically built using a uniform dielectric, microstripline has a dielectric $(K > 1)$ material in only part of the cross section. Construction of the line is easy to visualize from the picture of the line. The microstripline is different from the other lines shown because it has a nonuniform cross section and its properties will require some special consideration.

Both stripline and microstriplines typically have sidewalls of the outer conductor. The reasons for including these sidewalls are to totally contain the line's fields, prevent *incursion* of other fields, and provide structural support. In many cases these sidewalls are far enough away that they don't contribute to the electrical properties of the lines — that's why it is common to omit them in drawings. The approximation "far enough away," however, should be examined when calculating the properties of these lines.

At frequencies low enough for waveguide modes to be cut off, the only propagating mode on a transmission line is the *transverse electromagnetic* (TEM) mode. In this mode, the electric field lines are transverse to the line length (i.e., in the plane of a cross section) and the magnetic field lines wrap around the conductors and are also in the plane of a cross section of the line. This mode propagates at all frequencies down to direct current (DC) (0 frequency). At DC, the electrostatic properties of the line are the properties of the TEM mode, the properties that we are interested in.

A uniform lossless transmission line operating in TEM mode has a capacitance C and an inductance L per unit length. Finding C is a two-dimensional (2d) problem even though C is a three-dimensional (3d) parameter; all electric field lines are in the cross-sectional plane. Without proof or derivation here, we'll assert that the wave propagation properties of the line arise completely from L and C. These properties are the characteristic impedance of the line

$$Z_0 = \sqrt{\frac{L}{C}} \qquad (2.30)$$

and the wave speed

$$v = \frac{1}{\sqrt{LC}} \qquad (2.31)$$

The *characteristic impedance* is an impedance that, when connected to one end (terminating) of the line, will be the impedance measured at the other end of the line. This is unique because most terminating impedances will result in an impedance at the other end of the line that is a function of the terminating impedance, the frequency of the applied voltage, and the length of the line — at all frequencies greater than zero (DC). Also, terminating a line in its characteristic impedance is equivalent, insofar as the input impedance and the voltage and current along the line are concerned, to having an infinite length of line.

A lossless transmission line may be modeled entirely using the distributed capacitance and inductance (per unit length) of the line. The characteristic impedance of a lossless line is, however, entirely resistive. At first blush this seems to be a mistake — a resistor dissipates energy when a voltage is placed across it, so how can a structure consisting entirely of non-dissipative capacitors and inductors be described by a resistance? The answer to this is that the characteristic impedance of a line is what is "seen" when you apply a voltage to the beginning of a line that extends to infinity. In this case the traveling wave along the line is always advancing toward (but of course never reaching) infinity, and the voltage source at the beginning of the line "sees" energy disappearing. Terminating the line at a finite length with a resistor whose value is equal to the line's characteristic resistance results in actual energy dissipation (the resistor becomes hot) but the voltage source at the beginning of the line cannot distinguish between an infinite length line and a terminated finite length line.

Maxwell's equations tell us that a wave on a line with an air (or vacuum) dielectric will propagate at the speed of light v_0. From equation (2.31), we obtain

$$L = \frac{1}{v_0^2 C} \qquad (2.32)$$

$$Z_0 = \frac{1}{v_0 C} \qquad (2.33)$$

This means that C is the only parameter needed to determine everything that we need to know about a (lossless) transmission line.

If a transmission is uniformly filled with (or embedded in) a dielectric of relative constant K, then the capacitance is scaled (from the value with an air dielectric) by K. The inductance, however, doesn't change; it is calculated from equation (2.32) for the same structure assuming that $K = 1$.

The microstripline doesn't strictly operate in a TEM mode. At low frequencies (starting at 0 frequency, DC) however, the line will operate in a very TEM-like mode known as the *quasi-TEM mode*. The capacitance is calculated for the structure as given, and the inductance is calculated for the same structure with no dielectric material present. Since the capacitance value will be somewhere between the all-air capacitance value and an all-dielectric capacitance value, the wave velocity will also be between the all-air and all-dielectric values. A dielectric value K that, uniformly filling the space, would produce this wave velocity (and line capacitance), is referred to as the *effective dielectric constant* value of the structure.

Currents along the conductors of a transmission line will show the same distributions as the charge densities found in the examples in this book. In the case of a coaxial cable composed of concentric round lines, the symmetries of the geometry dictate that there be no current, or charge density, variations. In all other geometries, however, there definitely are variations. This is an important parameter to study because Ohmic (resistive) losses along a line depend (in the cross section) on the square of the current density. In other words, two different geometry lines can have very different losses (per unit length) even though they are created using the same materials and have the same cross-sectional conductor dimensions and conductor thicknesses.

Another parameter that enters into the loss calculation is the *radiofrequency (RF) skin effect*, which is a frequency-dependent effect that prevents currents from penetrating deeply into a real conductor. The RF skin effect is definitely not an electrostatic parameter and is outside the scope of this book.

Transmission lines are important for carrying RF signals. It is necessary to calculate the capacitance of transmission line structures to design transmission lines. When nonenclosed, or *open* lines are placed near one another, signals will couple. Again, a capacitance calculation is necessary to describe this coupling.

As digital signaling speeds increase, the pulses shorten and surprisingly short connection paths behave as transmission lines. An example of this is the multiline circuitboards used in, say, portable cellphones. These circuit boards must be modeled as complex transmission line structures and properly designed so that the shape of the digital pulses is not compromised and interline coupling is understood and controlled.

2.3 VACUUM TUBES AND CATHODE RAY TUBES

The first 70 years or so of electronics technology contain many examples of generating and controlling the flow of electrons in a vacuum. These include the use of vacuum tubes for general electronics and cathode ray tubes (CRTs) for displays. Both of these devices have been phased out by solid-state devices. Some electron beam devices, such as the scanning electron microscope (SEM), remain in use.

Design of electron beam control systems is a classical electrostatics project. Beam currents are typically low enough that distortion of the electric field by the electrons is ignored. This means that the overall problem decouples into two separate problems: the electrostatic problem of describing the potentials and fields for a given geometry and electrode voltages, and then description of electron trajectories as a function of launch conditions and the electric field. In all cases a high vacuum is maintained in an enclosed volume. This makes electron trajectory calculation an exercise in *Newton's laws*, circumventing the need to consider scattering by a background gas, and at the same time avoids gas breakdown.

One technique for launching electrons into a vacuum is called *thermionic emission*. A thin wire filament is heated (as in an incandescent lightbulb) hot enough to glow.[5] When a metal is heated, the conduction band electrons (the electrons that are free to move — i.e., the electrons

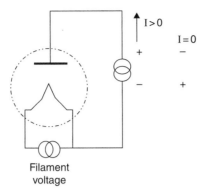

FIGURE 2.7   The vacuum tube diode.

responsible for electric current flow in metals) become very energetic. Some of them will acquire enough energy to "boil off" the surface of the metal into the surrounding vacuum. When Thomas Edison placed a metal electrode in the vacuum enclosed along with the filament, he collected these electrons with a positive potential; this is now called the *Edison effect.*[6]

Figure 2.7 shows Edison's structure. Because there are two elements, the filament and the anode (*plate*), this device is called a *vacuum tube diode.* Edison noticed that when the anode is positive (*biased positive*) with respect to the filament (as shown in Figure 2.7), current (I) flows. However, if the polarity of the plate voltage is reversed, so that the plate is biased negative with respect to the filament, no current flows. This was early evidence that current in metals is carried by *some sort of particle* that carries only a negative charge. For historical reasons, since *current flow* was defined as the movement of positive charge, even though the electrons flow from the filament to the plate, the electric current flow is from the plate to the filament. This subtlety often causes consternation in beginning engineering students, but is a perfectly workable system with no contradictions.

Since this diode will pass current in only one direction, it can be used to convert alternating current to direct current. This application is known as *rectification*, and the vacuum tube diode is often called the *vacuum tube rectifier.*

Figure 2.7 is a schematic representation of a vacuum tube diode. In practice, the filament is usually a length of wire mounted at (and connected to) either end. The anode evolved from being the simple plate of Edison's early experiments to being an open cylinder surrounding the filament. The term *plate* lived on, however, and is synonymous with the term *anode* in vacuum tube electronics.

Around 1907, Lee De Forest heralded in the era of electronics with the invention of the vacuum tube triode.[7,8]

De Forest added a third element to the structure: a metallic cylinder made of porous mesh between the filament and the plate. When this cylinder, called the *grid* because of its appearance, is biased slightly negative with respect to the filament, it lowers the local electric field and reduces the current flow between the filament and the plate. The grid itself would not collect any electrons because of its bias polarity; hence there is no *grid current.* If the grid is made negative enough, it would actually reduce the plate current flow to zero, or *cut off* the current. For small variations in grid voltage about a nominal reduced-current level (the *bias point*), variations in grid voltage would cause corresponding variations in plate current. The grid and its connections are shown symbolically in Figure 2.8.

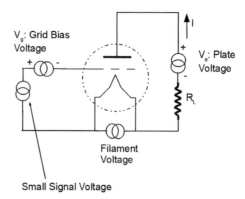

FIGURE 2.8   The vacuum tube triode.

If a *load resistor* (R_L in Figure 2.8) is placed in series with the plate voltage supply, the plate current is reduced somewhat by the Ohmic voltage drop across this resistor. Small voltage variations in the grid voltage cause variations about the nominal value of the plate current, which, in turn, cause voltage variations across the load resistor.

What is happening here is that a small signal voltage superimposed on the grid bias voltage, which requires zero power because there is no grid current, is causing a corresponding voltage variation in the load resistor — which in turn is dissipating power. A power amplifier has been built, and the age of electronics has begun.

There have been many advances in vacuum tubes, including (1) the indirectly heated cathode to reduce the effects of having an AC voltage on the filament; (2) a second grid between the first (*control*) grid and the anode to reduce anode–cathode capacitance that limits frequency response; and (3) a grid near the anode to reduce effects of secondary emission (electrons crashing into the anode and causing new electrons to be emitted by the anode). This final structure, the *vacuum tube pentode*, was the workhorse of the electronics industry for decades.

The *cathode ray tube* (CRT) is a vacuum tube with a filament, a cathode, a control grid, and an anode but also has other elements and a purpose entirely different from that of the vacuum tubes described above.

Figure 2.9 shows, schematically, an electrostatic CRT. Connecting wires are not shown; they are typically brought in at the end of the *neck* of the tube (the left end in Figure 2.9). An exception to this is the anode voltage, which is typically brought in near the anode, or screen, end (the right end in Figure 2.9).

The labeled components and their functions are as follows:

1 The *glass bottle* defines the structure and holds the vacuum.
2 The *heated cathode* is a source of electrons.
3 The *control grid* controls the current flow in the electron beam. Since the CRT is an electronic display, the control grid controls the brightness of the spot of light that will be created on the screen.
4 The *focusing electrodes* are cylindrical electrodes biased so as to cause the beam of electrons passing through them to bunch up, thus minimizing the cross-sectional area of the electron beam at the anode.
5 Two or more focusing electrodes ensure better focusing control.

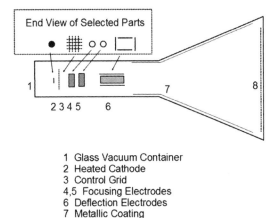

1 Glass Vacuum Container
2 Heated Cathode
3 Control Grid
4,5 Focusing Electrodes
6 Deflection Electrodes
7 Metallic Coating
8 Phosphor Coating

FIGURE 2.9   The electrostatic cathode ray tube.

6   There are two pairs of *deflection electrodes*. A potential on the vertical (in Figure 2.9) pair causes a horizontal deflection of electrons; a potential on the horizontal pair causes a vertical deflection of electrons. By setting these potentials, any point on the screen could be *illuminated* by electrons.

7   A *metalized coating* on the inside of the glass that is set to the anode voltage and creates an accelerating field for the electrons toward the screen.

8   A *phosphor coating* on the inside of the glass. Certain ceramic materials called *phosphors* emit light when struck by energetic electrons. This phosphor layer is also set at anode potential.

The CRT structure as shown was used for oscilloscopes and small television (TV) screens. Over the years, as TV screens became larger, anode voltages higher and manufacturing techniques more mature, the focusing and deflection electrodes were replaced by a magnetic structure (the *yoke*), a compound doughnut-shaped set of coils that is mounted around (on the outside of) the neck of the tube.

Manufacturing of the CRT became so well controlled and inexpensive that even the extension of the technology to the three-electron-gun shadow-mask tube for color TVs allowed this device to become a household item.[9]

2.4   FIELD EMISSION AND THE SCANNING ELECTRON MICROSCOPE

When the electric field at the surface of a metal then is high enough, and the field is directed so as to sweep electrons away from the metal, electrons are emitted from the metal by a process called *tunneling*. There is no way to explain tunneling in classical physics concepts. Tunneling is a quantum mechanical phenomenon. The electrons do not literally "tunnel through" the surface of the metal; they simply appear outside the surface. Historically, an accurate description of electron tunneling, called the *Fowler–Nordheim equations* after the developers of this description, was one of the early successes of the then new quantum theory early in the twentieth century.[10,11]

In practice, the very high electric field needed is obtained by the combination of a very sharp thin tip of metal and the use of high potentials. Metal field emission tips are typically made of a very hard, high-melting-temperature, metal such as molybdenum. Since the 1990s there has been considerable development of very thin carbon nanotubes as electron emitters.[12]

While at first using field emission for electron sources seems easier than using filaments, the extreme sharpness of a field emitter tip causes many issues in that molecule-level contamination and/or erosion can significantly modify a tip's electron emission properties.

The advantage of using a field emitting tip over using a hot filament is that the energy spread of emitted electrons is much narrower for a field emitter than it is for a filament. This means that it is much easier to produce an extremely narrow beam of electrons from a field emitter when such a beam is necessary.

An application that utilizes this property of an electron field emitting tip is the *scanning electron microscope* (SEM). The SEM has a structure similar, in a schematic sketch, to the electrostatic CRT structure of Figure 2.8 except that the column does not widen at the screen. The phosphor-coated screen itself is replaced by a metallic sample holder, at anode potential, onto which is placed a specimen that has a very thin (~10-nm) metallic coating. (The coating is just thick enough to provide some electrical conductivity while not distorting the surface features of the specimen. Near the specimen holder is an electrode biased so as to attract low energy electrons that are emitted by the specimen.

By the process of *secondary electron emission*,[12] when the electron beam is scanned across the specimen, electrons are knocked off the surface of the specimen and are, in turn, collected at the collection electrode. An image created on a display device by synchronizing position with the scanning of the specimen and modulating brightness with the collected secondary electrons from the specimen can result in a much higher magnification image of the specimen than is obtainable with an optical microscope. Despite the circuitous procedure of coating a specimen, placing it in the SEM, and pumping the high vacuum required, SEM images are so good that the SEM is a workhorse of the semiconductor and many other industries and research fields.

In the 1990s there was considerable interest in building field emission displays (FEDs) with arrays of field emission tips mass-produced using semiconductor industry photolithographic and vacuum deposition techniques.[13] Commercialization of these displays was overshadowed by the rapid advancements in liquid crystal displays (LCDs), and this type of display has almost totally disappeared.

2.5 ELECTROSTATIC FORCE DEVICES

Historically, electrical energy was converted to mechanical work using magnetic forces. For motors, solenoids, and other components, an electromagnet was the preferred structure. If we were constrained to build an electromagnet with a one-turn loop of wire, electromagnets would not be so useful. Fortunately, we are not so constrained — we can wind an electromagnet "coil" using hundreds or even thousands of turns (loops) of wire. The resulting magnetic forces are sufficient to perform many impressive tasks.

When we try to design a device using electric (Coulomb) forces, we do not have an analogy to the multiturn electromagnet. The only way to provide a sufficient force over a given surface area is to (1) increase the electric field by increasing the voltage or (2) decrease the gap between the electrodes. The resulting designs were not very attractive.

The advent of the microelectromechanical machining (MEMM) techniques has significantly improved the scenario described above.[14] Using processing techniques derived from semiconductor industry photolithographic techniques, small (micrometer-size) devices are readily batch-produced with repeatable characteristics.

An example of using Coulomb force to move an electrode is the *digital light projector* (DLP) invented at Texas Instruments in the late 1980s.[15] The heart of a DLP projector is a semiconductor chip on which millions of small mirrors are hinge mounted, each located above a transistor that is part of a matrix display addressing system. Each mirror can be moved far enough to switch between reflecting incoming light (from a bright source) out through a projection lens system or away from the lens system. The devices are small enough that the mirrors can be switched thousands of times per second. This allows for excellent grayscale and (with suitable colored light sources) color gamut. At the microscopic level, there is virtually no fatigue mechanism in the mirrors' motion and the devices are very longlasting.

Other applications for MEMM electrodes that can move include utilization of the natural resonant frequencies of any mechanical system to build electromechanical resonators for frequency reference,[16] small valves for microfluidic control,[17] and acceleration sensors.[18] Also, an accelerometer built with (or mounted) on a large thin diaphragm can be used as a microphone.

In all of these applications, the underlying physical mechanism is simply the Coulomb attraction between two electrodes with a potential (voltage) between them. For a microphone or a resonator or similar devices, it is necessary to bias the structure by applying a voltage that moves one of the electrodes a small percentage of the distance from its rest position toward the other electrode. As a result of this bias, sinusoidal variations in (movable) electrode position couple to sinusoidal variations in voltage, resulting in near-linear AC device performance.

There are so many types of MEMM structures that it is difficult to select a "typical" device to describe. The various references in this chapter barely scratch the surface of the list of existing and proposed devices. The analysis in Chapter 17 shows how Coulomb forces interact with mechanical (Hooke's law) spring forces to produce electrically controlled motion.

2.6 GAS DISCHARGES AND LIGHTING DEVICES

When a gas atom is struck by an energetic particle (typically an electron), an electron may be "knocked off" the gas atom, leaving a positively charged ion behind, causing the atom to be *ionized*. If a high electric field is present, the negatively charged electron and the positively charged gas ion will separate. The electron, which is much lighter than the ion, will gather more speed than the ion and may then crash into another gas atom, causing further ionization. If the gas pressure and electric field conditions are right, a steady-state condition representing a statistical balance between gas atoms ionizing and gas ions and electrons recombining can be achieved, and the gas will have *broken down*. This state of matter, composed of neutral atoms, electrons, and ionized atoms, is still electrically neutral and is called a *plasma*.[19]

If the high electric field necessary to sustain a plasma is produced by electrodes immersed into the gas, electrons will flow to the positive electrode and (positive) gas ions will flow to the negative electrode, where they combine with electrons supplied by the electrode. The net result is a steady flow of current. Energy is supplied by the source connected to these electrodes. When a neutral atom is ionized, it absorbs energy from the source. When an electron and an ion recombine, energy is released. The frequency of the released energy is a characteristic of the atom involved. Certain atoms, such as neon, release energy in the visible light spectrum. The *neon bulb* is a simple device – two electrodes are inserted through the walls of a small sealed glass bulb that has been evacuated and then backfilled with neon.

Some gases, such as mercury vapor, emit radiation at frequencies above the visible spectrum when ions recombine. If the glass bulb containing the gas is coated with a phosphor that emits visible light when struck by this radiation, then we have a very efficient light source known as a *fluorescent lamp*. If millions of small red, blue, and green fluorescent lamps are arrayed on a sheet in small chambers and proper addressing circuitry is provided, we have a *plasma display*.

In all the examples presented above, a static electric field, interacting with gas atoms under the right conditions, creates the gas plasma.

REFERENCES

1. J. Bird, *Electrical Circuit Theory and Technology*, 4th ed., Taylor & Francis, New York, 2010.

2. M. Ghausi, *Principles and Design of Linear Active Circuits*, McGraw-Hill, New York, 1965

3. L. N. Dworsky, *Modern Transmission Line Theory and Applications*, Wiley, New York, 1979

4. S. Frankel, *Multiconductor Transmission Line Analysis*, Boston, Artech House, 1977.

5. http://en.wikipedia.org/wiki/Cathode_heater.

6. http://www.merriam-webster.com/dictionary/edison%20effect.

7. http://en.wikipedia.org/wiki/Vacuum_tube#Triodes.

8. http://www.angelfire.com/electronic/funwithtubes/Basics_04_Triodes.html.

9. http://en.wikipedia.org/wiki/Shadow_mask.

10. http://ecee.colorado.edu/~bart/book/msfield.htm.

11. R. Gomer, *Field Emission and Field Ionization*, Harvard University Press, Cambridge, MA, 1961.

12. http://www.research.ibm.com/nanoscience/nanotubes.html.

13. J. Lee, D. Liu, and S. Wu, *Introduction to Flat Panel Displays*, Wiley, Hoboken, NJ, 2008.

14. http://www.mems.sandia.gov/.

15. http://www.dlp.com/technology/how-dlp-works/default.aspx.

16. http://www.electroiq.com/articles/stm/print/volume-9/issue-1/features/mems-resonators-vs-crystal-oscillators-for-ic-timing-circuits.html.

17. http://faculty.washington.edu/yagerp/microfluidicstutorial/tutorialhome.htm.

18. http://www.silicondesigns.com/tech.html.

19. http://en.wikipedia.org/wiki/State_of_matter#Plasma_.28ionized_gas.29.

3

Introduction to the Method of Moments Technique for Electrostatics

3.1 FUNDAMENTAL EQUATIONS

Our plan in this chapter is to approximate the charge distribution over a small region of an electrode as a constant and then to combine many of these approximations in a manner that results in a useful approximation to the exact solution of Poisson's equation over all of space. The method of moments (MoM) technique breaks conductor surfaces into small planar regions, assumes a constant charge distribution on each region, approximates Poisson's equation by a set of algebraic equations, and then creates an approximate solution by (exactly) solving these equations.

The term *method of moments* is derived from the use of *moment*, as in *momentum*, as a weighting function for multiplying a set of variables by before adding (or integrating) them. Sadiku traces the term back to a Russian literature origin.[1] It was popularized in the United States by Harrington, who used the name in the title of his seminal text on the subject.[2] The method of moments is a much more general technique than its application here portrays; the interested reader is encouraged to follow up with either of the above mentioned references.

Consider an unbounded space with a finite total area of ideal conductors present. Practically, of course, *unbounded* means everywhere, but the conductors that we are not considering are so far away as to be irrelevant to our discussion. The conductors can be either ideally thin (two-dimensional) sheets or actual volumes. Since all charge will reside on the surface of the conductors, we could consider a three-dimensional conductor to be made up of only its skin.

Each conductor is set at a voltage, or potential, referenced to another conductor. This means that there are electrical charge distributions present on these conductors. In the case of thin conductor sheets, the charge distribution is two-dimensional, in (on) the sheet.

Introduction to Numerical Electrostatics Using MATLAB, First Edition. Lawrence N. Dworsky.
© 2014 John Wiley & Sons, Inc. Published 2014 by John Wiley & Sons, Inc.

If the system was electrically neutral before voltages were applied, it remains electrically neutral, with charge moved from one conductor to another to support these voltages. Our problem is, given the geometry and the voltages, to find the charge distributions, which, in turn, will enable us to find the electric field and voltage distribution in all space, and, if desired, the stored energy and the (interelectrode) capacitances.

The electrode regions used for examples in this chapter will be rectangles, all in an X–Y plane and with sides parallel to the axes. These constraints are not fundamental, as will be shown in later chapters. They are established here so that meaningful and useful example problems can be formulated and while circumventing the complications of general three-dimensional problems.

When a capacitor is charged, a certain amount of charge is removed from one electrode and placed onto the other electrode. Overall charge neutrality of the system is maintained. The ratio of the charge moved to the voltage difference between the electrodes is a function of geometry (and dielectric materials—We will consider only linear homogeneous dielectric materials) only, and capacitance is defined by the ratio

$$C = \frac{Q}{V} \tag{3.1}$$

As an example, consider the parallel plate capacitor shown in Figure 3.1. Two rectangular electrodes, both in X–Y planes, are separated by a distance h. The top electrode is located directly over the bottom electrode.

Each rectangle is shown subdivided into subrectangles, each of area A_i (not necessarily all the same). If the total charge on each subrectangle is Q_i, and we assume that the charge density on each subrectangle is uniform and of value σ_i, then

$$\sigma_i = \frac{Q_i}{A_i} \tag{3.2}$$

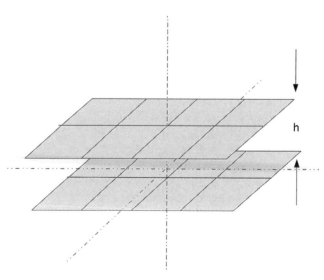

FIGURE 3.1 A parallel plate capacitor.

The voltage of subrectangle i is V_i. The contribution to this voltage of the charge on subrectangle j, in free space, is

$$V_{i,j} = \frac{\sigma_j}{4\pi\varepsilon_0} \iint \frac{dx\,dy}{\sqrt{(x_{0,i} - x_j)^2 + (y_{0,i} - y_j)^2 + h^2}}$$

$$= \frac{Q_j}{4\pi\varepsilon_0 A_j} \iint \frac{dx\,dy}{\sqrt{(x_{0,i} - x_j)^2 + (y_{0,i} - y_j)^2 + h^2}}$$

(3.3)

where the integration is over subrectangle j and is measured at the center of rectangle i.

Voltage V_i is the sum of the contributions from all of the rectangles (including rectangle i itself):

$$V_i = \sum_j \frac{Q_j}{4\pi\varepsilon_0 A_j} \iint \frac{dx\,dy}{\sqrt{(x_{0,i} - x_j)^2 + (y_{0,i} - y_j)^2 + h^2}}$$

(3.4)

Everything except the charge on the RHS of equation (3.4) is either a physical constant or a function of the geometry. We can therefore define

$$L_{i,j} = \frac{1}{4\pi\varepsilon_0 A_j} \iint \frac{dx\,dy}{\sqrt{(x_{0,i} - x_j)^2 + (y_{0,i} - y_j)^2 + h^2}}$$

(3.5)

and then

$$V_i = \sum_j L_{i,j} Q_j$$

(3.6)

We now have a set of i equations describing the voltage at the center of each subrectangle i in terms of the coefficients $L_{i,j}$ and the total charge on each sub-rectangle:

$$V_1 = L_{1,1}Q_1 + L_{1,2}Q_2 + \cdots + L_{1,n}Q_n$$
$$V_2 = L_{1,2}Q_1 + L_{2,2}Q + \cdots + L_{2,n}Q_n$$
$$\vdots$$
$$V_n = L_{1,n}Q_1 + L_{2,n}Q_3 + \cdots + L_{n,n}Q_n$$

(3.7)

Equivalently, in matrix notation,

$$\begin{bmatrix} V_1 \\ V_2 \\ \vdots \\ V_n \end{bmatrix} = \begin{bmatrix} L_{1,1} & L_{1,2} & \cdots & L_{1,n} \\ L_{2,1} & L_{2,2} & \cdots & L_{2,n} \\ & & \vdots & \\ L_{n,1} & L_{n,2} & \cdots & L_{n,n} \end{bmatrix} \begin{bmatrix} Q_1 \\ Q_2 \\ \vdots \\ Q_n \end{bmatrix}$$

(3.8)

This seems to be a complete set of n equations in n unknowns for the geometry described. The voltages and L coefficients are all known; we can solve the set of equations for all of the

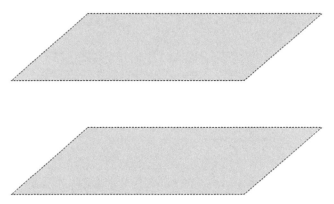

FIGURE 3.2   A parallel plate capacitor made up of two subelectrodes.

Q_i. For a two-electrode system, all the subrectangles on one electrode are at V_1 and all the rectangles on the other electrode are at V_2; the interelectrode capacitance is simply the (magnitude of the) the sum of the charges on either plate divided by the voltage difference between the plates.

Figure 3.2 shows a parallel plate capacitor consisting of two subelectrodes. The plates are rectangular and face each other. Since we haven't discussed calculating the L coefficient values yet, for now we'll simply make up some plausible values. From equation (3.5) we can see that all of the $L_{i,j}$ values are positive. Also, it is reasonable to assume that a diagonal term of the L matrix — the voltage contribution due to a uniform charge distribution of 1 on that same subrectangle — will be greater than the voltage contribution due to a uniform charge distribution of 1 on the other subrectangle (see Figure 3.2):

$$L_{1,1} > L_{1,2} > 0 \tag{3.9}$$

The symmetry of the capacitor in Figure 3.2 tells us that the diagonal terms as well as the off-diagonal terms are equal to each other. Therefore, in some set of units, a physically plausible set of values is

$$\begin{bmatrix} V_1 \\ V_2 \end{bmatrix} = \begin{bmatrix} 1.0 & 0.5 \\ 0.5 & 1.0 \end{bmatrix} \begin{bmatrix} Q_1 \\ Q_2 \end{bmatrix} \tag{3.10}$$

We wish to set the voltage difference between the two electrodes (which in this case are also the two subelectrodes) to be 1.0 V. Therefore we set $V_1 = -V_2 = 0.5$:

$$\begin{bmatrix} +0.5 \\ -0.5 \end{bmatrix} = \begin{bmatrix} 1.0 & 0.5 \\ 0.5 & 1.0 \end{bmatrix} \begin{bmatrix} Q_1 \\ Q_2 \end{bmatrix} \tag{3.11}$$

From the symmetries shown, we expect $Q_1 = -Q_2$, and indeed, solving equation (3.11) yields $Q_1 = +1.0$ and $Q_2 = -1.0$.

The values chosen for V_1 and V_2 were somewhat arbitrary as our only constraint was that the difference between them be 1.0. Repeating the solution of equation (3.10) using the choices $V_1 = 1.0$ and $V_2 = 0.0$, we obtain $Q_1 = +\frac{4}{3}$ and $Q_2 = -\frac{2}{3}$.

Clearly there is a problem here. While $Q_1 - Q_2 = 2.0$ as before, the second solution doesn't demonstrate charge neutrality in the system. We can still calculate capacitance, but we are certainly not looking at a charge-neutral system.

3.2 A WORKING EQUATION SET

The first solution *worked right* because we have a physically symmetric system to which we applied (anti)symmetric boundary conditions. Unfortunately we can't always simply set the voltage to 0.5 and −0.5 V (multielectrode systems), and what if the physical system isn't symmetric, as shown in Figure 3.3?

We would be well served to understand the root of this situation and to develop a general-purpose solution.

If we can get an infinite number of different solutions to this problem based on the choices V_2 and V_1 while keeping $V_2 - V_1 = 1$, then there must be another variable lurking somewhere. This missing variable is an unknown reference voltage V_r. We really only are specifying

$$\begin{bmatrix} V_1 - V_r \\ V_2 - V_r \end{bmatrix} = \begin{bmatrix} L_{1,1} & L_{1,2} \\ L_{2,1} & L_{2,2} \end{bmatrix} \begin{bmatrix} Q_1 \\ Q_2 \end{bmatrix} \tag{3.12}$$

Now we seem to have one more variable than we have equations. Fortunately, we do have available another important equation, the equation that insists on total charge neutrality, that is, the sum of all $Q_i = 0$. Rearranging equation (3.12) to show V_r as one of the unknown variables and including the charge neutrality condition as an additional equation, we get

$$\begin{bmatrix} V_1 \\ V_2 \\ 0 \end{bmatrix} = \begin{bmatrix} L_{1,1} & L_{1,2} & 1 \\ L_{2,1} & L_{2,2} & 1 \\ 1 & 1 & 0 \end{bmatrix} \begin{bmatrix} Q_1 \\ Q_2 \\ V_r \end{bmatrix} \tag{3.13}$$

Solving these equations formally yields

$$Q_1 = \frac{V_1 - V_2}{2(L_{1,2} - L_{1,1})} \tag{3.14}$$

$$Q_2 = \frac{V_2 - V_1}{2(L_{1,2} - L_{1,1})} \tag{3.15}$$

$$V_r = \frac{-(V_1 + V_2)}{2} \tag{3.16}$$

As can be seen by these three equations, $Q_1 = -Q_2$ always; but $V_r = 0$ only when $V_1 = -V_2$.

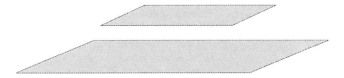

FIGURE 3.3   An asymmetric parallel plate capacitor.

The reference voltage V_r is the voltage with respect to a reference point at infinity; that is, it represents the work required to bring all the charge to the electrodes from infinity and distribute it properly. An asymmetric voltage distribution and/or an asymmetric structure can cause this voltage to differ from 0. When this happens, the net charge on the structure will not be zero unless we properly account for V_r.

For a problem with more than two subrectangles (but with all the subrectangles part of two electrodes), some of the rectangles will have voltage V_1 applied and the rest will have V_2 applied:

$$
\begin{bmatrix} V_1 \\ V_1 \\ V_2 \\ V_2 \\ \vdots \\ 0 \end{bmatrix} = \begin{bmatrix} L_{1,1} & L_{1,2} & L_{1,3} & L_{1,4} & \cdots & 1 \\ L_{2,1} & L_{2,2} & L_{2,3} & L_{2,4} & \cdots & 1 \\ L_{3,1} & L_{3,2} & L_{3,3} & L_{3,4} & \cdots & 1 \\ L_{4,1} & L_{4,2} & L_{4,3} & L_{4,4} & \cdots & 1 \\ \vdots & \vdots & \vdots & \vdots & \vdots & \cdots & 1 \\ 1 & 1 & 1 & 1 & 1 & 0 \end{bmatrix} \begin{bmatrix} Q_1 \\ Q_2 \\ Q_3 \\ Q_4 \\ \vdots \\ V_r \end{bmatrix} \tag{3.17}
$$

In the four-subrectangle example shown in equation (3.17), we obtain for example

$$
Q_1 + Q_2 = -[Q_3 + Q_4] \tag{3.18}
$$

$$
|Q_1 + Q_2| = |Q_3 + Q_4| = Q_{\text{total}} \tag{3.19}
$$

and then the interelectrode capacitance is

$$
C = \frac{Q_{\text{total}}}{|V_2 - V_1|} \tag{3.20}
$$

Our next task is to evaluate the $L_{i,j}$ coefficients.

3.3   THE SINGLE-POINT APPROXIMATION FOR OFF-DIAGONAL TERMS

As part of solving a real problem, we need to evaluate all of the L coefficients, as defined by equation (3.5).

This expression was written assuming that rectangle j is in the X–Y plane at $Z = 0$. It is still a general expression, however, because the L coefficient would be the same if we rotated and translated the coordinate system of any geometry so that rectangle j *is* in the X–Y plane at $Z = 0$.

If rectangles i and j are far apart (this approximation will require some further discussion), then a very simple approximation for $L_{i,j}$ can be derived. Assume that all of the charge on rectangle j has been concentrated at the center of rectangle j. The integration disappears, leaving us with the simple expression

$$
L_{i,j} \approx \frac{1}{4\pi\varepsilon_0} \frac{1}{\sqrt{\left(x_{0,i} - x_{0,j}\right)^2 + \left(y_{0,i} - y_{0,j}\right)^2 + h^2}} \tag{3.21}
$$

In other words, $L_{i,j}$ is approximately the voltage at the center of rectangle i due to a unit point charge at the center of rectangle j.

3.4 EXACT SOLUTIONS FOR THE DIAGONAL TERM AND IN-PLANE TERMS

For the self (diagonal) terms, equation (3.21) is impossible to evaluate, with $x_{0,i} = x_{0,j}$, and so on. Fortunately, in the case of rectangular regions, as well as many other region shapes, L may be calculated exactly.

For the self-(diagonal) terms, $x_{0,i} = y_{0,i} = h = 0$, leaving us with

$$L_A \equiv \frac{1}{4\pi\varepsilon_0 A_j} \int\int \frac{dx\, dy}{\sqrt{x_j^2 + y_j^2}} \tag{3.22}$$

The simplest case to evaluate (although not very useful for real problems) is the case of a circular electrode. Let area j be a circle of radius a. Equation (3.22), with the integral transposed to polar coordinates, is

$$L_A \equiv \frac{1}{4\pi\varepsilon_0(\pi a^2)} \int_0^{2\pi}\int_0^a r\, dr\, \frac{d\theta}{r} = \frac{1}{2\pi\varepsilon_0 a} \tag{3.23}$$

As this equation demonstrates, the self-L integrals are very well behaved. While this behavior doesn't guarantee that a closed-form solution can always be found, it does suggest that if a numerical procedure encounters difficulty at (0,0), the problem is with the numerical procedure, not the integral itself.

Evaluating L_a for a rectangle as shown in Figure 3.4, we obtain

$$L_A = \frac{1}{4\pi\varepsilon_0} \frac{1}{4ab} \int_{-b}^{b}\int_{-a}^{a} \frac{dx\, dy}{\sqrt{x^2 + y^2}} \tag{3.24}$$

Once again converting to polar coordinates, and noting the four-quadrant symmetry, we get

$$L_A(a,b) = \frac{1}{4\pi\varepsilon_0 ab} \int_0^{(\pi/2)}\int_0^{r(\theta)} dr\, d\theta \tag{3.25}$$

This integral is broken into two parts

$$L_A(a,b) = \frac{1}{4\pi\varepsilon_0 ab} \left[\int_0^{\arctan(b/a)}\int_0^{a/\cos((\theta))} dr\, d\theta + \int_{\arctan(b/a)}^{\pi/2}\int_0^{b/\sin(\theta)} dr\, d\theta \right] \tag{3.26}$$

and finally, after some algebra,

$$L_A(a,b) = \frac{1}{4\pi\varepsilon_0} \left(\frac{1}{b}\ln\frac{b + \sqrt{a^2 + b^2}}{a} - \frac{1}{a}\ln\frac{-a + \sqrt{a^2 + b^2}}{b} \right) \tag{3.27}$$

Equation (3.27) can be used as the basis for exact solutions to many L values for cross terms between rectangles in the same plane.

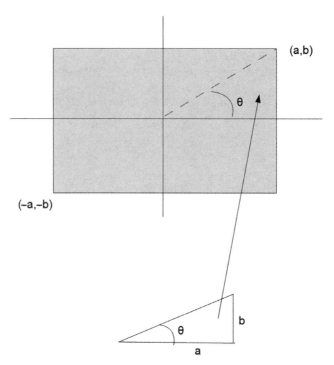

FIGURE 3.4   Rectangle of dimensions ($2a,2b$) showing triangle used in developing equation (3.26).

Consider first Figure 3.5a. Define L_b as the voltage at (0,0) due to a uniform charge distribution (having a total charge of 1) on the rectangle shown. Remember that setting the voltage point at (0,0) does not restrict the generality of this expression; the coordinate system origin can be translated and/or rotated as necessary — it is only the relative positions of the voltage point and the charged rectangle that matter:

$$L_{i,j} = \frac{1}{4\pi\varepsilon_0} \frac{1}{2b(d-a)} \int\limits_{-b}^{b} \int\limits_{a}^{d} \frac{dx\,dy}{\sqrt{x^2+y^2}} \tag{3.28}$$

If a second charged rectangle, having the same charge as the first rectangle, is added as shown in Figure 3.5b, the voltage would be doubled. Therefore, we may write

$$L_{i,j} = \frac{1}{4\pi\varepsilon_0} \frac{1}{4b(d-a)} \left[\int\limits_{-b}^{b} \int\limits_{-d}^{a} \frac{dx\,dy}{\sqrt{x^2+y^2}} + \int\limits_{-b}^{b} \int\limits_{a}^{d} \frac{dx\,dy}{\sqrt{x^2+y^2}} \right] \tag{3.29}$$

Multiplying both sides of equation (3.29) by $4b(d-a)$ and referring to equation (3.24) (Figure 3.5c), we obtain

$$4b(d-a)L_{i,j} = \frac{1}{4\pi\varepsilon_0} \left[4dbL_a(d,b) - 4abL_a(a,b) \right] \tag{3.30}$$

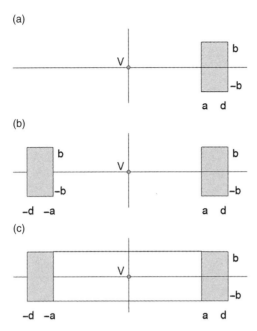

FIGURE 3.5 Steps in evaluating in-plane coefficient for a rectangle along the X axis.

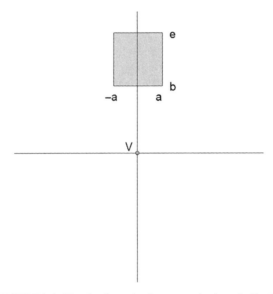

FIGURE 3.6 Notation for an in-plane rectangle along the Y axis.

or

$$L_b(a,b,d) = \left[\frac{dL_a(d,b) - aL_a(a,b)}{d-a} \right] \tag{3.31}$$

Figure 3.6 shows the same situation except with the rectangle centered on the Y axis. In this case

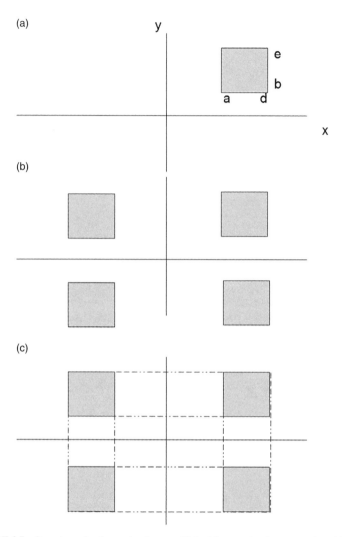

FIGURE 3.7   Steps in evaluating an in-plane coefficient for a rectangle not crossing either axis.

$$L_c = \frac{eL_a(e,a) - bL_a(b,a)}{e - b} \qquad (3.32)$$

Figure 3.7a shows the derivation of a more general solution.

This solution is exact for in-plane L coefficients where any rectangle j, which, when rotated to be parallel to the X–Y axes and translated so that the field (measurement) point is at $(0,0,0)$, does not cross either of these axes.

First we note that the desired L term will be just $\frac{1}{4}$th of the L term for four identical rectangles distributed symmetrically about the axis in Figure 3.7b. Then, by combining combinations of large rectangles as shown in Figure 3.7c, following the same process as described above, we get

$$L_c(a,b,d,e) = \frac{deL_a(d,e) + abL_a(a,b) - aeL_a(a,e) - dbL_a(d,b)}{2(e-b)(d-a)} \qquad (3.33)$$

This approach may be extended to many other relative rectangle locations; the only limit is the ingenuity of the extender. In practice, these exact solutions are very useful. Unfortunately, equation (3.5) for rectangles in different planes ($h \neq 0$) is an elliptic integral that cannot be evaluated exactly.

3.5 APPROXIMATING $L_{i,j}$

For off-diagonal terms ($i \neq j$), $L_{i,j}$ cannot be calculated exactly. The general expression [equation (3.5)] cannot be integrated analytically, and a numerical approximation is needed. MATLAB function `integral2` does the job very well. However, an electrode divided into a 50×50 subelectrode pattern contains 2,500 subelectrodes. A simple parallel plate capacitor consisting of two such electrodes therefore has 5000 subelectrodes. The (fully populated) $L_{i,j}$ array has 25,000,000 values to calculate. This is a significant number of numerical integrations (although some inspection of symmetries could reduce this number considerably).

Figure 3.8 shows $1/L_{i,j}$ versus z_i for the parameters given in the figure. $1/L_{i,j}$ is shown rather than $L_{i,j}$ because the region near $z_i = 0$ is of primary interest, and the inverse function provides clearer plots in this situation. The function was calculated using `integral2`. As may be seen, $L_{i,j}$ is identically $L_{i,i}$ at $z_i = 0$.

Also shown in Figure 3.8 is the single-point approximation [equation (3.21)]. As expected, this is an excellent approximation for large values of z_i but it is useless near $z_i = 0$, going to infinity at $z_i = 0$.

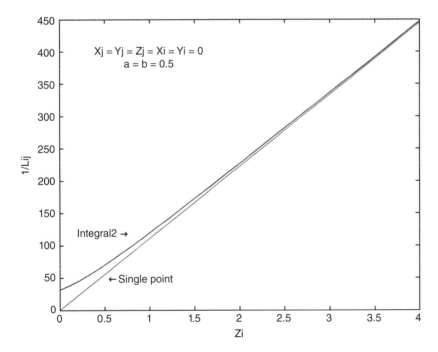

FIGURE 3.8 $1/L_{i,j}$ and the single-point approximation versus z_i.

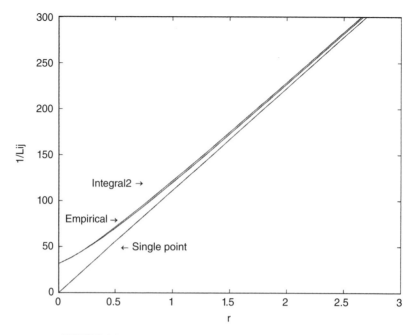

FIGURE 3.9 Figure 3.8 with the empirical function superimposed.

Figure 3.9 is almost a repeat of Figure 3.8, with the empirical function

$$L_{i,j} \approx \frac{1}{4\pi\varepsilon_0} \left[\frac{1}{r + \{L_a[1 + (r/a)]\}^{-1}} \right] \tag{3.34}$$

(for $a = b$) superimposed, and

$$r = \sqrt{(x_i - x_j)^2 + (y_i - y_j)^2 + (z_i - z_j)^2} \tag{3.35}$$

It should be pointed out that the formulation of the original problem using $L_{i,j}$ is a carefully stated approximation of the original problem; using numerical integration to evaluate the coefficients is a rigorous approach to getting the right numbers. Using the single-point approximation for points far enough away is a limiting approximation that can be derived. Equation (3.34), however, has no physical basis other than its resemblance to the actual function. It is, however, exact at $r = 0$ and approaches the single-point function for large r.

Equation (3.34) gives us a simple and rapidly calculated function that will make it easy to write programs that handle large numbers of coefficients. The results will be good although not as accurate as those of a program that actually performed the integration for each coefficient.

L_empirical_tester.m is a MATLAB program that calculates the values for Table 3.1 and for Figures 3.8 and 3.9:

TABLE 3.1 Accuracy of Empirical Formula versus Accuracy of MATLAB
`Integer2` **Calculation**

x_i	y_i	z_i	$a = 0.5$	Errors (%) $a = 0.05$	$a = 0.005$
0	0	10	−0.05	0	0
0	0	1	−1.60	−0.05	0
0	0	0.1	0.66	−1.60	−0.05
0	0	0.010	0.23	0.66	−1.60
3	0	0.1	−1.80	−0.02	0
2	0	0.1	−3.70	−0.04	0
1	0	0.1	−11.80	−0.17	0
0.1	0	0.1	−5.70	−3.80	−0.06
1	0	0	−12.00	−0.18	0

```
% test script for the various Lij calls and approximations

Lself = @ (a,b) (log((b + sqrt(a^2 + b^2))/a)/b ....
        - log((-a + sqrt(a^2 + b^2))/b)/a);

eps0 = 8.854;
a = .5; b = a;
xj = 0; yj = 0; zj = 0;
xi = 0; yi = 0.; zi = .5;

Lmat = integral2(@ (x,y) 1./sqrt((xi - x).^2 + (yi - y).^2 ...
        + zi.^2), xj-a, xj + a, yj-b, yj + b);
Lmat = Lmat/(4*pi*eps0)/(4*a*b);
fprintf ('matlab routine integral2 %f \n', Lmat)

La = Lself(a,b)/(4*pi*eps0);
fprintf ('La (self) integral %f \n', La)

r = sqrt((xi - xj)^2 + (yi - yj)^2 + (zi - zj)^2);
Lempir = r + 1/Lself(a,b)/(1 + r/a);
Lempir = 1/(Lempir*4*pi*eps0);
err = (Lempir - Lmat)/Lmat;
fprintf ('empirical f & err: %f %f\n', Lempir, err)

% plotting

xax = []; exact = []
for zi = 0:.1:4
  xax = [xax,zi];
  Lmat = integral2(@ (x,y) 1./sqrt((xi - x).^2 + (yi - y).^2 ...
        + zi.^2), xj-a, xj + a, yj-b, yj + b);
  Lmat = Lmat/(4*pi*eps0)/(4*a*b);
  f = 1/Lmat;
  exact = [exact,f];
end

pt = 4*pi*eps0*xax;
empir = (xax + 1/Lself(a,b)./(1 + xax/a))*4*pi*eps0;
```

```
plot (xax, exact, xax, pt, xax, empir)

axis ([0,3,0,300])
xlabel ('r')
ylabel ('1/Lij')
text(.4,120,'integral2 \rightarrow')
text(.55,50,'\leftarrow single point')
text (.5,400,'Xj = Yj = Zj = Xi = Yi = 0')
text (.7,375,'a = b = 0.5')
text (.07,80,'empirical \rightarrow')
```

We now have all of the tools necessary to construct and solve real problems.

PROBLEMS

3.1 Find L_a for a square electrode 1 cm on each side. Find L_b for an identical electrode placed next to the first electrode, using both the exact formula and the numerical approximations that were examined.

3.2 For a single square electrode 1 cm on each side "alone in the universe," find the capacitance. In this situation, do not add the extra row and column as described in equation (3.34). Can you interpret what this capacitance means?

3.3 Repeat Problem 3.2 for the three-subelectrode-structure shown in Figure P3.3. Take advantage of the symmetry so only two equations in two unknowns must be solved. Calculate the capacitance. Look at the charge distribution (a three-point plot). Can you interpret these?

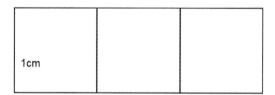

FIGURE P3.3 Diagram for Problem 3.3.

3.4 Calculate the capacitance of the simple parallel plate capacitor (square electrodes, 1 cm on each side, 1 cm separation) shown in Figure P3.4. By proper use of symmetries and boundary conditions (electrode voltages), this can be solved with only one equation in one unknown.

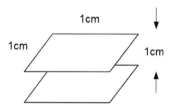

FIGURE P3.4 Diagram for Problem 3.4.

3.5 Compare the MATLAB integration, single-point, and empirical calculations for several trajectories in the plane and out of the plane about a rectangle of size $a = 3$ cm, $b = 1$ cm.

3.6 Follow up on the above mentioned point — if the empirical calculation isn't an improvement over the single-point calculation, why is it of any use?

REFERENCES

1. M. N. O. Sadiku, *Numerical Techniques in Electromagnetics* (*with MATLAB*), 3rd ed., CRC Press, New York, 2011. Note that the current edition is the 3rd edition (with MATLAB), 2011; the second edition (published in 2002, without MATLAB) is available online at http://www.scribd.com/doc/83690179/Numerical-Techniques-in-Electromagnetics.

2. R. F. Harrington, *Field Computation by Moment Methods*, Electromagnetic Waves (series), IEEE Press, 1992 (original text, 1998).

4

Examples Using the Method of Moments

4.1 A FIRST MODELING PROGRAM

We are now in a position to actually write a complete modeling program. We will use it to learn some more electrostatics, and then we will look at its limitations and inaccuracies and plan how we might improve it.

The basic components of this first program are as follows:

1 Data input — we have to be able to "tell" the program what structure and what boundary conditions we want to examine.

2 Data input processing — the program must be able to take the input data information and convert it into the subrectangles (cells) as described by the data input.

3 Calculation of all the $L_{i,j}$ coefficients.

4 Adding the extra row and column to the $L_{i,j}$ matrix (and the extra column to the V vector) as described in Chapter 3 to ensure that charge neutrality is maintained.

5 Solving the linear equation set.

6 Postprocessing and presentation of results — do we want capacitance, voltage, or electric field at some specific points?

7 Warnings of possible inaccuracies from the program.

8 Automatic improvements of the accuracy of the calculations.

Components 1 and 2 are in general a very sophisticated technology. We want our modeling software to be able to interact with our computer-aided design (CAD) software and extract necessary modeling information from the drawings as accurately and automatically as

Introduction to Numerical Electrostatics Using MATLAB, First Edition. Lawrence N. Dworsky.
© 2014 John Wiley & Sons, Inc. Published 2014 by John Wiley & Sons, Inc.

possible. This is a very important component of electrostatic (or general electromagnetic, mechanical, etc.) modeling. On the other hand, it is not a fundamental component of electrostatics or numerical electrostatic approximations. Our first modeling program will handle only very limited geometry descriptions, so we'll begin with the brute-force path to data input—we'll describe our geometry in a handwritten data file.

4.2   INPUT DATA FILE PREPARATION FOR THE FIRST MODELING PROGRAM

We shall create a data file in which each line presents the minimum amount of information necessary to describe an electrode. The *minimum amount* criterion is intended to make life as simple as possible for the program user while letting the computer do the (necessary) repetitive work.

This program's limitation is that our basic — and only — structure is a rectangle in the *X–Y* plane as shown in Figure 4.1a. It must be made up of square subsections. The side of

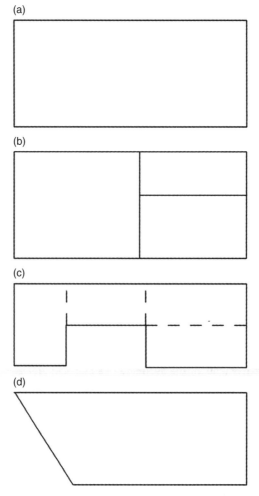

FIGURE 4.1   Basic rectangular modeling elements.

each square, $2a$, is the same for every rectangle in the dataset. The rectangle can be described as a single rectangle, or the same result can be obtained by describing several rectangles which combine to the same original rectangle, as in Figure 4.2b. This might seem like a waste of time and effort, but it is sometimes useful when a parameter study described by varying a dimension of one of the component rectangles is planned. Similarly, the electrode shown in Figure 4.1c is possible because it consists of rectangles.

On the other hand, the structure of Figure 4.1d is not allowed; it cannot be described by a combination of rectangles. (*Note*: This is not a fundamental limitation of the method — only an imposed limitation in this chapter.)

It would not be difficult to add the ability to include rectangles in the X–Z and Y–Z planes. The purpose of this first program, however, is to introduce all the aspects of building a modeling program and to learn some electrostatics from the results. Discussion of this capability will therefore be postponed to later chapters.

The default orientation of the input rectangles will be with their axes parallel to the X and Y axes. However, there is much to learn from structures with one or more electrodes rotated (in the X–Y plane), as shown in Figure 4.2.

The most common rotation will be about the rectangle's own center (Figure 4.2a), although the more general case of a rotation about an arbitrary point will be considered later in this chapter.

Rotation capability will be anticipated by including the required information in each data line, but our first program will ignore this information.

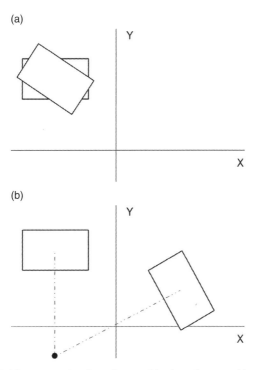

FIGURE 4.2   Rectangular electrode rotated in plane about an arbitrary point.

The first line (row) in the file shown in Table 4.1 contains the value of a. The program's internal constants will use meters for the measure of length, so a should be specified in meters.

Each remaining line in the program (an arbitrary number of lines) contains the description of a rectangular electrode. This description assumes that the electrode's sides are parallel to the axes, even though rotation will be considered later on.

For each line, the entries, in the order shown, are as follows:

X_0, Y_0 = center of electrode

N_x, N_y = number of cells in x and y that make up this rectangle

  h = height (z value) of rectangle in X–Y plane

  V = voltage applied to electrode

  θ = rotation information, not used yet

The total structure, as shown in Figure 4.3, is of two identical square plates, each subdivided into 16 square plates. The plates are one unit apart (the individual Z values don't matter here;

TABLE 4.1   Complete Data File for a Parallel Plate Capacitor

0.125						
0.0	0.0	4	4	0	0.5	0
0.0	0.0	4	4	1	−0.5	0

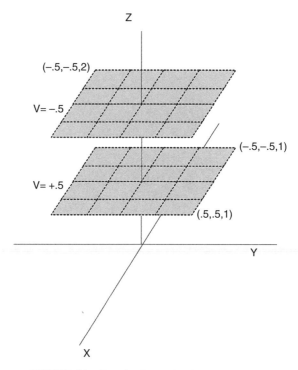

FIGURE 4.3   Capacitor layout described by Table 4.1.

only the difference between them) with a voltage difference of one volt between them (the individual *V* values don't matter here; only the difference between them.

The data file may be prepared by any ASCII text editor and saved to any convenient file name. A typical file extension for text files such as `.txt` is recommended but not required.

4.3 PROCESSING THE INPUT DATA

The MATLAB function `get_data.m` reads the input file and produces two arrays, an array that contains the information needed to generate the $L_{i,j}$ array and an array containing the *V* data. This modeling program will use only the corrected single-point approximation (the empirical formula) for the off-diagonal $L_{i,j}$ terms. The information needed is therefore the location of the center of each rectangle and its size. In keeping with prior definitions, a subrectangle's dimensions are $2a \times 2b$; the *a* and *b* values are calculated and stored.

The following `get_data.m` program returns the arrays `data_prep` and volts:

```
function [a, data_prep, volts] = get_data()
% get_data gets the user created data file and converts it to
% a file of cells
% getdata2 assumes a uniorm nx and ny so that there is only 1 a and 1 b

filename = uigetfile('*.txt');  % read the file
fid = fopen(filename);
all_data = (fscanf(fid, '%g'))';
nr_lines = (length(all_data) - 1)/7;
closeresult = fclose(fid)

a = all_data(1);
data_in = [];                     % parse the data
for i = 1:nr_lines
  j = 7*(i-1) + 2;
  data_in = [data_in; all_data(j:j + 6)];
end

volts = [];
data_prep = [];
for i = 1:nr_lines                % loop through data file and build output
  x0 = data_in(i,1); y0 = data_in(i,2); nx = data_in(i,3); ny =
data_in(i,4);
  x_ll = x0 - a*nx;   x_ur = x0 + a*nx;
  y_ll = y0 - a*ny;   y_ur = y0 + a*ny;

% x_ll = data_in(i,1);   x_ur = data_in(i,2);
% y_ll = data_in(i,3);   y_ur = data_in(i,4);
h = data_in(i,5); V = data_in(i,6); theta = data_in(i,7)*pi/180;
  for j = 1:nx
    xj = x_ll - a + j*2*a;
    for k = 1:ny
      yj = y_ll - a + k*2*a;
      if theta == 0
```

```
        line = [xj, yj, h];
    else
        % These 3 lines are for rotated electrode calculations
        x_new = xj*cos(theta) - yj*sin(theta);
        y_new = xj*sin(theta) + yj*cos(theta);
        line = [x_new, y_new, h];
    end
    data_prep = [data_prep; line];
    volts = [volts; V];
    end
  end
end
end
```

Table 4.2 shows these arrays for the dataset of Table 4.1.

The order of the cells generated is arbitrary and inconsequential. The volts array and the data_prep array, however, must both have the cells in the same order. Each line in data_prep presents a cell; its center (x_j, y_j) and its height above $z = 0$ (z_j). The volts array presents the applied voltage at each subrectangle. The two arrays could have easily been

TABLE 4.2   Arrays for Dataset in Table 4.1

X_0	Y_0	h	V
−0.375	−0.375	2	0.5
−0.375	−0.125	2	0.5
−0.375	0.125	2	0.5
−0.375	0.375	2	0.5
−0.125	−0.375	2	0.5
−0.125	−0.125	2	0.5
−0.125	0.125	2	0.5
−0.125	0.375	2	0.5
0.125	−0.375	2	0.5
0.125	−0.125	2	−0.5
0.125	0.125	2	0.5
0.125	0.375	2	0.5
0.375	−0.375	2	0.5
0.375	−0.125	2	0.5
0.375	0.125	2	0.5
0.375	0.375	2	0.5
−0.375	−0.375	1	−0.5
−0.375	−0.125	1	−0.5
−0.375	0.125	1	−0.5
−0.375	0.375	1	−0.5
−0.125	−0.375	1	−0.5
−0.125	−0.125	1	−0.5
−0.125	0.125	1	−0.5
−0.125	0.375	1	−0.5
0.125	−0.375	1	−0.5
0.125	−0.125	1	−0.5
0.125	0.125	1	−0.5
0.125	0.000	1	−0.5
0.375	−0.375	1	−0.5
0.375	−0.125	1	−0.5
0.375	0.125	1	−0.5
0.375	0.375	1	−0.5

combined, but ultimately the matrix solution setup will require them to be separate arrays, so they were constructed separately from the start.

4.4 GENERATING THE $L_{i,j}$ ARRAY

The following is the L_fit.m program:

```
function L = L_fit(a, data_prep)
%function L = L_pt_corr(a, b, xi, yi, zi)
% This is the empirical function fit

% xj, yz, zj is the center of the source rectangle in the xyplane
% xi, yi, zi is the field point

global eps0

xi = data_prep(:,1);
yi = data_prep(:,2);
zi = data_prep(:,3);

n = length(xi);      % all the lengths are the same
xs = (repmat(xi,1,n) - repmat(xi',n,1)).^2;
ys = (repmat(yi,1,n) - repmat(yi',n,1)).^2;
zs = (repmat(zi,1,n) - repmat(zi',n,1)).^2;

p = 1.414*a;
La = (log((a+p)/a)/a - log((-a+p)/a)/a);

r = sqrt(xs+ys+zs);

L = r + 1/La./(1+r/a);
L = 1./L/(4*pi*eps0);

end
```

The MATLAB function L_fit.m, with the data_prep array for its information, produces the $L_{i,j}$ coefficients using the empirical derived in Chapter 3. The function is vectorized so that no explicit loops are seen. Since the empirical formula returns L_a when r the field point is in the center of the source rectangle, the entire matrix of points can be calculated using this formula.

The main script for this program, mom1.m, assumes the task of adding the extra row and column to the $L_{i,j}$ array and the extra column to the volts array as specified in Chapter 3.

4.5 SOLVING THE SYSTEM AND EXAMINING SOME RESULTS

The mom_1.m program is as follows:

```
% This is the first MOM program with refining added

global eps0
```

```
eps0 = 8.854;
% Read the input file and process it into cell data
% Data Format:

[a, data_prep, volts] = get_data;

% Generate the L data
  L = L_fit(a, data_prep);

  nr_vars = length(L);
  new_col = L(1,1)*ones(nr_vars,1);
  L = [L, new_col];
  new_row = [new_col',0];
  L = [L; new_row];

  volts = [volts;0];
  q = L\volts;

  cap = 0;
  for i = 1:length(q) - 1
    if q(i) > 0
      cap = cap + q(i);
    end
  end

  q
  cap
```

When mom_1.m solves the set of linear equations, the result is the vector q of the charges on each subelectrode. Remember that the last entry in q is actually a scaled voltage reference number—it is not the charge on any subelectrode. Ignoring this last entry, there are 32 entries, 16 for the first electrode and 16 for the second electrode.

Figure 4.4 shows the charge on the top electrode. The charge on the bottom electrode is a mirror image with the sign reversed.

Since the voltage difference between the two electrodes is 1 V, the capacitance is the sum of all the charges shown in Figure 4.4, $C = 28.5$ pF. The parallel plate capacitance for this structure is

$$C_{pp} = \frac{\varepsilon_0 A}{d} = \frac{8.854(1)(1)}{(1)} = 8.854\,\text{pF} \tag{4.1}$$

Is a capacitance of >3 times the parallel plate capacitance reasonable? In other words, does this simulation make sense? Also, we see that the charge distribution is lowest in the center of the electrodes, higher at the edges, and highest at the corners. Is this physically possible and reasonable?

Before answering these questions, let's get some more data. By reducing a but increasing n_x and n_y, we can increase the number of subelectrodes uniformly.

Figure 4.5 shows the result of this exercise. As n increases, the capacitance increases, leveling out at approximately 29.74 pF.

3.55	2.57	2.57	3.55
2.57	1.54	1.54	2.57
2.57	1.54	1.54	2.57
3.55	2.57	2.57	3.55

FIGURE 4.4   Charge distribution on a simple parallel plate capacitor.

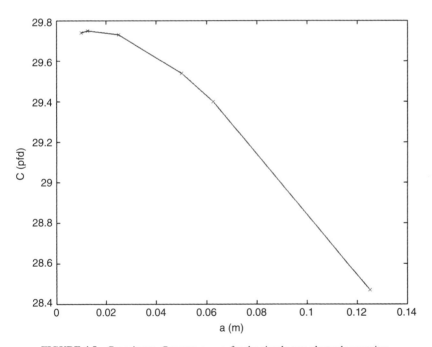

FIGURE 4.5   Capacitance C versus $n_x = n_y$ for the simple two-electrode capacitor.

Figure 4.6 is a contour plot of the electrode charge for a fairly high resolution ($n = 50$) calculation.

Figure 4.6 shows in much greater detail the story that Figure 4.4 was trying to tell. The highly mobile charge on an electrode pushes charge out to the edges because there is no counterbalancing (repulsive) charge at the edge to push back. This effect is much more

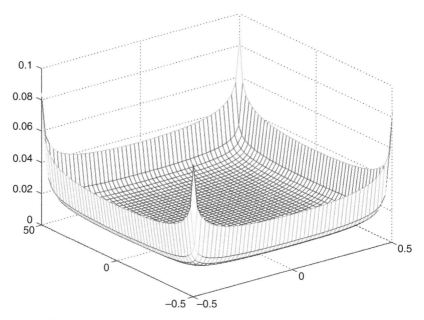

FIGURE 4.6   High-resolution plot of charge distribution on a parallel plate capacitor.

significant in the corners because there is a lack of counterbalancing charge in two dimensions rather than in only one dimension, as there is away from the corners along an edge. The corner of an electrode in this capacitor forms a right angle.

4.6   LIMITS OF RESOLUTION

Figure 4.6 makes a clear case for higher-resolution calculations, the result is more accurate.

Making all of the cells as small as possible is not the only way to achieve good accuracy. There is no inherent reason why all the cells must be of the same size. We could start with all equal-size cells, examine our results, and then subdivide cells where the charge is the highest.

Figure 4.7 shows an example of the results of such a calculation. Using a threshold of some multiple of the minimum charge on the electrode, cells with a sufficiently high charge are simply subdivided into four smaller squares. The existing program must first be augmented so that each line in the data_prep array carries its own value of a, and then the chosen lines in data_prep each are replaced by four lines describing the new squares.

Figure 4.7 graphically shows what the new dataset would look like for this example. In this figure each dot represents the center of a (square) cell.

Figure 4.8 shows the program results for the $a = 0.01$ dataset as h is reduced from 1.

Figure 4.8 shows the ratio of the calculated capacitance to the ideal parallel plate capacitor capacitance. The figure shows that this ratio approaches 1 as the distance between the two electrode plates decreases. What is happening here in terms of the electric field deserves some discussion.

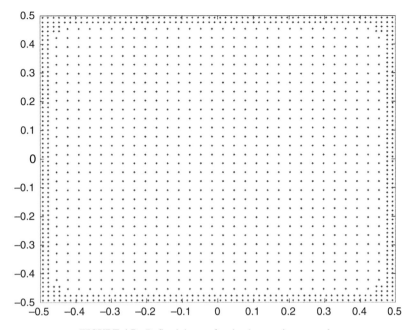

FIGURE 4.7 Refined dataset for simple capacitor example.

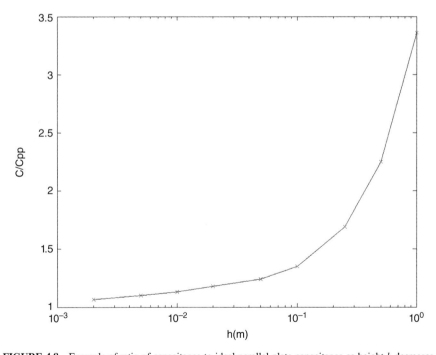

FIGURE 4.8 Example of ratio of capacitance to ideal parallel plate capacitance as height h decreases.

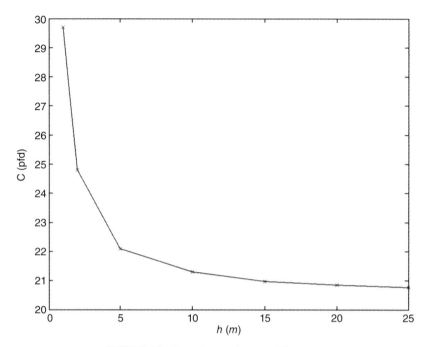

FIGURE 4.9 Example capacitance as h increases.

First, however, consider the situation when the plates are pulled farther and farther apart (see Figure 4.9).

As is expected, C decreases as h increases. However, C does not fall toward zero but levels off at ~20.5 pF. At some point each electrode loses its effect on the charge on the other electrode. This is the *self-capacitance* discussed in the problem set for Chapter 3.

4.7 VOLTAGES AND FIELDS

Once the charge distribution in a structure is known, we can approximate the voltage anywhere in space by assuming that the charge in each cell is concentrated at the center of the cell. Then, the voltage V_i at point (x_i, y_i, z_i) is given by

$$V_i = \sum_j \frac{q_j}{4\pi\varepsilon_0 r_{i,j}} \tag{4.2}$$

where

$$r_{i,j} = \sqrt{(x_i - x_j)^2 + (y_i - y_j)^2 + (z_i - z_j)^2} \tag{4.3}$$

Figure 4.10 shows the equipotential surfaces at the center of the electrodes for the capacitor (Figure 4.3) with square electrodes of side 1, electrode separation = 1. These data were

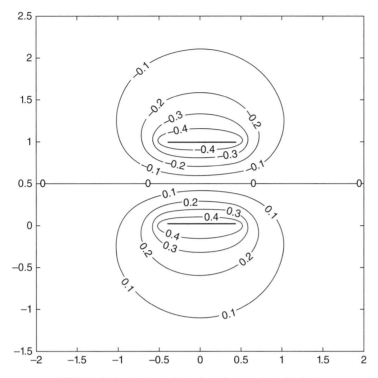

FIGURE 4.10 Equipotential surfaces for capacitor with $h = 1$.

taken with a program run at $n_x = n_y = 50$. As the symmetry requires, the plane $z = 0.5$ is an equipotential at $V = 0$. This will prove important in subsequent discussions.

Equation (4.2) is reasonably accurate provided the field point (i) is not too close to one of the charges. If this occurs, the voltage will approach infinity as point i approaches point j. This problem can be avoided by replacing the problematic term in equation (4.2) with a numerical integral (assume that the charge is distributed uniformly over its cell) or alleviated by using an empirical formula such as the one used in Chapter 3 for calculating $L_{i,j}$.

The magnitude of the electric field at point i, with the same assumptions and caveats as described above, is

$$| \vec{E} | = \sum_j \frac{q_j}{4\pi\varepsilon_0 r_{i,j}^2} \tag{4.4}$$

The X component of E is

$$E_x = | \vec{E} | \frac{x_i - x_j}{r_{ij}} \tag{4.5}$$

and therefore

$$E_x = \sum_j \frac{q_j (x_i - x_j)}{4\pi\varepsilon_0 r_{i,j}^3} \tag{4.6}$$

with corresponding expressions for E_y and E_z.

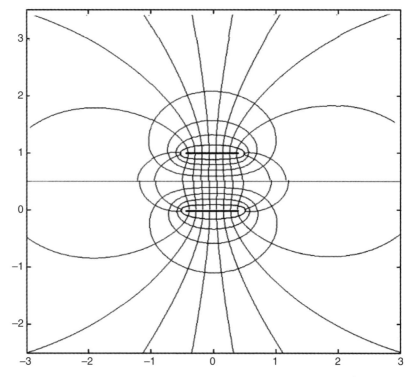

FIGURE 4.11 Electric field magnitude lines superimposed on equipotentials of Figure 4.10.

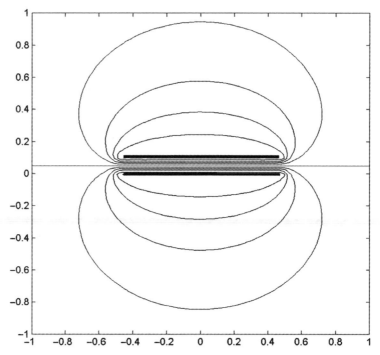

FIGURE 4.12 Equipotentials for a tightly spaced parallel plate capacitor.

Figure 4.11 is a repeat of Figure 4.10 with electric field lines superimposed on the equipotential surfaces. Since this is the $X–Z$ plane at the center (in y) of square electrodes, $E_y = 0$, so these field lines are showing the full picture.

Note that there is significant field outside the region between the electrodes. This is why the capacitance is greater than the ideal parallel plate capacitance. The majority of the electric field energy not stored between the plates is due to field lines terminating at or near the edges of the electrodes, referred to as *fringe*, or *fringing fields*.

Figure 4.12 shows the equipotential surfaces for the same capacitor but with h (the plate separation height) decreased from 1.0 to 0.1. The electric field lines are not shown in this figure, but this omission is really inconsequential because the tightly bunched equipotentials between the electrodes and the widely spaced equipotentials elsewhere tell us that the electric field energy is concentrated principally between the electrodes and we should expect a capacitance close to the parallel plate capacitance.

Figure 4.5 showed that as the capacitor plates are separated, at $h = 5$ the capacitance has dropped to ≈ 22 pF and on further plate separation, the capacitance asymptotically approaches ≈ 20.5 pF. At $h = 5$ we would therefore expect the equipotential surfaces to show us electrodes that still see the effects on each other, but only slightly. Figure 4.13 illustrates this point exactly. For a single electrode we would expect approximately elliptically shaped curves near the electrode, with the curves becoming circular farther away from the electrode.

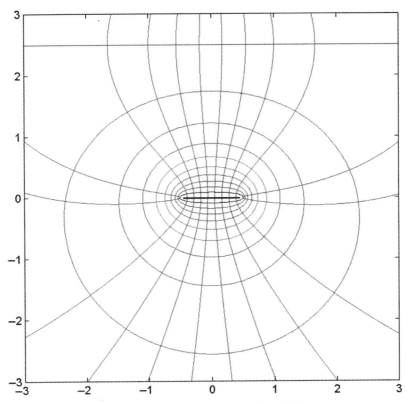

FIGURE 4.13 Equipotentials for widely spaced parallel plate capacitor.

4.8 VARYING THE GEOMETRY

Various characteristics of the simple capacitor structure can be examined quite easily simply by changing numbers in the data file. For example, Figure 4.14 shows a cross section of the capacitor with one electrode offset (in X) with respect to the other. This offset is accomplished just by changing the first number in either the second or the third line of the file.

Figure 4.15 shows the results of the electrode offset for two values of h (the electrode separation). When $h = 0.1$, the capacitance falls almost linearly with x_{offset}, falling to approximately 25% of its original value, until $x_{offset} = 1$. At this point the two electrodes are no longer facing each other at all, and the capacitance decline with increasing x_{offset} levels off, falling to approximately 21 pF at $x_{offset} = 3$. This latter capacitance is, of course, the same

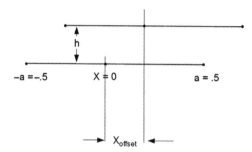

FIGURE 4.14   Cross section of capacitor with one electrode offset.

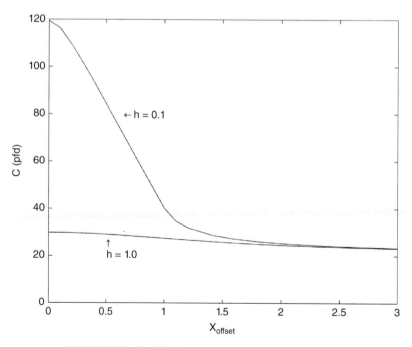

FIGURE 4.15   Parallel plate capacitance versus electrode offset.

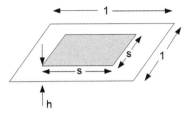

FIGURE 4.16 Capacitor with different-size square electrodes.

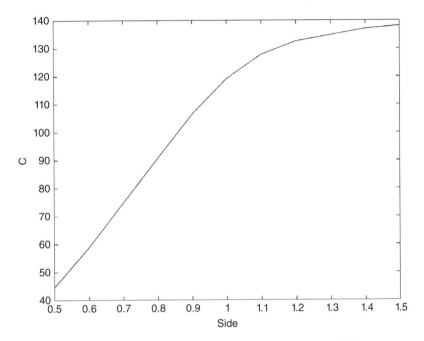

FIGURE 4.17 Capacitance of the capacitor shown in Figure 4.16.

capacitance that was approached in Figure 4.9, which showed the capacitance as the two electrodes were pulled apart (h increased). This make physical sense because electrodes that are too far apart to "see" each other don't know in which direction their greatest separation distance occurs.

When $h = 1.0$, the capacitance was already close to its maximum plate separation distance, so we see very little change as x_{offset} is increased. At $x_{offset} = 3$, we can't see the difference between the $h = 1.0$ and the $h = 0.1$ cases because, again, each electrode no longer knows where the other one is.

Figure 4.16 shows another possible variation. In this case both electrodes are kept square and centered, but the size of one of the electrodes is varied. Figure 4.17 shows the results of these variations when $h = 0.1$.

There is no reason why either of the electrodes has to be square, or why different-size electrodes and misorientation cannot be combined.

Figure 4.18 shows a capacitor with connecting leads brought out. In this example the electrodes are rectangular and the two connecting leads do not have the same dimensions.

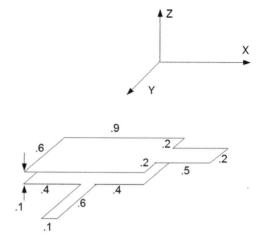

FIGURE 4.18   Capacitor with two different connecting leads.

TABLE 4.3   Data File for Capacitor Shown in Figure 4.18

0.01						
0	0	45	30	0	−0.5	0
0	0.6	5	30	0	−0.5	0
0	0	45	30	0.1	0.5	0
0.7	0	50	10	0.1	0.5	0

Table 4.3 shows the data file for `mom1.m` that creates this capacitor.

The format of the file is the same as in the previous examples. All dimensions are in meters. The first line in the data file is the cell $\frac{1}{2}$ side, $a = b = 0.01$. The second line describes the electrode of the lower section. The third line describes the connecting lead to this (lower) electrode. The geometry description (center and lengths) render it contiguous with the lower electrode, but this is not necessary for the program to operate correctly—the two pieces could be separate in space. The fourth and fifth lines describe the upper electrode and its connecting lead, respectively. The calculated capacitance is 76.7 pF. If we remove the connecting lines (lines 3 and 5 in the data file), the calculated capacitance drops to 69.2 pF. This is approximately a 10% drop—this may or may not be important, depending on the application.

The final variation to be presented in this section is the calculation for a rotated electrode. The calculation is not difficult or involved; it was omitted in the original program because it might have been a distraction.

The following equations rotate a point (x_j, y_j, z_j) about the z axis by an angle θ:

$$x_{new} = x_j \cos(\theta) - y_j \sin(\theta) \tag{4.7}$$

$$y_{new} = x_j \sin(\theta) + y_j \cos(\theta) \tag{4.8}$$

This code is implemented in `get_data.m`. It is not executed until a nonzero rotation is encountered

Figure 4.19 shows an example of an electrode rotating about its center. The data file used is shown in Table 4.4. Figure 4.20 shows the result of this rotation.

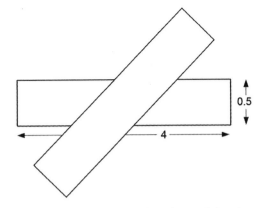

FIGURE 4.19 Capacitor with an electrode rotated about its center.

TABLE 4.4   Data File for Capacitor Shown in Figure 4.19

0.01						
0	0	25	200	0	−0.5	45
0	0	25	200	0.1	0.5	0

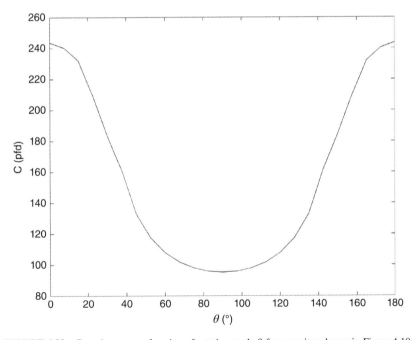

FIGURE 4.20 Capacitance as a function of rotation angle θ for capacitor shown in Figure 4.19.

To rotate an electrode about an arbitrary point (either inside our outside of the electrode), it is necessary only to define that point as the center of the coordinate system.

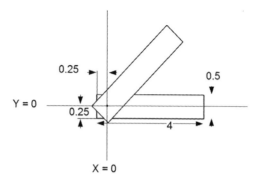

FIGURE 4.21   Capacitor with an electrode rotated about an arbitrary point.

TABLE 4.5   Data File for Capacitor Shown in Figure 4.22

0.01						
−0.25	−0.25	20	160	0	−0.5	45
−0.25	−0.25	20	160	0.1	0.5	0

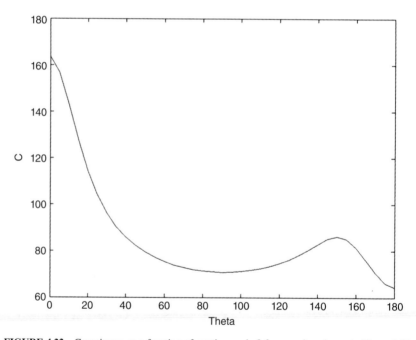

FIGURE 4.22   Capacitance as a function of rotation angle θ for capacitor shown in Figure 4.21.

Figure 4.21 shows an example of this situation; with respect to the origin, the lower left corner of the electrodes is (−0.25,−0.25). The data file is shown in Table 4.5, and the capacitance as a function of θ is shown in Figure 4.22.

Figure 4.22 shows a curious capacitance behavior—while it is expected that the capacitance will fall drastically as θ is increased from 0, the capacitance variation near $\theta = 180$ degrees might be unexpected. Examination of Figure 4.22, however, shows that this behavior is perfectly reasonable because of the electrodes' overlay area as θ is varied.

The capacitor shown in Figures 4.21 and 4.22 is clearly a poor choice as a variable capacitor for tuning a radio or as a rotation angle encoder; monotonic capacitance versus rotation angle behavior is mandatory in these applications. This type of information is a very valuable result of simulation; we usually can forgive an absolute capacitance error of a few percent as long as this error repeats in a production environment, but surprises such as the behavior shown Figure 4.22, while educational in simulation, are often very costly once a design has been committed.

PROBLEMS

4.1 Create a dataset to analyze a square parallel plate capacitor with identical square electrodes 1 cm on a side and 1 cm apart. Use a resolution of $a = 0.0002$ m. Find the capacitance.

4.2 Modify the dataset of Problem 4.1 so that one of the electrodes has a square hole in the center, 0.2 cm on a side. Find the capacitance.

4.3 Show the charge density profile on both electrodes for the results of Problem 4.2.

4.4 Repeat Problems 4.2 and 4.3 with the same hole in both electrodes.

4.5 Calculate voltages and electric fields for $(0,0,Z)$, that is, along the z axis for all of the cases described above.

4.6 Modify `mom1.m` to locate the cells with the largest $|q|$, subdivide these cells so as to improve resolution, and recalculate the charge distribution and capacitance. Use the dataset from Problem 4.4 to demonstrate the results.

5

Symmetries, Images, and Dielectrics

5.1 SYMMETRIES

When there are physical symmetries in a geometry description, the same symmetries must be present in L, q, and V. These symmetries can be used to reduce the number of variables required for a given resolution model.

Consider the 12-cell parallel plate capacitor shown in Figure 5.1.

All 12 cells are rectangular — not necessarily square but all aligned the same. The top and bottom electrodes are in the $(z = -h/2)$ and $(z = +h/2)$ planes. They are at $V = -0.5$ and $V = +0.5$ V.

Because of the physical and voltage symmetries of the geometry description, there is no need to include the extra equation to guarantee charge neutrality.

From the physical symmetries of the geometry, we know that

$$q_1 = q_4 = q_3 = q_6 = -q_7 = -q_9 = -q_{10} = -q_{12} \tag{5.1}$$

and also that

$$q_2 = q_5 = -q_8 = -q_{11} \tag{5.2}$$

The first two equations describing this structure are

$$L_{1,1}q_1 + L_{1,2}q_2 + \cdots + L_{1,12}q_{12} = \frac{V}{2}$$

$$L_{2,1}q_1 + L_{2,2}q_2 + \cdots + L_{2,12}q_{12} = \frac{V}{2} \tag{5.3}$$

Introduction to Numerical Electrostatics Using MATLAB, First Edition. Lawrence N. Dworsky.
© 2014 John Wiley & Sons, Inc. Published 2014 by John Wiley & Sons, Inc.

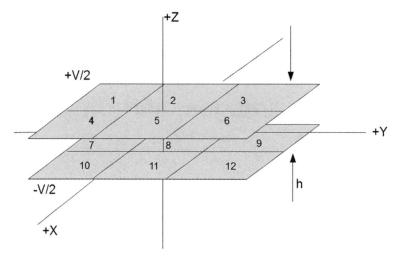

FIGURE 5.1 Twelve-cell parallel plate capacitor.

Substituting all the symmetry–based equalities yields

$$\left(L_{1,1}+L_{1,3}+L_{1,4}+L_{1,6}-L_{1,7}-L_{1,9}-L_{1,10}-L_{1,12}\right)q_1 + \left(L_{1,2}+L_{1,5}-L_{1,8}-L_{1,11}\right)q_2$$
$$\left(L_{2,1}+L_{2,3}+L_{2,4}+L_{2,6}-L_{2,7}-L_{2,9}-L_{2,10}-L_{2,12}\right)q_1 + \left(L_{2,2}+L_{2,5}-L_{2,8}-L_{2,11}\right)q_2$$

$$(5.4)$$

In other words, this structure is accurately described by only two equations in two unknowns. Unfortunately, since it is impossible to predict a priori what geometry somebody wants to study, a computer program that identifies and then automatically utilizes "any symmetry that might come along" is very difficult to write.

There is, however, one type of symmetry that occurs so often that it deserves special mention.

5.2 IMAGES

Suppose that, in the previous example, we considered symmetries only across the $Z=0$ plane (i.e., $q_1 = -q_7$, $q_2 = -q_8$, etc.). The original set of equations becomes two separate but identical sets of equations:

$$\begin{bmatrix} L_{1,1}-L_{1,7} & L_{1,2}-L_{1,8} & \cdots & L_{1,6}-L_{1,12} \\ L_{2,1}-L_{2,7} & L_{2,2}-L_{2,8} & \cdots & L_{2,6}-L_{2,12} \\ & & \cdots & \\ \vdots & \vdots & \cdots & \vdots \\ & & \cdots & \\ L_{6,1}-L_{6,7} & L_{6,2}-L_{6,8} & \cdots & L_{6,6}-L_{6,12} \end{bmatrix} \begin{bmatrix} q_1 \\ q_2 \\ q_3 \\ q_4 \\ q_5 \\ q_6 \end{bmatrix} = \begin{bmatrix} +0.5 \\ +0.5 \\ +0.5 \\ +0.5 \\ +0.5 \\ +0.5 \end{bmatrix} \quad (5.5)$$

FIGURE 5.2 Edge view of structure shown in Figure 5.1.

$$\begin{bmatrix} L_{1,1}-L_{1,7} & L_{1,2}-L_{1,8} & \cdots & L_{1,6}-L_{1,12} \\ L_{2,1}-L_{2,7} & L_{2,2}-L_{2,8} & \cdots & L_{2,6}-L_{2,12} \\ & & \cdots & \\ \vdots & \vdots & \cdots & \vdots \\ & & \cdots & \\ L_{6,1}-L_{6,7} & L_{6,2}-L_{6,8} & \cdots & L_{6,6}-L_{6,12} \end{bmatrix} \begin{bmatrix} -q_1 \\ -q_2 \\ -q_3 \\ -q_4 \\ -q_5 \\ -q_6 \end{bmatrix} = \begin{bmatrix} -0.5 \\ -0.5 \\ -0.5 \\ -0.5 \\ -0.5 \\ -0.5 \end{bmatrix} \qquad (5.6)$$

This breakdown of a two-electrode problem into two smaller, identical problems can be applied to any structure that has a mirror symmetry about a plane (in this case the $z = 0$ plane). There is another way to describe this situation that is mathematically identical to the above that has not yet been introduced.

Figure 5.2 shows the same structure as in Figure 5.1, but in cross section. Because of the symmetries, $Z = 0$ is an equipotential at 0 V. To clarify, the entire $Z = 0$ *plane* is an equipotential surface, not just the region between or in any sense near the electrodes.

If we were to place an actual conductor at the $Z = 0$ plane (often called a *ground plane*) and remove the lower electrode entirely, the charge (and charge distribution) on the upper electrode would not change. Viewing this another way, if we were to replace an actual ground plane by a lower electrode, as in Figure 5.2, then we could model the charge distribution, voltage distribution, electric field, and capacitance of the upper electrode over an infinite ground plane. The charge distribution on this imaginary lower electrode is called the *image charge*, as it is a reverse (in sign) image of the charge distribution on the upper electrode.

What we now have is a tool for modeling any electrode(s) located in a Z plane above a ground plane. With a modest number of cells we can simulate the response of a conductor (the ground plane) that is infinite in extent. While we obviously cannot build such a ground plane, we often work with structures having ground planes that are of sufficient extent in comparison to the electrode(s) above them to be considered infinite in extent.

One small caveat: The original capacitor in Figure 5.2 had 1.0 V between the two electrodes, that is, 0.5 V between either electrode and the ground plane. The charge, voltage, and field responses (for $Z > 0$) will be the same in both cases. There will be a factor of 2 correction, however, for the capacitance of the two situations.

Since the number of cells, for the same resolution, is reduced by 50%, the number of $L_{i,j}$ matrix entries is cut by a factor of 4. This improves both execution time and solution accuracy. Consider the following programs:

- Program mom_image.m:

```
% This is the first MOM program modified for ground plane image
% modeling
```

```
global eps0
eps0 = 8.854;

clear

% Read the input file and process it into cell data
% Data Format:

  [a, data_prep, volts] = get_data_image;

% Generate the L data
L = L_fit_image(a, data_prep);

nr_vars = length(L);
q = L\volts;

cap = 2*sum(q)
```

- Program `get_data_image.m`:

```
function [a, data_prep, volts] = get_data_image()
% get_data gets the user created data file and converts it to
% a file of cells
% Note that the rotation angle capability of the previous program
% is now moot and the parameter has been removed.
% file format:
% Line 1: a, h
% lines 2 and on: x0, y0, nrx, nry, V

filename = uigetfile('*.txt'); % read the file
fid = fopen(filename);
all_data = (fscanf(fid, '%g'))';
nr_lines = (length(all_data) - 2)/5;
closeresult = fclose(fid)

a = all_data(1);
data_in = []; % parse the data
for i = 1:nr_lines
  j = 5*(i-1) + 2;
  data_in = [data_in; all_data(j:j + 5)];
end

volts = [];
data_prep = [];
for i = 1:nr_lines % loop through data file and build output
  x0 = data_in(i,1); y0 = data_in(i,2); nx = data_in(i,3); ny =
  data_in(i,4);
  x_ll = x0 - a*nx; x_ur = x0 + a*nx;
  y_ll = y0 - a*ny; y_ur = y0 + a*ny;

  h = data_in(i,5); V = data_in(i,6);
  for j = 1:nx
  xj = x_ll - a + j*2*a;
```

```
    for k = 1:ny
    yj = y_ll - a + k*2*a;
    line = [xj, yj, h];
    data_prep = [data_prep; line];
    volts = [volts; V];
    end
    end

  end

  end
```

- Program L_fit_image.m:

```
function L = L_fit_image(a, data_prep)

%  This is the empirical function fit

%  xj, yz, zj is the center of the source rectangle in the xyplane
%  xi, yi, zi is the field point

global eps0

xi = data_prep(:,1);
yi = data_prep(:,2);
zi = data_prep(:,3);

n = length(xi); % all the lengths are the same
xs = (repmat(xi,1,n) - repmat(xi',n,1)).^2;
ys = (repmat(yi,1,n) - repmat(yi',n,1)).^2;
zs = (repmat(zi,1,n) - repmat(zi',n,1)).^2;

p = 1.414*a;
La = (log((a+p)/a)/a - log((-a+p)/a)/a);

r = sqrt(xs+ys+zs); % real cells
L = r + 1/La./(1+r/a);
L = 1./L/(4*pi*eps0);

zs = (repmat(zi,1,n) - repmat(-zi',n,1)).^2; % image cells
r = sqrt(xs+ys+zs);
L2 = r + 1/La./(1+r/a);
L2 = 1./L2/(4*pi*eps0);

L = L - L2;

end
```

The MATLAB program mom_image.m, along with the necessary functions get_data_image.m and L_fit_image.m, are a rewrite of the mom1.m program from Chapter 4. This program images every cell described in the data file at some height *h* over the ground plane with an image cell at height *h* below the ground plane.

The data file format is the same as for the `mom1.m` program. For example, the simple data file `data_image.txt`

```
.025
0. 0. 120 20 .5 .5
```

describes a 6 × 1 m rectangular electrode situated 0.5 m over an infinite ground plane, with 0.5 V applied (with respect to the ground plane). The program's output is that the capacitance of this structure is 251.3 pF.

Figure 5.3 shows this same structure with the 6-m dimension replaced by the variable *L*. This structure may be regarded as a microstrip transmission line (as described in Chapter 2) with an air dielectric. Microstrip lines are almost never constructed this way; typically a dielectric layer sits on the ground plane and supports the rectangular electrode. However, since we haven't discussed dielectric interfaces yet, we'll just treat this as an example calculation of a microstrip having a dielectric of (relative) constant 1.

Figure 5.4 shows the calculated capacitance for *L* varying from 1 to 6 m. Superimposed on the data points is a best-fit line generated by the MATLAB function `polyfit`. As may be seen, the fit to the straight line is excellent. The slope of the line is 38.3 pF/m.

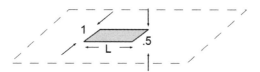

FIGURE 5.3   Air dielectric microstripline with length variable *L*.

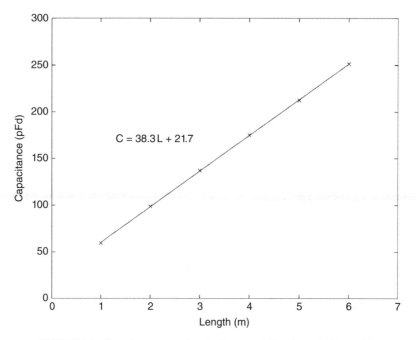

FIGURE 5.4   Capacitance versus length of microstripline shown in Figure 5.3.

What these data and their excellent straight-line fit are telling us is that the microstrip line has a constant capacitance per unit length (the standard transmission line capacitance), as described by the slope of the line. Using equation (2.33), the characteristic impedance of this transmission line is

$$Z_0 = \frac{1}{v_0 C} = \frac{1}{(3 \times 10^8)(38.3 \times 10^{-12})} = 87.0\,\Omega \tag{5.7}$$

The published calculation[1] give this characteristic impedance as $88.6\,\Omega$. The calculated value is within 2% of the published value.

The `polyfit` line has an $L = 0$ intercept of 21.7 pF. This does not mean that a zero-length line would still have capacitance. Indeed, if we were to reduce the length (L) of the line from 1 to 0, we would find that the capacitance of the line does, indeed, go to zero at zero length. The 21.7 pF intercept means that there is an excess, or fringing, capacitance at each end of the line of $21.7/2 \sim 10.9$ pF.

5.3 MULTIPLE IMAGES AND THE SYMMETRIC STRIPLINE

Figure 5.5a shows the cross section of a conventional stripline: a metal center conductor sandwiched, centered, between two ground planes. In this example a thin center conductor of width W is sandwiched between two very wide and infinitely long ground planes, $h/2$ away from the center conductor.

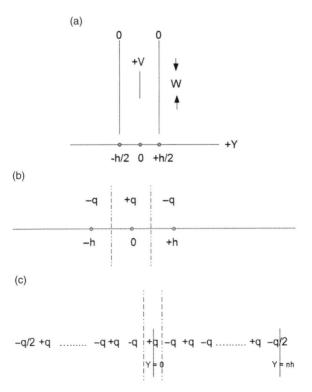

FIGURE 5.5 Multiple-image placement for symmetric stripline: (a) Stripline cross section; (b) First-pass image charge placement; (c) Final image charge placement.

Following the logic of the last section, we would like to replace the (equipotential surface) ground planes with image charges. We therefore replace the ground plane at $z = +h/2$ with an image charge of $-q$ at $z = +h$ and the ground plane at $-h/2$ with an image charge of $-q$ at $z = -h$ (Figure 5.5b).

This doesn't quite do the job, however; the image charge at $+h$ distorts the equipotential at $-h/2$ and the image charge at $-h$ distorts the equipotential at $+h/2$. We can correct for this by giving the image charges their own separate image charges; that is, charges of $+q$ at $z = +2h$ and $-2h$.

You can see where this is going—these latter image charges, in turn, need their own $-q$ separate image charges at $z = +3h$ and $-3h$, and so on, indefinitely (Figure 5.5c). Fortunately, this series converges. For example, at $x = 0$ and $y = +h/2$ (the $+h/2$ ground plane directly in the line of the charges), we have

$$V = \frac{q}{4\pi\varepsilon_0} \frac{2}{h} \left[1 - \frac{2}{2} + \frac{2}{3} - \frac{2}{4} + \cdots + -\frac{12}{2n} \right] \tag{5.8}$$

Note that the outermost charges have been set to a value of $\frac{1}{2}$ rather than 1. This gives the entire system charge neutrality without disturbing the equipotential surfaces noticeably (for n large enough).

The following MATLAB program generates $v = 0$ equipotential surfaces for an arbitrary number of charge points and plots the results. Note that the dimensions are normalized to h because only the dimension ratios are relevant. Also, there are no electrostatic units present—zero voltage is zero voltage in any set of units:

- Program `multiple_image_equipotentials.m`:

```
% Viewing of equipotential surfaces for stripline

h = 1;   % outer conductor - outer conductor separation

equip = zeros(1,2); % initialize final data array
y1 = .4*h; % starting point for dumb search
for x = 0 : .1 : 5 % scan x values
  for y = y1 : .001 : 10*h % search y values
  v = mult_images(x,y,h); % get the voltage at a test point
  if y == y1 % initialize the comparison
  v_old = v;
  end
  if v*v_old < 0 % look for a sign change
  equip = [equip; x,y];
  break
  end
  end
end

  equip = equip(2:end,:); % strip nonsense starting point
  xp = equip(:,1)/h;
  yp = equip(:,2)/h;
  plot(xp,yp,'k', -xp,yp,'k', xp,-yp,'k', -xp,-yp,'k')
```

```
xlabel ('w/h')
ylabel ('y/h')
```

- Program multiple_images.m:

```
function v = mult_images (x,y,h)
%    generates voltage for a string of -n to n multiple image
%    charges at point x,y for charge separation h

n = 35;    % nr of image points on each side
v = 0;

for i = 1 : 2*n + 1
y_src = (i - 1 - n)*h; % y location of each point
rad = sqrt (x^2 + (y - y_src)^2) ; % distance to test point
term = (-1)^i/rad; % voltage and alternating charge signs
if i == 1 | i == 2*n + 1 % preserving charge neutrality
term = term/2;
end
v = v + term;
end

end
```

Note that the numerical search for $v = 0$ in these programs is a very "dumb" search; the program simply steps y in small increments and looks for a sign change in v. If we were writing a more generalized program to calculate equipotential surfaces at arbitrary voltage levels, and to be smart enough to follow contours, we could either write a more sophisticated program or simply calculate a grid of voltages and use MATLAB's internal contour plotting capabilities as was demonstrated in Chapter 1.

Figure 5.6 shows the $v = 0$ equipotential surfaces calculated for $n = 35$. The dimensions of the figure are in units of h since it is only the width : height ratios that matter. As the figure shows, for widths $< 2h$ these equipotential surfaces are barely distinguishable from the original ground planes.

The equation set to be solved is a direct extension of equation (5.5), replacing the one real and one image charge with the series of charges

$$L_{i,j} = \sum_{k=-n}^{k=+n} L_{i,k}(r) \tag{5.9}$$

where

$$r = \sqrt{(x_i - x_j)^2 + (y_i - y_j)^2 + z_k^2} \tag{5.10}$$

$$z_k = kh \tag{5.11}$$

Remember that regardless of the number of image charges inserted, the size of the set of equations to solve depends only on the number of cells in the electrode (the stripline center conductor); program execution time will be dominated by the generation of $L_{i,j}$.

MATLAB program mom.stripline calculates the capacitance of stripline structures:

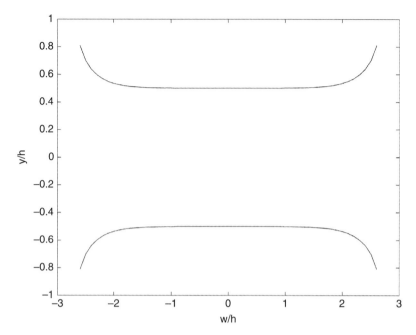

FIGURE 5.6 Equipotential surface approximation for multiple-image charges.

- Program mom_stripline.m:

```
% This is the first MOM program modified for ground plane image(s)
% Note alternative function calls for microstrip or stripline

global eps0
eps0 = 8.854;
clear

% Read the input file and process it into cell data
% Data Format:
[a, h, data_prep, volts] = get_data_stripline;

% Generate the L data
%L = L_fit_image(a, data_prep); % for air microstrip calculations
tic
L = L_fit_stripline(a, h, data_prep); % for stripline calculations
toc

%nr_vars = length(L);
tic
q = L\volts;
toc

cap = sum(q)
%output = [];
%for i = 1: length(q)
% output = [output;  [data_prep(i,:),  q(i)]];
%end
%save output.txt output -ascii
```

- Program get_data_stripline.m:

```
function [a, h, data_prep, volts] = get_data_image()
% get_data gets the user created data file and converts it to
% a file of cells
% Note that the rotation angle capability of the previous program
% is now moot and the parameter has been removed.
% file format:
% Line 1: a, h
% lines 2 and on: x0, y0, nrx, nry, V

filename = uigetfile('*.txt'); % read the file
fid = fopen(filename);
all_data = (fscanf(fid, '%g'))';
nr_lines = (length(all_data) - 2)/5;
closeresult = fclose(fid)

a = all_data(1); h = all_data(2);
data_in = []; % parse the data
for i = 1:nr_lines
  j = 5*(i-1) + 3;
  data_in = [data_in; all_data(j:j + 4)];
end

volts = [];
data_prep = [];
for i = 1:nr_lines % loop through data file and build output
  x0 = data_in(i,1); y0 = data_in(i,2); nx = data_in(i,3); ny =
  data_in(i,4);
  x_ll = x0 - a*nx; x_ur = x0 + a*nx;
  y_ll = y0 - a*ny; y_ur = y0 + a*ny;

  V = data_in(i,5);
  for j = 1:nx
  xj = x_ll - a + j*2*a;
  for k = 1:ny
  yj = y_ll - a + k*2*a;
  line = [xj, yj];
  data_prep = [data_prep; line];
  volts = [volts; V];
  end
  end

end

end
```

- Program L_fit_stripline.m:

```
function L = L_fit_mult_image(a, h, data_prep)

%function L = L_pt_corr(a, b, xi, yi, zi)
% This is the one point approximation with empirical function fit
```

```
% xj, yz, zj is the center of the source rectangle in the xyplane
% xi, yi, zi is the field point

global eps0
x_list = data_prep(:,1);
y_list = data_prep(:,2);
n = length(x_list); % all the lengths are the same
p = 1.414*a;
La = (log((a + p)/a)/a - log((-a + p)/a)/a);
nr_images = 35; % number of image points on each side
L = zeros(n);

for i = 1 : n
  xi = x_list(i); yi = y_list(i);
    for j = 1 : n
    xj = x_list(j); yj = y_list(j);
    for k = -nr_images : nr_images
    zj = k*h;
    r = sqrt((xi - xj)^2 + (yi - yj)^2 + zj^2);
    term = (-1)^k/(r + 1/La/(1 + r/a));
    if k == - nr_images | k == nr_images
    term = term/2.;
    end
    L(i,j) = L(i,j) + term;
    end
    end
  end
L = L/(4*pi*eps0);

end
```

In these programs, the data file format has been updated slightly. A sample data file
(data_image_stripline.txt) is

```
.05 1.
0. 0. 10 6 1.
```

The first line is the values for a and h. The second (and any subsequent) lines are
X_0, Y_0, n_x, n_y, and V. In this example the line is 0.6 m wide and 1.0 m long; $a = 0.05$,
$h = 1$, and $V = 1$.

Figure 5.7 shows the results of a repeat of the microstripline calculation, but for stripline.
Two cases are shown: $W = 1.0$ m and $W = 0.6$ m. In both cases the center ground plane–
ground plane distances are 1.0 m.

For both examples the fit to a straight line is excellent. The calculated Z_0 for the $W = 1.0$-m
line is 65.4 Ω and for the $W = 0.6$-m line, 90.5 Ω. Both numbers are within 2% of published
values.[2]

One last point—stripline is usually built using a uniform solid dielectric rather than an air
dielectric. The calculations in this case do not change; we simply replace the permittivity of
free space with the permittivity of the dielectric material.

Figure 5.8 shows a junction of two 3-m-long pieces of each transmission line described

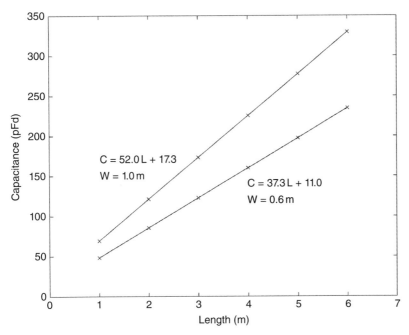

FIGURE 5.7   Two air–dielectric striplines with length variable *L*.

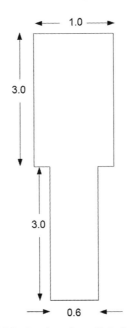

FIGURE 5.8   Junction of two dissimilar striplines.

above. The dataset for calculating the capacitance of this structure is

```
.05 1.
0. 0.30 6 1.
3. 0. 30 10 1.
```

The total capacitance of this structure as calculated by `mom_stripline.m` is 283.1 pF. Referring to Figure 5.7, a 3 m length of the wider line would have a line capacitance of 3(52.0) pF and an edge capacitance of (17.3)/2 pF. Similarly, a 3 m length of the narrower line would have a line capacitance of 3(37.3) and an edge capacitance of (11.0)/2. The total of these four capacitances is 282.1 pF. The excess capacitance, ~1.0 pF, is therefore the fringing capacitance (to ground) at the junction of the two lines. This junction fringing capacitance must be factored into any RF circuit modeling of transmission line structures.

5.4 DIELECTRIC INTERFACES

When there are more than one dielectric materials (with different dielectric constants) in a structure, then the simple relationship between Q and V as described in Chapter 1 is no longer correct and must be modified.

Consider, for example, Figure 5.9, where plane $Y = 0$ is an interface between two dielectric materials. For $Y \leq 0$, $K > 1$; for $Y > 0$, $K = 1$ (e.g., air). Designate these regions as regions 1 and 2, respectively. There is a fixed charge Q at some point $(x_0, y_0, -a)$, in region 1—the dielectric material. We want to know the voltage V both in the dielectric material and the air.

At the interface, we will require two boundary conditions to be satisfied: (1), the potential V must be continuous across the dielectric interface

$$V_1(x,0,z) = V_2(x,0,z) \tag{5.12}$$

and (2) the normal electric flux density (D) must be continuous:

$$K \frac{\partial V_1}{\partial y}\bigg|_{y=0} = \frac{\partial V_2}{\partial y}\bigg|_{y=0} \tag{5.13}$$

Our goal here is to allow calculation of V from inside the dielectric (in region 1) as if the entire space were dielectric and also to allow calculation of V from outside the dielectric (in region 2) as if the entire space were dielectric.

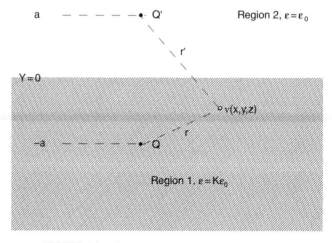

FIGURE 5.9   Dielectric interface shown in cross section.

As shown in Figure 5.9, we begin by postulating an image charge Q' "mirrored" across the dielectric interface from Q, that is at (x_0, y_0, a), and then we find Q' so that both boundary conditions are satisfied.

At an arbitrary point in the dielectric ($Y \le 0$), referring to Figure 5.9, we obtain

$$V_1 = \frac{1}{4\pi K \varepsilon_0} \left[\frac{Q}{r} + \frac{Q'}{r'} \right] \tag{5.14}$$

In region 2, we only see an effective charge Q'' at the location of Q:

$$V_2 = \frac{Q''}{4\pi K \varepsilon_0 r} \tag{5.15}$$

Q' and Q'' are presently unknown. At $Y = 0$ (the interface), $r = r'$. Therefore, from the first boundary condition [equation (5.12)], we have

$$Q + Q' = Q'' \tag{5.16}$$

To satisfy the second boundary condition [equation (5.13)], we begin with

$$r = \sqrt{x^2 + (y-a)^2 + z^2} \tag{5.17}$$

and then

$$\frac{\partial}{\partial y} \frac{1}{r} = \frac{-1}{r^2} \frac{\partial r}{\partial y} = -\frac{y-a}{r^3} \tag{5.18}$$

Similarly

$$\frac{\partial}{\partial y} \frac{1}{r'} = -\frac{y+a}{r'^3} \tag{5.19}$$

At $Y = 0$, we obtain

$$\frac{\partial}{\partial y} \frac{1}{r} = \frac{a}{r^3} \tag{5.20}$$

$$\frac{\partial}{\partial y} \frac{1}{r'} = -\frac{a}{r'^3} = -\frac{a}{r^3} \tag{5.21}$$

Using these relations in the second boundary condition, we obtain

$$K(Q - Q') = Q'' \tag{5.22}$$

Combining equations (5.16) and (5.22), we have

$$Q' = Q\frac{K-1}{K+1} \tag{5.23}$$

In the dielectric; therefore

$$V = \frac{Q}{4\pi K\varepsilon_0}\frac{1}{r} + \frac{K-1}{K+1}\frac{1}{r'} \tag{5.24}$$

In the case of microstripline, the actual charge Q is on the dielectric interface surface. Referring to Figure 5.9, for any point on the surface, $r = r'$, and then

$$V = \frac{Q}{4\pi K\varepsilon_0 r}\left(1 + \frac{K-1}{K+1}\right) \tag{5.25}$$

Figure 5.10 shows a microstrip transmission line section. This is sometimes referred to as an *open* microstrip line because there is no enclosing metallic box. A dielectric slab is fully metalized on one side (the bottom in this figure). The small comblike symbol in the lower left of the figure is the standard electric circuit designation for a conductor set to zero, or *ground* voltage. The dielectric slab thickness is h. The relative dielectric constant of the slab is K, or sometimes ε_r. The width of the line is w, and the length is L.

The microstrip line is modeled using multiple images, as shown in Figure 5.11.

At $y = h$ and $y = -h$ (the top and bottom of the dielectric) are the real charges Q and $-Q$. At the same points in space are the image charges described in equation (5.25), $+\varepsilon$ and $-\varepsilon$, where

$$\varepsilon = \frac{K-1}{K+1} \tag{5.26}$$

Following the logic of the multiple-image stripline model, the image charges $+\varepsilon$ and $-\varepsilon$ each require their own image charges. Unlike the image charges for the stripline model, however, these image charges do not have the same magnitude as the first image charges. As derived above, the magnitude of these charges is the original charge multiplied by ε. This logic is recursive, again following the logic of the stripline model, leading to an infinite number of image charges. In this case, however, not only does the effect of each charge diminish because of the increased distance (from $Y = 0$); the charges themselves decrease with increased distance. When $K = 1$, then $\varepsilon = 0$, and, of course, there are no image charges.

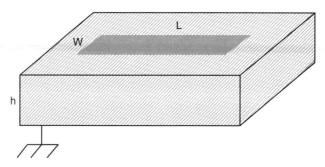

FIGURE 5.10   An open microstrip transmission line section.

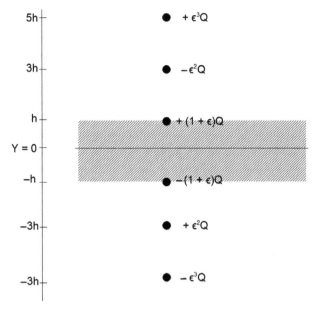

FIGURE 5.11   Multiple-image pattern for the open microstripline.

When $K = 5$ (for example), then $\varepsilon = \frac{2}{3}$ and the 10th image charge will have a magnitude of $\left(\frac{2}{3}\right)^{10} \approx 0.02$.

The MATLAB program `mom_k_microstrip.m` implements the analysis of the structure in Figure 5.11:

- Program `mom_k_microstrip.m`:

```
% This is the open microstrip line program with a dielectric

clear
global eps0
eps0 = 8.854;

% Read the input file and process it into cell data
% Data Format:

[a, K, h, data_prep, volts] = get_data_k_microstrip;

% Generate the L data
L = L_fit_k_microstrip(a, K, h, data_prep);

q = L\volts;

cap = sum(q)
```

- Program `L_fit_k_microstrip.m`:

```
function L = L_fit_k_microstrip(a, er, h, data_prep)
```

```
% This is the one point approximation with empirical function fit set
% up for the open microstrip line with rel dielectric er

% xj, yz, zj is the center of the source rectangle in the xyplane
% xi, yi, zi is the field point

global eps0

x_list = data_prep(:,1);
y_list = data_prep(:,2);
n = length(x_list); % all the lengths are the same
p = 1.414*a;
La = (log((a + p)/a)/a - log((-a + p)/a)/a);
L = zeros(n);
k_max = 51;

for i = 1 : n
 xi = x_list(i); yi = y_list(i);
 for j = 1 : n
 if i > j
 L(i,j) = L(j,i);
 else
 xj = x_list(j); yj = y_list(j);
 zp = 0; zm = -2*h;
 rp = sqrt((xi - xj)^2 + (yi - yj)^2 + zp^2);
 rm = sqrt((xi - xj)^2 + (yi - yj)^2 + zm^2);
 L(i,j) = (1/(rp + 1/La/(1 + rp/a)) ....
 - 1/(rm + 1/La/(1 + rm/a)));
 sign = -1;
 for k = 1 : 2 : k_max
 z = k*h; zp = z - h; zm = -z -h;
 eps = ((er-1)/(er + 1))^((k + 1)/2);
 sign = -sign; %(-1)^((k + 3)/2);
 rp = sqrt((xi - xj)^2 + (yi - yj)^2 + rp^2);
 rm = sqrt((xi - xj)^2 + (yi - yj)^2 + rm^2);
 L(i,j) = L(i,j) + sign*eps*(1/(rp + 1/La/(1 + rp/a)) ....
 - 1/(rm + 1/La/(1 + rm/a)));
 end
 end
 end
end

L = L/(4*pi*eps0*er);

end
```

• Program `get_data_k_microstrip.m`:

```
function [a, K, h, data_prep, volts] = get_data_k_microstrip()
% get_data gets the user created data file and converts it to
% a file of cells

% file format:
```

```
% Line 1: a, K, h
% lines 2 and on: x0, y0, nrx, nry, V

filename = uigetfile('*.txt'); % read the file
fid = fopen(filename);
all_data = (fscanf(fid, '%g'))';
nr_lines = (length(all_data) - 3)/5;
closeresult = fclose(fid)

a = all_data(1); K = all_data(2); h = all_data(3);
data_in = []; % parse the data
for i = 1:nr_lines
j = 5*(i-1) + 4;
data_in = [data_in; all_data(j:j + 4)];
end

volts = [];
data_prep = [];
for i = 1:nr_lines % loop through data file and build output
x0 = data_in(i,1); y0 = data_in(i,2); nx = data_in(i,3); ny =
data_in(i,4);
x_ll = x0 - a*nx; x_ur = x0 + a*nx;
y_ll = y0 - a*ny; y_ur = y0 + a*ny;

V = data_in(i,5);
for j = 1:nx
xj = x_ll - a + j*2*a;
for k = 1:ny
yj = y_ll - a + k*2*a;
line = [xj, yj];
data_prep = [data_prep; line];
volts = [volts; V];
end
end

end

end
```

The data file format for this program is

```
0.25 5 .5
0. 0. 40 20 1.
```

The first line contains the values a, K, and h for the structure. The second line (and possible subsequent lines) contains the values x_0, y_0, n_x, n_y, and V for each rectangular electrode on the surface of the dielectric.

In the sample data file shown above, the line is 2 m long and 1 m wide. Repeating the procedure of previous examples, that is, running the program for several lines from 1 to 6 m and fitting the resulting capacitances to a straight line (again, the fit is excellent), the slope of the line gives us a transmission line capacitance C of 150.8 pF/m. The Wheeler model[1] for this structure predicts $C = 140.2$ pF/m. These two numbers differ by $< 8\%$; this is a good, but not excellent, result. We would like to improve the accuracy of this calculation.

5.5 TWO-DIMENSIONAL CROSS SECTIONS OF UNIFORM THREE-DIMENSIONAL STRUCTURES

The modeling procedure used for the stripline and microstripline transmission lines is useful in that it produces numbers for the characteristic impedance and the line end fringing capacitance and allows us to model junctions of dissimilar lines. In many cases, however, only the characteristic impedance is of interest. We would like an accurate method for finding the characteristic impedance, hopefully a method that does not require curve fitting of multiple program runs.

In more general terms, we would like a procedure for calculating the per-unit-length capacitance (and field, voltage distribution, and conductor charge distribution) of infinitely long uniform structures.

Consider an infinitely long line of charge running along the z axis of a coordinate system. In cross section this line is simply a point.

Because of the symmetry of this simple structure, all field lines must emanate from the line and be in X–Y planes. A (mathematical) cylinder surrounding the line, centered at the line, is therefore a Gaussian surface with electric field lines normal to it everywhere. Also, because of the symmetry, the field strength is the same at every point on the cylinder. In cross section (see Figure 5.12) the cylinder is simply a circle.

Given Q as the charge per unit length and r as the radius of the cylinder, using Gauss's law, we may write

$$Q = \int_s \vec{D} \cdot \vec{ds} = 2\pi r \varepsilon_0 E_r \tag{5.27}$$

or

$$E_r = \frac{Q}{2\pi \varepsilon_0 r} \tag{5.28}$$

Integrating this, we obtain

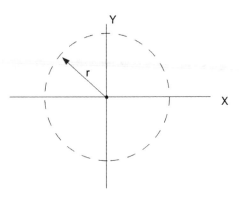

FIGURE 5.12   Cross section of an infinite line of charge and a Gaussian surface.

$$V = \frac{-Q}{2\pi\varepsilon_0}\ln(r) + V_0 \tag{5.29}$$

where V_0 is mathematically an arbitrary constant of integration and physically the arbitrary voltage reference level. If we pick a reference radius r_0, we may write

$$V = \frac{-Q}{2\pi\varepsilon_0}\ln(r) \tag{5.30}$$

and then

$$V = \frac{-Q}{2\pi\varepsilon_0}\ln\frac{r}{r_0} \tag{5.31}$$

The two forms of the expression are equivalent; both appear in the literature. Setting $r_0 = 1$ or $V_0 = 0$, we arrive at

$$V = \frac{-Q}{2\pi\varepsilon_0}\ln(r) \tag{5.32}$$

This expression is often puzzling when it is initially encountered. While the expression for a point charge, specifically

$$V = \frac{Q}{4\pi\varepsilon_0 r} \tag{5.33}$$

is monotonic, equation (5.32) changes sign, passing through 0 at $r = 1$.

Fortunately, there is no issue here. Remember that equation (5.33) came from

$$V = \frac{Q}{4\pi\varepsilon_0 r} + V_0 \tag{5.34}$$

and by setting V_0 to, say, -1, we can make it also go through 0 and change sign.

Returning to equation (5.32), for setting up a method of moments problem (and solution), the sign of the charge is arbitrary provided we are consistent. Therefore, although it is not physically correct, we can safely drop the minus sign in equation (5.32), leaving us with

$$V = \frac{Q}{2\pi\varepsilon_0}\ln(r) \tag{5.35}$$

to work with.

Figure 5.13 shows the cross section of a strip of width $2a$, centered at (x_j, y_j). For calculating the diagonal term $L_{i,i}$, we simply set $x_i = x_j = y_i = y_j = 0$, and then

$$L_{i,i} = \frac{1}{2\pi\varepsilon_0}\frac{1}{2a}\int_{-a}^{a}\ln(x)dx = \frac{1}{2\pi\varepsilon_0}[\ln(a) - 1] \tag{5.36}$$

For the off-diagonal terms $(i \neq j)$, we obtain

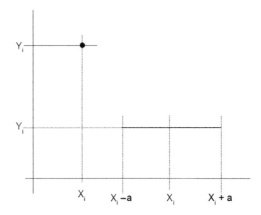

FIGURE 5.13   Layout for calculating $L_{i,j}$.

$$L_{i,j} = \frac{1}{2\pi\varepsilon_0} \frac{1}{2a} \int\limits_{x_j-a}^{x_j+a} \ln[r(x)]dx \qquad (5.37)$$

Referring to Figure 5.13, letting $x_p = x_i - x_j$, $y_p = y_i - y_j$, we obtain

$$L_{i,j} = \frac{1}{4\pi\varepsilon_0 a}\left[\frac{x_p+a}{2}\ln\left((x_p+a)^2+y_p^2\right) - \frac{x_p-a}{2}\ln\left((x_p-a)^2+y_p^2\right) - 2a \right.$$

$$\left. + y_p\arctan\frac{x_p+a}{y_p} - y_p\arctan\frac{x_p-a}{y_p} \right] \qquad (5.38)$$

This expression is messy, which won't bother a computer at all. The important point is that it is an exact integration, unlike the approximations used previously.

Figure 5.14 shows the results of using these formulas to calculate the air dielectric microstripline capacitance (the X symbols) overlaid on the Wheeler calculation results[1] for the same structure. As may be seen, the agreement is excellent (<1% error) for all points.

The MATLAB program which produced these results is shown below

- Program mom2d.m:

```
% This is the open microstrip line program using infinitely long
% strips of conductor each 2a wide

global eps0
eps0 = 8.854;
clear

% Read the input file and process it into cell data

  [a, data_prep, volts] = get_data_2d;
% Generate the L data
L = L_2d(a, data_prep);
```

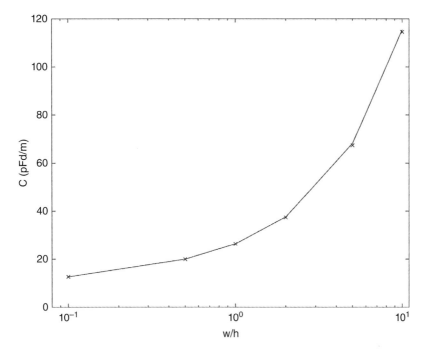

FIGURE 5.14   Results of air dielectric microstripline calculation.

```
q = L\volts;

cap = 0; cap2 = 0;
for i = 1 : length(q)
 if q(i) > 0
 cap = cap + q(i);
 else
 cap2 = cap2 + q(i);
 end
end
disp ([cap, cap2])
```

- Program `get_data2d.m`:

```
function [a, data_prep, volts] = get_data_2d()
% get_data gets the user created data file and converts it to
% a file of cells

% file format:
% Line 1: a
% lines 2 and on: x0, nrx, h V

filename = uigetfile; % read the file
fid = fopen(filename);
all_data = (fscanf(fid, '%g'))';
nr_lines = (length(all_data) - 1)/4;
closeresult = fclose(fid)
```

```
a = all_data(1);
data_in = []; % parse the data
for i = 1:nr_lines
 j = 4*(i-1) + 2;
 data_in = [data_in; all_data(j:j+3)];
end

volts = [];
data_prep = [];
for i = 1:nr_lines % loop through data file and build output
 x0 = data_in(i,1); nx = data_in(i,2); h = data_in(i,3);
 x_l = x0 - a*nx; x_r = x0 + a*nx;
 V = data_in(i,4);

 for j = 1:nx
 xj = x_l - a + j*2*a;
 data_prep = [data_prep; [xj,h]];
 volts = [volts; V];
 end

end

end
```

- Program L_2d.m:

```
function L = L_2d(a, data_prep)

% calculates the Lij set for the 2d structure

global eps0

x_list = data_prep(:,1); h_list = data_prep(:,2);
n = length(x_list);
L = zeros(n);

for i = 1 : n
 xi = x_list(i);
 hi = h_list(i);
 for j = 1 : n
 if i == j
 L(i,j) = log(a) - 1;
 elseif i > j
 L(i,j) = L(j,i);
 else
 xj = x_list(j);
 hj = h_list(j);
 xd = xi - xj; hd = hi - hj;
 L(i,j) = ((xd+a)/2*log((xd+a)^2+hd^2) ...
 - (xd-a)/2*log((xd-a)^2+hd^2) ...
 -2*a + hd*atan2(xd+a,hd) - hd*atan2(xd-a,hd) )/a/2;
 end
```

```
  end
end
L = L/(2*pi*eps0);
end
```

A sample data file for these programs is

```
.01
0. 100 0.4 1.
0. 0. 100 0. -1.
```

The first data line contains a. Each subsequent line contains x_0, n_x, h, and V. The example above specifies two 2-m-wide lines, 0.4 m apart with +1 V on one line and −1 V on the other. This corresponds to a microstripline 2 m wide that is 0.2 m above the ground plane, with 1 V on it above a 0-V ground plane.

For a three-dimensional (3d) calculation, an 80×40 rectangle electrode, for example, contains 3200 cells. If there are two electrodes, there are 6400 equations to solve. In the preceding example there are only 200 equations to solve, but the resolution is much finer than in the case of the 3d example.

Extending the preceding analysis to include image calculations for stripline and microstripline (both without and with a dielectric) is straightforward and mimics the 3d analyses of these structures. The accuracy of the results of these extensions is excellent.[3]

5.6 CHARGE PROFILES AND CURRENT BUNCHING

Any of the stripline or microstripline models described in this chapter can be used to examine the charge profile on the line(s). The very high resolution of the two-dimensional (2d) cross-sectional calculation of Section 5.5, however, makes it the model of choice for looking at charge density.

Figure 5.15 shows the charge profile on the upper conductor of the air microstripline when $w/h = 2/2 = 1$. As may be seen, the charge is significantly bunched at the outer edges of the conductor. In the case of the conductor pair structure used in Section 5.5 this is the profile that is found on both conductors. But what about the charge profile on the ground plane? We never actually modeled a structure with a ground plane.

The point of the image charge calculation(s) is that (in the microstrip case), nothing would change if an actual ground plane were placed between the two conductors. Furthermore, if the lower conductor were then removed (and the ground plane set at 0 V), nothing on the upper conductor side of the ground plane would change. In other words, if the ground plane is at $y = 0$, then $D = D_y$ at $y = 0$ (symmetry forces this conclusion), D_y is the same as it was when there were two conductors and no ground plane, and Poisson's equation tells us that the charge on the ground plane is proportional to D_y.

From equation (5.28), the contribution to D at point X_i due to a charge on the upper conductor Q_j at point x_j is

$$\left|\overrightarrow{D_{i,j}}\right| = \varepsilon_0 \left|\overrightarrow{E_{i,j}}\right| = \frac{Q_j}{2\pi r_{ij}} \tag{5.39}$$

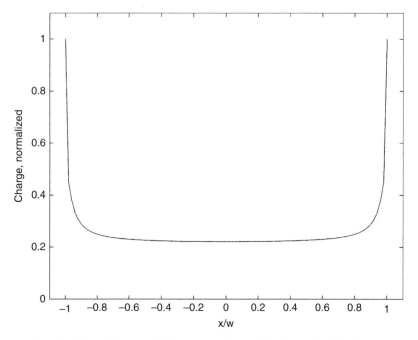

FIGURE 5.15 Charge profile in upper conductor of air microstrip, with $w/h = 1$.

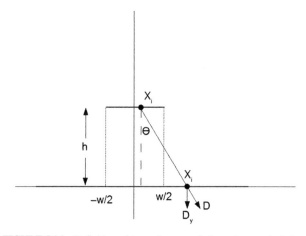

FIGURE 5.16 Definition of terms for ground-plane charge calculation.

where (see also Figure 5.16)

$$r_{ij} = \sqrt{\left(x_i - x_j\right)^2 + h^2}$$ (5.40)

$$\theta = \arctan \frac{h}{x_j - x_i}$$ (5.41)

$$D_{y;i,j} = |\vec{D}_{i,j}| \cos(\theta)$$ (5.42)

and finally, D_y is the sum of the contributions of all the charges Q_j:

$$Q_i \sim D_{y,i} = \sum_j D_{y,i,j} \tag{5.43}$$

Figure 5.17 shows the charge distribution on (a cross section of) the microstripline ground plane (the same microstripline as in Figure 5.15). Clearly these distributions look nothing like each other. While the center conductor line charge distribution is very uniform across most of the line and grows rapidly (bunches) at the edges, the ground-plane charge distribution resembles a flattened Gaussian distribution function, peaking at the center and falling off with distance from the center. The ground plane seems almost to be ignoring the very high charge distributions at the edges of the center conductor.

A study of the losses in transmission lines is beyond the scope of this book. It is, however, worth noting that losses in the conductors is due to the Ohmic resistance of the conductors. The power loss at a point is proportional to the resistivity of the material and the current density *squared* at that point. In the microstripline example we see that there is tremendous loss at the edges. For a given material system and a given allowed cross-sectional area of the line, the power loss is kept at a minimum by minimizing current bunching. Circular coaxial cable is best because there is no current bunching; unfortunately, sometimes other practical issues dominate and microstripline or stripline must be used.

There are other practical applications of the information gleaned from these charge distributions. First, it is clear that the highest electric field is at the edges of the center conductor. This is important if we are worrying about material breakdown or arcing under high voltage conditions. Next, we can see that (using the preceding example) a ground plane wider than approximately 20 m will be carrying virtually no current at its edges. This gives us a lower limit on how narrow we can make the ground plane without increasing the resistance of the line or disturbing the characteristics of the line with mounting screws, for instance.

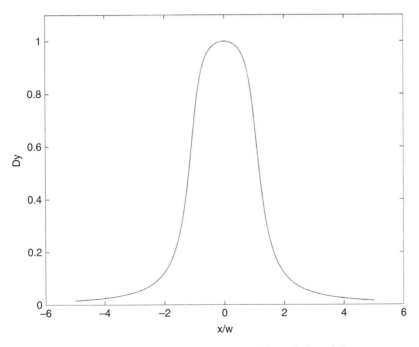

FIGURE 5.17   Charge distribution on ground plane of microstripline.

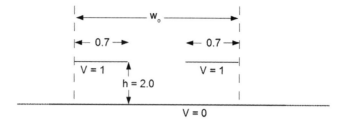

FIGURE 5.18 Cross section of two-line microstrip.

If it is necessary to connect a small component between the center conductor of the line and the ground plane, it should be connected using a small hole in the top of the upper conductor (going down through the dielectric layer). Since there is very little charge at the center of the upper conductor, there will be almost no change in the line's capacitance or resistance (and hence its overall performance) because of this hole. As we will show in Chapter 6, there is almost no field in the center of a hole in the upper conductor and a surprisingly large hole (as compared to the upper conductor's width) can be drilled before the effects of this hole are noticed. The other side of this observation is that drilling a series of centered holes in order to decrease the capacitance of microstripline or stripline (when that is a desired result) is a futile effort.

This same characteristic of microstriplines can be seen from a different perspective using the following example. We know from the preceding calculations that the capacitance of an air microstripline when $w = h = 2$ m is 26.3 pF/m. The same program predicts that the capacitance of this structure with $h = 2$ and $w = 0.7$ (35% of the previous value) is 17.7 pF/m. If we calculate the sum of the capacitances of two such lines spaced very far apart, we get 35.4 = 2 (17.7) pF/m, as we would expect.

However, when we start bringing the lines closer together, we see that the capacitance begins to decrease. Figure 5.18 shows a cross section of the structure. W_0 is the total outside width of the two lines, as shown. The minimum value W_0 can have is 1.4 m; at this value the two lines touch, forming a single 1.4 m-wide line.

Figure 5.19 shows the total capacitance of the two lines as they are moved closer together. The interesting point is the capacitance at W_0 (the outside width) = 2.0. At this point, $C = 25.7$ pF.

What we have created here is an effective upper conductor for an air microstripline composed of two 0.7 m-wide lines and a 0.6 m gap between them. The capacitance is only ~2% lower than the capacitance of a full solid 2 m-wide line even though we have removed 30% of the center of the line.

5.7 CYLINDER BETWEEN TWO PLANES

This section describes a calculation method that does not use the method of moments. It is, however, an interesting use of image calculations and approximations based on naturally occurring equipotential surface shapes. Also, a slight variation on the calculation of 2d cross sections of infinitely long uniform structures is presented.

Starting with Coulomb's law, the potential at a point $p = (x_p, y_p, 0)$ due to a charge q at $(0,0,0)$ is (dropping the $z = 0$ index since point p will always be in the X–Y plane) as follows:

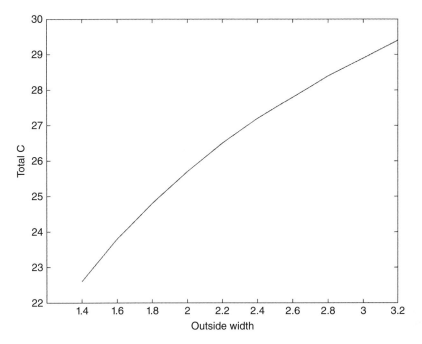

FIGURE 5.19 Capacitance of two-line microstrip versus Outside Width.

$$V(x_p,y_p) = \frac{q}{4\pi\varepsilon_0}\frac{1}{\sqrt{x_p^2+y_p^2}} \tag{5.44}$$

If we want to replace q with a line of charge distribution σ (in coulombs per unit length), running from $(0,0,-s)$ to $(0,0,s)$ (i.e., running along the z axis through the origin), we can replace equation (5.44) by

$$4\pi\varepsilon_0 V(x_p,y_p) = \int_{-s}^{s} \frac{dz}{\sqrt{x_p^2+y_p^2+z^2}} = 2\int_{0}^{s} \frac{dz}{\sqrt{x_p^2+y_p^2+z^2}} \tag{5.45}$$

which integrates to

$$4\pi\varepsilon_0 V(x_p,y_p) = -\ln\left(x_p^2+y_p^2\right) + 2\ln\left(s+\sqrt{x_p^2+y_p^2+s^2}\right) \tag{5.46}$$

In order to consider a 2d cross section (e.g., a transmission line description) of some geometry, we would be interested in the voltage at (x_p,y_p) due to an infinitely long line of charge; that is, we would want to let s grow very large — actually, to approach infinity. As a increases, however, the second term above approaches

$$2\ln(2s) \tag{5.47}$$

which itself goes to infinity as a goes to infinity. Needless to say, this is not a useful result. Now consider the structure shown in Figure 5.20.

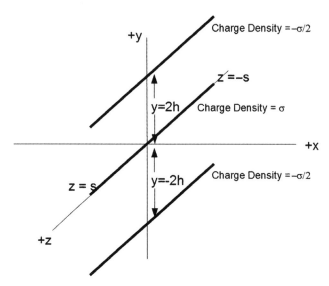

FIGURE 5.20   Three-line neutral charge structure.

We have added two more lines of charge, parallel to the first line, at $y = 2h$ and at $y = -2h$. Each of these new lines carries a charge density of $-\sigma/2$. This structure is electrically neutral for all values of s.

The voltage at any point in the X–Y plane is, by extension of the above, as follows:

$$
4\pi\varepsilon_0 V\left(x_p, y_p\right) = 2\int_0^s \frac{dz}{\sqrt{x_p^2 + y_p^2 + z^2}} - 2\int_0^s \frac{dz}{\sqrt{x_p^2 + \left(y_p + 2h\right)^2 + z^2}}
$$
$$
- 2\int_0^s \frac{dz}{\sqrt{x_p^2 + \left(y_p - 2h\right)^2 + z^2}}
$$
(5.48)

Each of these three integrals integrates to two terms as above; the second terms of each integral are the only terms that contain the parameter s. These terms are

$$
2\ln\left(s + \sqrt{x_p^2 + y_p^2 + s^2}\right) - \ln\left(s + \sqrt{x_p^2 + \left(y_p + 2h\right)^2 + a^2}\right) - \ln\left(s + \sqrt{x_p^2 + \left(y_p - 2h\right)^2 + s^2}\right)
$$
(5.49)

As s increases, this approaches

$$
2\ln(2s) - \ln(2s) - \ln(2s)
$$
(5.50)

which, in the limit as a goes to infinity, goes to zero.

The remaining terms are

$$
4\pi\varepsilon_0 V\left(x_p, y_p\right) = -\ln\left(x_p^2 + y_p^2\right) + \ln\left(x_p^2 + \left(y_p + 2h\right)^2\right) + \ln\left(x_p^2 + \left(y_p - 2h\right)^2\right)
$$
(5.51)

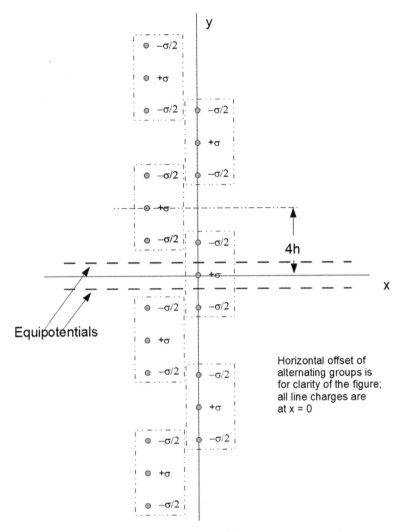

FIGURE 5.21 Multiline structure stacking to create equipotentials.

This structure and the accompanying expression for V_p are not very useful by themselves. However, if we stack many of these structures, at $z = 0$, $z = \sigma 4h$, $z = \sigma 8h$, ..., as shown in Figure 5.21, we get a structure that is reminiscent of the image charge structure presented earlier in this chapter. Since each structure is charge-neutral, the entire assembly must be charge-neutral. When there are many of these stacked structures, the lines $y = 2h$ and $y = -2h$, near $x = 0$, are essentially equipotentials.

The Following MATLAB program can be used to calculate the capacitance of a thin cylinder between two plates:

```
function [Cap] = wbp(r)
%wbp calculates the Capacitance of a thin wire between two parallel plates
% This is the cross section of an infinitely long structure
% r is the radius of the wire in units of h. 2 h is the plate separation.
```

```
% i.e. we are simulating equipotential surfaces at yp = h and -h.
% r is at (x,y) = (0,0)

eps0 = 8.854; % pFd/m

h = 1; % This is arbitrary since everything scales
yp = h; % upper equipotential surface

% equipotential line
yc = [-250:250]*4*h;
V = [];
for xp = 0:25
    charge_line = -log(xp^2 + (yp - yc).^2) ....
    + .5*log(xp^2 + (yp - yc + 2.*h).^2) ....
    + .5*log(xp^2 + (yp - yc - 2.*h).^2);
    V = [V, sum(charge_line)];
    V1 = mean(V)/4/pi/eps0;
    sigma = std(V);
end
fprintf('Line Average: %f, sigma: %f %f\n', V1, sigma, sigma/V1)

% equipotential circle
yc = [-250:250]*4*h;
V = [];
for t = 0: 2*pi/100: 2*pi
    xp = r*cos(t); yp = r*sin(t);
    charge_line = -log(xp^2 + (yp - yc).^2) ....
    + .5*log(xp^2 + (yp - yc + 2.*h).^2) ....
    + .5*log(xp^2 + (yp - yc - 2.*h).^2);
    V = [V, sum(charge_line)];
    V2 = mean(V)/4/pi/eps0;
    sigma = std(V);
end
fprintf('Circle Average: %f, sigma: %f %f\n', V2, sigma, sigma/V2)

Cap = 1./abs(V2 - V1);

end
```

Using a total of 251 of the three-line structures, arranged as shown in Figure 5.21, we calculated the voltage along $y = h$ and $y = -h$ from $x = 0$ to $x = 25\,h$ in steps of h for $r = 0.1\,h$. The mean of the voltage is 2×10^{-5}, with a standard deviation of 2% of the mean. This is an excellent equipotential plane.

Although thousands of lines of charge density are present, the equipotential surface near one of these charge densities should still appear circular. For the same parameters as given above, the mean voltage on a circle of radius 0.1, taken at 100 evenly spaced points, is 0.0458 V with a standard deviation of 6.4% of the mean. This is a good approximation to a circle.

This is a natural structure for calculating the capacitance of a thin round wire sandwiched between two parallel plates. Since the charge enclosed by the circle is 1 C/m, the total charge crossing the r surface must also be 1 C/m, and

$$C = \frac{1}{V_{\text{line}} - V_{\text{circle}}} = 21.87\,\text{pF/m} \qquad (5.52)$$

For $r < 0.1\,h$, the circular equipotential approximation improves. For example, at $r = 0.05\,h$ the standard deviation of the voltage around the circle is only 1.2% of the mean. On the other hand, for $r > 0.1\,h$, it quickly deteriorates. For $r = 0.15\,h$, for example, the standard deviation is 17% of the mean.

As noted at the beginning of this discussion, this is an example of some properties of charge distributions that just happened to come together to create equipotential surfaces that approximate a useful physical structure. The accuracy of the result is excellent (for $r < 0.1\,h$). Unfortunately, it is very difficult to generalize this technique to more than a "keep your eyes open" recommendation.

PROBLEMS

5.1 Create a square electrode parallel plate capacitor file for a 1×1 cm square electrode (and its image). Use a resolution of $a = 0.02$ cm. Calculate the capacitance for various values of h, then plot the capacitance and the ideal parallel plate capacitance.

5.2 Simulate a parallel plate capacitor using $a = 0.0002$, $h = 0.003$, and vary $n_x = n_y$ from 15 to 100. Plot the results and from these results to estimate the parallel plate and edge capacitances of this structure.

5.3 When transmission lines such as microstriplines are made with very thin center conductors, only the width : height ratio of these lines matters in determining the characteristic impedance. What is the characteristic impedance of a microstripline built on a material with a relative dielectric constant of 10 and a center conductor width : thickness ratio of 1 : 1?

5.4 Assume that the center conductor and ground-plane metals in the example for `mod_2d.m` are very thin. (This is an RF skin effect issue. This approximation ensures that the current density is uniform through the thickness of the conductors. It is not the best way to build these lines, but it is an accurate way of avoiding RF issues in this problem). Estimate what percentage of the Ohmic losses occur in the upper conductor of the line.

5.5 The efficiency of the 2d calculation process enables useful "thick" conductors to be reasonably simulated by adding layers of conductors in the X–Y plane. Build an air–microstrip transmission line with $w = h = 1$ and calculate how the characteristic impedance varies as the line thickness varies from 0 to 0.1. Compare the results to published calculations.

5.6 Again using the `mod_2d.m` calculation, create a file for two identical horizontal thin conductors adjacent at the same value of y. The conductors are 10 units long and are separated by two units. In this case the image ground plane is the vertical line between the ground planes. Set up this solution, and then find the electric field at the tip of the conductor facing the ground plane and along a line from the conductor to the ground plane. This problem simulates the electric field along a sharp edge near a counterelectrode plane.

REFERENCES

1. http://www.cepd.com/calculators/microstrip.htm.

2. http://www.ideaconsulting.com/strip.htm.

3. A. Farrar and A. T. Adams, Characteristic impedance of microstrip by the method of moments, *IEEE Trans. Microwave Theory Tech.* **MTT-18**: 65–66   (Jan. 1970).

6

Triangles

6.1 INTRODUCTION TO TRIANGULAR CELLS

All of the examples and calculations presented thus far have been based on the use of rectangular cells in an X–Y plane. Rectangular cells in an X–Y plane are very useful for many practical problems, so this wasn't a bad choice. However, rectangular cells are very awkward to work with in three-dimensional (3d) structures, especially when the surfaces involved are not planar. A good example of this problem is the surface of a sphere; there is simply no convenient way to approximate the surface of a (conducting) sphere using a collection of rectangular cells. Also, when a structure has regions where smaller cells are needed to resolve complex geometric features and/or accurate approximate regions where the electric field is changing rapidly but larger cells are satisfactory for much of the geometry, the programs presented in Chapters 1 and 3–5 seldom can do the job — they were written to be as simple as possible and used square cells. A much more useful choice is triangular cells of arbitrary size, location, and orientation. Also, and finally, readers by now must be wondering whether there isn't more to electrostatics than transmission line sections.

Restricting ourselves to uniform-size rectangular cells in X–Y planes let us develop basic method of moments (MoM) solution techniques for different situations (images, dielectric interfaces, rotated electrodes, etc.) and look at solutions without being distracted by complexities of processing geometric information. The `get_data_xxx.m` functions presented in Chapters 4 and 5 translated two- or three-line input files to all the necessary cell information in only a few lines of code. Moving on to triangles of arbitrary size location and orientation in 3d space brings with it a lot of programming baggage that must be layered on top of the basic algorithm of describe geometry — calculate $L_{i,j}$ — solve the linear equation set. Unfortunately, there is no way around this.

Introduction to Numerical Electrostatics Using MATLAB, First Edition. Lawrence N. Dworsky.
© 2014 John Wiley & Sons, Inc. Published 2014 by John Wiley & Sons, Inc.

Approximating a surface with triangular (or other) regions is called *tessellation*. Generating tessellations in two dimensions for MoM problems or in two or three dimensions for finite element or other techniques is a sophisticated field into itself, and there are many texts describing many techniques as well as many software packages available. A full discussion of this subject is outside the scope of this book and won't be delved into deeply here. Instead, in this chapter we'll work with existing MATLAB capability and a freely available package that, together, give us a very versatile and powerful capability. The calculations to be described are correct regardless of how the cell structure was created, so these discussions are not limiting in any sense.

6.2   RIGHT TRIANGLES

Limiting our mathematics to right triangles simplifies the work to be done. As shown in Figure 6.1a, any triangle can be subdivided into two right triangles by drawing a line from the vertex opposite the longest side to the longest side, at a 90° angle to the longest side.

In the case of a triangle with two equal longest sides or three equal sides, the choice is arbitrary.

Consider any right triangle, rotated so that its legs are parallel to the X and Y axes (Figure 6.1b). The leg parallel to the X axis is of length a, the leg parallel to the Y axis is of length b. It can easily be shown that the three lines drawn from each vertex to the midpoint of the opposite side intersect at the common point $(a/3, b/3)$. This point is the barycenter,[1] that is, the center of mass, of the right triangle—regardless of the values of a and b. If we assume that all the charge is concentrated at this point (the long-distance approximation for $L_{i,j}$), then the odd-order moments cancel,[2] making this the best choice for this point.

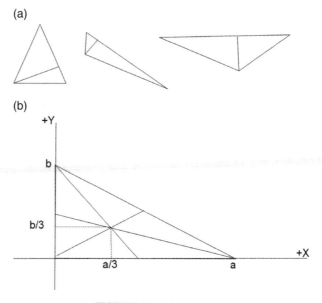

FIGURE 6.1   Right triangles.

Note that the center of a rectangle is its barycenter, due to the symmetries of the rectangle, so no problems arose with rectangles and there was no need to mention this consideration earlier.

Going forward, assume that every triangle (other than right triangles themselves) that is used to describe an electrode geometry will be broken into two right triangles and that there can then be up to twice as many cells (variables) in the numerical problem as there are in the geometric description. If the initial triangles (number, sizes, and positions) have been well chosen to describe the problem, then these subdivisions will not degrade the quality of the model.

6.3 CALCULATING $L_{i,i}$ (SELF) COEFFICIENTS

Figure 6.2 shows the right triangle of Figure 6.1 translated so that the barycenter of the triangle is at the origin of the axes. Note that there has been no rotation—sides a and b are still parallel to the X and Y axes, respectively. The triangles of Figures 6.2 and 6.1 *look* different; this was done deliberately to emphasize the point that the definitions of a and b are consistent in both figures and the calculations below will always be correct, regardless of the relative sizes of a and b. This is not the same as saying that values of a and b are not important.

Without any loss of generality, $L_{i,i}$ is the voltage at $(0,0)$ due to a uniform charge distribution on the triangle with a total charge of Q_i:

$$\sigma_i = \frac{Q_i}{A_i} = \frac{2Q_i}{ab} \tag{6.1}$$

$$L_{i,i} = \frac{1}{4\pi\varepsilon_0 A_i} \iint \frac{dx\,dy}{\sqrt{x^2 + y^2}} \tag{6.2}$$

where the integral is over the area of the triangle.

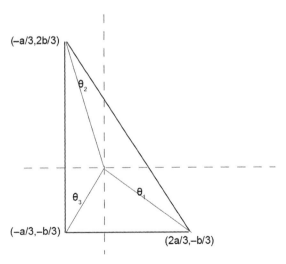

FIGURE 6.2 Layout for calculating $L_{i,i}$ for a right triangle.

Switching to polar coordinates and substituting the area of the triangle, we obtain

$$L_{i,i} = \frac{1}{2\pi\varepsilon ab} \int_0^{2\pi} \int_0^{r(\theta)} dr\, d\theta \tag{6.3}$$

Referring to Figure 6.2, the three angles shown are, in terms of a and b, as follows:

$$\theta_1 = -\arctan\frac{b}{2a} \tag{6.4}$$

$$\theta_2 = \pi - \arctan\frac{2b}{a} \tag{6.5}$$

$$\theta_3 = \frac{3\pi}{2} - \arctan\frac{b}{a} \tag{6.6}$$

The entire triangle is then divided into the three regions

$$\theta_1 \leq \theta < \theta_2, r = \frac{b}{3}\frac{1}{\sin(\theta) + (b/a)\cos(\theta)} \qquad \text{(region 1)} \tag{6.7}$$

$$\theta_2 \leq \theta < \theta_3, r = \frac{a}{3}\frac{1}{\cos(\pi-\theta)} \qquad \text{(region 2)} \tag{6.8}$$

$$\theta_3 \leq \theta < \theta_1, r = \frac{b}{3}\frac{1}{\cos\{[(3\pi)/2]-\theta\}} \qquad \text{(region 3)} \tag{6.9}$$

The integrals in these three regions are

$$\frac{b}{3}\frac{1}{\sqrt{1+\frac{b^2}{a^2}}}\ln\frac{\tan\left(\frac{1}{2}(\theta_2 + \arctan(b/a))\right)}{\tan\left(\frac{1}{2}(\theta_1 + \arctan(b/a))\right)} \qquad \text{(region 1)} \tag{6.10}$$

$$\frac{-a}{3}\ln\frac{\tan[(\theta_3/2) + (\pi/4)]}{\tan[(\theta_2/2) + (\pi/4)]} \qquad \text{(region 2)} \tag{6.11}$$

$$\frac{-b}{3}\ln\frac{\tan(\theta_1/2)}{\tan(\theta_3/2)} \qquad \text{(region 3)} \tag{6.12}$$

At this point it is straightforward to substitute the angle definitions into equations (6.10)–(6.12), sum the three results, and get the final expression for $L_{i,i}$. The resulting expression, however, is so messy that it would take at least half of the page to write it out. Since there is very little doubt that anyone needing this expression will generate the numbers using computer code, it is just as useful to leave things as they are. Starting with a and b, calculate the three angles using equations (6.4), (6.5), and (6.6); substitute these angles into equations (6.10), (6.11), and (6.12); and add the three resulting numbers together to obtain the value of the integral in equation (6.3).

6.4 CALCULATING $L_{i,j}$ FOR $i \neq j$

As in the case of rectangles, we cannot integrate the general expression for $L_{i,j}$. Following the same procedure as in the case of rectangles to develop an approximate expression, the following expression gives good results. Let j represent the charged (right) triangle with (x_j, y_j, z_j) as its barycenter. Let (x_i, y_i, z_i) be the field point at which we want the voltage (in the case, the barycenter of triangle i).

Let

$$r = \sqrt{\left(\left(x_i - x_j\right)^2 + \left(y_i - y_j\right)^2 + \left(z_i - z_j\right)^2 \right)} \tag{6.13}$$

Then

$$L_{i,j} \approx \frac{1}{4\pi\varepsilon_0 r + \dfrac{1}{L_{j,j}\left(1 + r/s\right)}} \tag{6.14}$$

where

$$s = \frac{a+b}{4} \tag{6.15}$$

6.5 BASIC MESHING AND DATA FORMATS FOR TRIANGULAR CELL MoM PROGRAMS

Our modeling program using triangular cells in principle parallels the rectangle program(s), but there are now many more tasks to complete. Some of these tasks are due to the use of triangles, and some of them are necessary because, unlike the previous constraint that all electrodes be in an X–Y plane, we're eliminating all constraints on electrode location and orientation.

A good user-oriented modeling program would take the issue of data validation very seriously. For example, are we accidentally specifying that one electrode, at some voltage, passes through or lies on top of another electrode at some other voltage? For our purposes this is extra programming; it contributes neither to understanding of the electrostatics or modeling nor to the results. In general, ignoring this task is a poor decision; it is being made here to try and keep a large task from becoming an overwhelming task for a newcomer.

Going forward, task 1 consists of generating triangular element surfaces that represent the geometry we wish to model. Since the results of this task are not unique, we must make some decisions in addition to the basic "how much resolution do we need — aka what is the value of a?" choice that came up in the previous models.

For example, Figure 6.3 shows a simple handmade right triangle mesh of a planar square electrode. There is a possible problem, however. While both the upper left and lower right triangles fill the corners the same way and the upper right and lower left triangles fill the corners the same way, all four corners are certainly not treated similarly. If we were to build

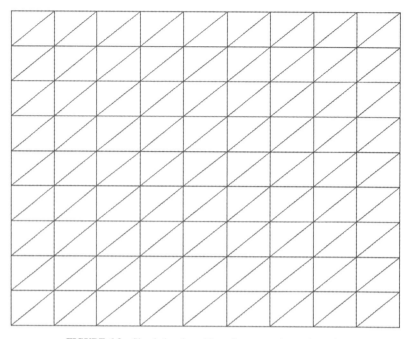

FIGURE 6.3   Simple hand meshing of a square planar electrode.

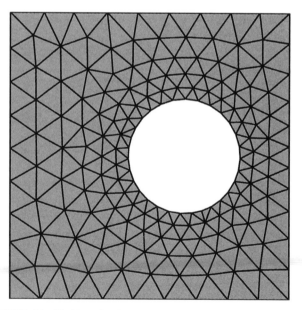

FIGURE 6.4   Meshing of a planar square electrode with an off-center hole.

a capacitor using two identical planar squares, we would probably get a good calculation of capacitance, but what about the fields and charge densities in the corners? Physically we know that they should be identical, but would they be identical?

Figure 6.4 shows a more complicated structure. This mesh was generated using a software package that will be presented later. The configuration of the drawing has been

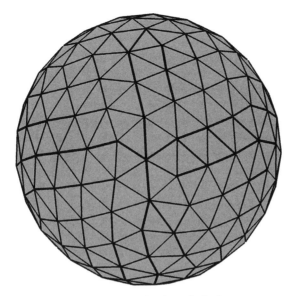

FIGURE 6.5   Uniformly meshed sphere.

changed from that in Figure 6.3. Because we might now consider electrode surfaces with holes (and/or irregular outer boundaries) the actual electrode region is shown in gray; this makes an internal hole very visible.

This electrode has been set up so that the hole is off-center in the square and the resolution near the hole is higher than elsewhere. Now we have to make not one, but several, decisions about resolution.

Figure 6.5 shows a meshed sphere surface. All of the triangles are not identical. Even when unwanted, this happens sometimes because the geometry is not amenable to identical triangular meshing and sometimes the generation software just can't figure out how to do it. In the case of concentric spherical electrodes, for example, we know that physically the charge density on both spheres must be absolutely uniform; however, the model will show small but noticeable variations. This means that we must consider cell size and the expected limits of resolution.

The program `mom_tri_1.m` is presented here to demonstrate a triangular cell script:

```
% mom_tri_1.m   script for solving triangular cell mom problems

clear

% This is a first pass test program to demonstrate all the routines
% necessary to model triangles

% Each row in the array all_tris is a triangle that is, going forward,
% brought in from the mesh generator(s).
% Cols 1:3 is the first node (arbitrary) point (nx1,ny1,nz1)
%      4:6         2nd                          (nx2,ny2,nz2)
%      7:9         3rd                          (nx3,ny3,nz3)
%      10 is the voltage the triangle is set to
```

```
% This data file is hand-generated and written into the routine

all_tris =                [-.5  0  0    1    0  0   1    2   0    .5];  % #1
all_tris = [all_tris;     [-.5  0  0    1    2  0  -.5   2   0    .5]]; % #2
all_tris = [all_tris;     [-1   0  1    1    0  1  1.5   1   1.  -.5]]; % #3
all_tris = [all_tris;     [-1   0  1   1.5   1  1   0    1   1   -.5]]; % #4

q = triangles_1(all_tris);

V_off = q(end)
q = q(1:end-1)
C1 = sum(q.*(q>0))
C2 = sum(q.*(q<0))
```

MATLAB program `mom_tri_1.m` is a simple demonstration script for setting up and solving a method of moments (MoM) problem using triangular cells. This program includes two input electrodes, each composed of two triangles. These triangles are shown in Figure 6.6.

In Figure 6.6 the triangles are numbered according to their order in the computer file. The lines in the file can appear in any order without affecting the results. The program `mom_tri_1.m` (a utility program for triangular cell MoM modeling) calls `triangles_1.m` and passes the triangle information (`all_tris`) to it.

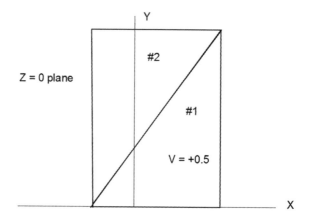

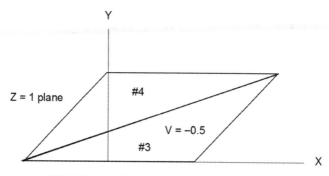

FIGURE 6.6   Triangles described in `mom_tri_1.m`.

```
function [q, rt_tris] = triangles_1(all_tris)

  %   triangles_1.m

  [nr_input_tris,p] = size(all_tris);
  fprintf ('Number of triangles in input file = %d \n', nr_input_tris)

  % INPUT DATA VERIFICATION SHOULD OCCUR HERE IN A REAL WORKHORSE PROGRAM

  % Generate right triangles from the data
  [rt_tris, volts] = make_right_tris(all_tris);

  tic
  L = get_L_2(rt_tris);    % vectorized
  nr_vars = length(L);
  new_col = ones(nr_vars,1);
  L = [L, new_col];
  new_row = [new_col',0];
  L = [L; new_row];
  toc

  % Solve and finish

  volts = [volts;0];
  q = L\volts;

end
```

Program `triangles_1.m` is a utility program for calling other programs and performing some minimal file processing. It immediately calls `make_right_tris.m`, passing the triangle data (`all_tris`) to it. Ultimately this returns an array of right triangles (`right_tris`) and a vector of triangle voltages (volts).

The `make_right_tris.m` program — which provides a function for translating a triangle data file to an all-right-triangle data file — is as follows:

```
function [right_tris, volts] = make_right_tris( all_tris )
% Reads the all_tris list line by line and then splits every
%   triangle into two right triangles. An incoming right
%   triangle is just passed along

  right_tris = [];    volts = [];
  [nr_input_tris,p] = size(all_tris);
  for i = 1: nr_input_tris
    % nodes and voltage of the ith input triangle
    nodes = [all_tris(i,1:3);  all_tris(i,4:6);  all_tris(i,7:9)];
    v = all_tris(i,10);

    [angs,nodes] = get_included_angles(nodes);   % get all the included
angles
    if abs(sum(angs) - pi) > .01*pi     % check for problems
      fprintf ('Sum of angles error in triangle %d \n', i)
    end
```

```
    % If triangle i already a right triangle, just add it to the list
    if (abs(angs(1) - pi/2) < .001*pi/2)
      right_tris = [right_tris; [nodes(1,:), nodes(2,:), nodes(3,:)]];
      volts = [volts;v];
    else
    % find the point for the new right angle vertices
      s21 = sqrt(sum((nodes(2,:) - nodes(1,:)).^2));
      s32 = sqrt(sum((nodes(3,:) - nodes(2,:)).^2));
      f = s21/s32*cos(angs(2));
    % create the new node and new triangles
      new_node = nodes(2,:) + f*(nodes(3,:) - nodes(2,:));
      right_tris = [right_tris; [new_node, nodes(2,:), nodes(1,:)]];
      right_tris = [right_tris; [new_node, nodes(3,:), nodes(1,:)]];
      volts = [volts;v]; volts = [volts;v];

    end
  end
end


% --------------------------------------------------------------


function [angs, nodes] = get_included_angles(nodes)

  angs = [];    % get the included angles, return biggest as first
  v12 = nodes(2,:) - nodes(1,:);
  s12 = sqrt(sum(v12.^2));
  v13 = nodes(3,:) - nodes(1,:);
  s13 = sqrt(sum(v13.^2));
  dot = sum(v12.*v13);
  angs = [angs,acos(dot/s12/s13)];
  v21 = -v12;
  s21 = s12;
  v23 = nodes(3,:) - nodes(2,:);
  s23 = sqrt(sum(v23.^2));
  dot = sum(v21.*v23);
  angs = [angs,acos(dot/s21/s23)];
  v31 = -v13;
  s31 = s13;
  v32 = -v23;
  s32 = s23;
  dot = sum(v31.*v32);
  angs = [angs,acos(dot/s31/s32)];
  [ang_max, index] = max(angs);
  if index == 2
    temp = nodes(2,:);
    nodes(2,:) = nodes(1,:);
    nodes(1,:) = temp;
    temp2 = angs(2); angs(2) = angs(1); angs(1) = temp2;
  elseif index == 3
    temp = nodes(3,:);
    nodes(3,:) = nodes(1,:);
    nodes(1,:) = temp;
```

```
  temp2 = angs(3); angs(3) = angs(1); angs(1) = temp2;
  end

end
```

The `make_right_tris.m` program performs several operations on the `all_tris` array. Each line of `all_tris` describes a triangular electrode and the voltage on it. For each line of `all_tris`, the `make_right_tris.m` program performs the following tasks:

1 Splitting the nine (9) geometry numbers into a 3×3–node array, each row containing the (x,y,z) information for one of the triangle's nodes (arbitrarily ordered at this point).

2 Calling the `get_included_angles` array, which calculates a 1×3 array of the included triangle angles, corresponding to the three nodes. It then rearranges the order of both the angles and node arrays so that the largest angle (and corresponding node) is first.

3 Adding up the angles and ensuring that they sum to π. (This was actually a debugging check during program development, but it can't hurt to leave it in.)

4 Creating a row in `all_tris` [if `angs(1)` = $\pi/2$, this is already a right triangle] with the node information and an entry in the volts vector with the voltage information. It then loops to the next triangle.

5 Finding the point on the longest side, which, when connected to vertex 1, creates two right triangles. With reference to Figure 6.6, using the notation $N_i = (x_i, y_i, z_i)$, we obtain

$$S_{12} \equiv |N_2 - N_1| \tag{6.16}$$

and so on. The new node, N_4, is given by

$$N_4 = N_2 + f(N_2 - N_1) \tag{6.17}$$

where $0 < f < 1$, f is yet to be determined. From Figure 6.7 and equation (6.17), we obtain

$$S_{24} = |N_4 - N_2| = f|N_3 - N_2| \tag{6.18}$$

Therefore

$$f = \frac{S_{24}}{|N_3 - N_2|} = \frac{S_{12} \cos(\theta_2)}{|N_3 - N_2|} = \frac{|N_2 - N_1| \cos(\theta_2)}{|N_3 - N_2|} \tag{6.19}$$

6 Adding the two new right triangles, (N_4, N_1, N_2) and (N_4, N_1, N_3), to the `rt_triangles` array and add the voltage (twice) to the volts vector.

The program `get_L_1.m`—a first-pass L calculation program—is as follows:

```
function [L] = get_L_1(rt_tris )
%Generate the L array from the right triangles
```

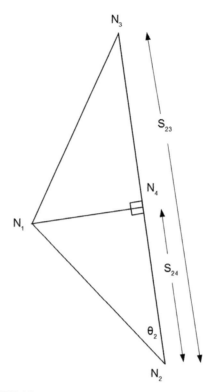

FIGURE 6.7 Arbitrary triangle split into two right triangles.

```
eps0 = 8.854;
[nr_rt_tris,junk] = size(rt_tris);
fprintf('Number of right triangles = %d \n', nr_rt_tris)
L = zeros(nr_rt_tris,nr_rt_tris); % allocate L array

for j = 1: nr_rt_tris % j triangles are the charged triangles
  a = sqrt(sum((rt_tris(j,4:6) - rt_tris(j,1:3)).^2));
  b = sqrt(sum((rt_tris(j,7:9) - rt_tris(j,1:3)).^2));
  scl = (a + b)/4.; % empirical fitting function parameter
  err_check = (a^2 + b^2 - sum((rt_tris(j,7:9) - rt_tris(j,4:6)).^2)) ...
        / (a^2 + b^2) ;
  if abs(err_check) > 1.e-3
    fprintf ('Right triangle file error, #%d %g\n', j, err_check)
  end
  % barycenter of jth rt tri
  bary_jth = (rt_tris(j,1:3) + rt_tris(j,4:6) + rt_tris(j,7:9))/3;

  th1 = -atan(b/2/a);
  th2 = pi - atan(2*b/a);
  th3 = 3*pi/2 - atan(b/a);

  t1 = b/3/sqrt(1 + b^2/a^2)*log(tan(.5*(th2 + atan(b/a)))/tan(.5*(th1
+ atan(b/a))) );
  t2 = -a/3*log(tan(th3/2 + pi/4.)/tan(th2/2 + pi/4.));
  t3 = -b/3*log(tan(th1/2)/tan(th3/2));
```

```
  L_self_tri = (t1 + t2 + t3)/(2.*pi*eps0*a*b);

  for i = 1: nr_rt_tris % this is the field point
    bary_ith = (rt_tris(i,1:3) + rt_tris(i,4:6) + rt_tris(i,7:9))/3;
    %  barycenter - barycenter distance
    r = sqrt(sum((bary_ith - bary_jth).^2));
    L(i,j) = 1/(4*pi*eps0*r + 1./L_self_tri./(1 + r/scl));
  end
 end
end
```

The `triangles_1_.m` program passes the `rt_tris` array to `get_L_1.m`, which generates the *L* matrix. The `get_L_1.m` program is straightforward and follows the `get_L` routines presented in previous chapters. Calculating the self- (diagonal) *L* term is much more involved than it was for rectangles, but this is just a bookkeeping task for the computer. The `get_L_1.m` program has been written with explicit looping through all of the *L* matrix terms. This was done for clarity, certainly not for runtime efficiency. This issue will be addressed in Section 6.6.

The rest of the program is identical to the rectangle programs — the extra row and column needed to ensure charge neutrality are added to the *L* array, the set of linear equations is solved for *q*, and capacitance is calculated. For this simple dataset, the result is that $C = 40.7$ pF.

6.6 USING MATLAB TO GENERATE TRIANGULAR MESHINGS

The example in Section 6.5 was complete in that it started with a description of the desired electrode geometry and ended up with a completed calculation of the charge distribution on the electrodes. Generating the triangle descriptions by hand, however, does not allow for interesting geometries. Although not an impossible task, manual generation of triangular meshings very quickly becomes an overwhelming task when the electrode geometries and desired resolutions are not trivial.

Fortunately, MATLAB has built-in capabilities that perform this task very well. Consider the `mom_tri_2.m` program, which presents a triangular cell script for basic MATLAB data generation:

```
% mom_tri_2.m script for solving triangular cell mom problems

close all; clear

all_tris = mesh_prob_6_5;
[q, rt_tris] = triangles_1(all_tris);

V_off = q(end)
q = q(1:end-1);
C1 = sum(q.*(q > 0))
C2 = sum(q.*(q < 0))

figure(2)
```

```
hold on
y = 0;

for z = 1 : .1 : 1.51
  x_plot = []; v_plot = [];
  for i = 0 : 100
    x = (i/100);
    x_plot = [x_plot, x];
    pt = [x,y,z];
    v_plot = [v_plot, get_V(pt,rt_tris,q)];
  end
  plot(x_plot, v_plot)
end

axis ([0 1 -.05 .55])
H = line ([.5 .5], [0, .5]);
set(H, 'Linestyle', '--')
xlabel ('(X,Y,Z) = (X,0,0)')
ylabel ('Volts')
```

The `mom_tri_2.m` program is a simple modification of `mom_tri_1.m`. The explicit definition of the `all_tris` array has been replaced by a call to the function `mesh_gen_1`.

The `mesh_gen_1.m` program — which provides a function for generating and displaying the basic meshing procedure — is as follows:

```
function all_tris = mesh_gen_1
% mesh_gen_1.m first pass generated triangles data from Matlab routine

  test = [-0.5 : .07 : .5];
  [x,y] = meshgrid(test);
  %h = 1.0;   z = 0*x + h;        % flat electrode example
  h = 1.5;   z = 0*x + h;         % Gaussian Tip example

  tri = delaunay(x,y);    % generate triangles
  nr_tris = length(tri);

  trimesh(tri,x,y,z)

  all_tris = [];          % generate all_tris array
  V = 0.5;
  for i = 1:nr_tris
    n = tri(i,:);
    t1 = [x(n(1)), y(n(1)), z(n(1))];
    t2 = [x(n(2)), y(n(2)), z(n(2))];
    t3 = [x(n(3)), y(n(3)), z(n(3))];
    all_tris = [all_tris; [t1, t2, t3, V]];

  end

  %C = [.5,.5,.5; .5,.5,.5; .5,.5,.5];   % insert for b&w graphics
  %colormap(C)                           % insert for b&w graphics
```

```
axis ([-.75 .75 -.75 .75 0. 1.6])
hold on

% ---- bottom electrode --------------------------------------
%h = 0.0; z = 0*x + h; % flat electrode example

% improved tip resolution ------------------------------
rad = sqrt (x.^2 + y.^2);
x = x.*rad; y = y.*rad;
% ------------------------------------------------------
%z = 0*x;
% Gaussian Tip example -----------------------
z = exp (-50*(x.^2 + y.^2));
trimesh (tri, x, y, z);
% ------------------------------------

V = -.5;
for i = 1:nr_tris
  n = tri(i, :);
  t1 = [x(n(1)), y(n(1)), z(n(1))];
  t2 = [x(n(2)), y(n(2)), z(n(2))];
  t3 = [x(n(3)), y(n(3)), z(n(3))];
  all_tris = [all_tris; [t1, t2, t3, V]];
 end
end
```

(*Note*: The mesh_gen_1.m program included several lines that are commented out. These lines will be brought in, introducing new capabilities, as this section proceeds.)

The mesh_gen_1.m program begins by defining a square (x,y) grid array. Then a grid $z(x,y)$ is described. In this first example z is simply a constant of value h. Treating this as a three-dimensional (3d) structure when there really is no z structure to describe is overkill, but it sets up the correct format for the next example, which will have true z structure.

The function delaunay (x,y) assigns a node number to each point in the grid (x,y) and then fills the space with triangles. Delaunay triangles fill a space with nonoverlapping triangles and strive to avoid "long, skinny" triangles.[3] This is a trivial task when working with a uniform rectangular grid of points (this example) but can become very challenging when working with an arbitrary set of points.

The function trimesh combines the triangle information with the node location information and produces 3d graphics of the meshed structure.

The mesh_gen_1 program creates the all_tris array described in Section 6.5. It then repeats the task for a second value of h, creating the simple two-electrode structure shown in Figure 6.8.

An important attribute of MoM structures is that "what happens in mesh_gen_1 stays in mesh_gen_1." The all_tris array has node location and voltage information organized by triangles, but there is no numbering of nodes. This means that different geometric structures may share common node locations, but the MoM analysis doesn't know whether these are common node locations or just very close node locations. A possible problem here is that it also doesn't know whether, say, two planes at different voltages are touching or even crossing through each other. Since we are concerned more with the

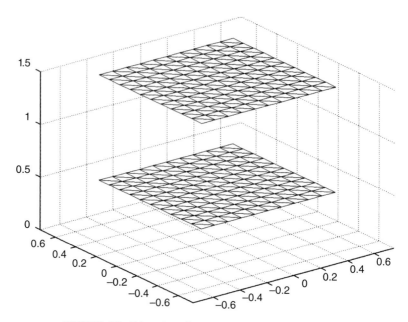

FIGURE 6.8   Triangular cell structure created by mesh_gen_1.

MoM analysis than with the grid creation subtleties, we shall leave this type of checking to visual inspection.

The mesh_gen_1 program uses different colors to represent the voltage of the different electrodes. This is, of course, not possible in a black-and-white publication. Two lines that suppress the color information are shown in the listing.

Once the all_tris array has been defined, mom_tri_2 is identical to mom_tri_1. The resulting calculation predicts a capacitance of 29.8 pF.

The MATLAB tic-toc timer function shows that virtually all of the execution time in this program is used by the get_L_1 function. The *L* array is of size (number of right triangles).[2] The number of elements in this array increases quickly according to the number of triangles in the structure. *L* is a fully populated array. The get_L_1 function addresses every element of this array using a pair of nested loops. While writing get_L_1 this way shows what the function is doing clearly, it does not fully utilize MATLAB's efficiency in processing *vectorized* (no explicit) loops.

The get_L_2 function is a direct replacement for get_L_1, but with no explicit loops. It could probably be written to operate more efficiently than it does, but the code was generated to parallel get_L_1 as closely as possible. Here is the get_L2_.m program — a vectorized version of get_L_1.m:

```
function [L] = get_L_2(rt_tris)
%  Generate the L array from the right triangles

  eps0 = 8.854;

  [n,junk] = size(rt_tris);
  fprintf('Number of right triangles = %d \n', n)

  a2 = rt_tris(:,4:6); a1 = rt_tris(:,1:3);
```

```
a_vec_sq = sum((a2 - a1)'.^2); a_vec = sqrt(a_vec_sq);
b2 = rt_tris(:,7:9); b1 = rt_tris(:,1:3);
b_vec_sq = sum((b2 - b1)'.^2); b_vec = sqrt(b_vec_sq);
c2 = rt_tris(:,7:9); c1 = rt_tris(:,4:6);
c_vec_sq = sum((c2 - c1)'.^2); c_vec = sqrt(c_vec_sq);
scl_vec = (a_vec + b_vec)/4;
err_vec = (a_vec_sq + b_vec_sq - c_vec_sq)./(a_vec_sq + b_vec_sq);
err_max = max(abs(err_vec));
fprintf ('Right triangle err_max = %g \n', err_max)
bary = (rt_tris(:,1:3) + rt_tris(:,4:6) + rt_tris(:,7:9))/3;

th1_vec = -atan(b_vec./2./a_vec); %-atan2(b,2*a);
th2_vec = pi - atan(2.*b_vec./a_vec); %pi - atan2(2*b,a);
th3_vec = 3*pi/2 - atan(b_vec./a_vec); %3*pi/2 - atan2(b,a);

t1_vec = b_vec/3./sqrt(1 + b_vec.^2./a_vec.^2).*log(tan(.5*(th2_vec ...
    + atan(b_vec./a_vec)))./tan(.5*(th1_vec + atan(b_vec./a_vec)))) );
t2_vec = -a_vec/3.*log(tan(th3_vec./2 + pi/4.)./tan(th2_vec./2
+ pi/4.));
t3_vec = -b_vec/3.*log(tan(th1_vec./2)./tan(th3_vec./2));

L_self_vec = (t1_vec + t2_vec + t3_vec)./(2*pi*eps0.*a_vec.*b_vec);

bx = bary(:,1); xs = (repmat(bx,1,n) - repmat(bx',n,1)).^2;
by = bary(:,2); ys = (repmat(by,1,n) - repmat(by',n,1)).^2;
bz = bary(:,3); zs = (repmat(bz,1,n) - repmat(bz',n,1)).^2;
r = sqrt(xs + ys + zs);
a_mat = repmat(a_vec',1,n);
b_mat = repmat(b_vec',1,n);
L_self_mat = repmat(L_self_vec',1,n);
scl_mat = repmat(scl_vec',1,n);
L = 1./(4*pi*eps0*r + 1./L_self_mat./(1 + r./scl_mat));
```

end

To switch to the vectorized function, `triangles_1.m` merely replaces the line

```
L = get_L_1(rt_tris)
```

with

```
L = get_L_2(rt_tris)
```

Since the calculations are identical, the results will be identical. The execution time should decrease by approximately a factor of 150.

6.7 CALCULATING VOLTAGES

Before proceeding to some more interesting geometries, let's consider the issue of finding the voltage and the electric field at arbitrary points in space, particularly near our structure. Returning to our basic definitions, we have

$$V(x_i, y_i, z_i) = \sum_j \lambda_{i,j} q_j \qquad (6.20)$$

where $\lambda_{i,j}$ is the voltage at point j due to all of the (charged) rectangles with barycenters (x_j, y_j, z_j). The notation λ is used to differentiate this term from the $L_{i,j}$ term that was used to calculate q. The methods for calculating the two terms are similar; the i in $L_{i,j}$ refers to the contribution to the voltage at the barycenter of the ith triangle due to the charge on the jth triangle, and the i in $\lambda_{i,j}$ refers to the contribution to the voltage at field point i due to the charge on the jth triangle. In the former case we had set the voltage on all triangles to their applied (boundary) values and found the charge distribution q_j, which supports these applied values. In the latter case we will use the charge distribution q_j, which we have used to find the voltage at an arbitrary point i. If i happens to be the barycenter of one of the electrode triangles, the voltage we find should be the original applied voltage (boundary condition) for that triangle. Here is the get_V.m program—a function used to calculate voltage at an arbitrary point:

```
function V = get_V(pt,rt_tris,q)
%  Get voltage at a given point
%  This is actually a reduced version of get_L_2

  xi = pt(1); yi = pt(2); zi = pt(3);

  eps0 = 8.854;

  a2 = rt_tris(:,4:6); a1 = rt_tris(:,1:3);
  a_vec_sq = sum((a2 - a1)'.^2); a_vec = sqrt(a_vec_sq);
  b2 = rt_tris(:,7:9); b1 = rt_tris(:,1:3);
  b_vec_sq = sum((b2 - b1)'.^2); b_vec = sqrt(b_vec_sq);
  c2 = rt_tris(:,7:9); c1 = rt_tris(:,4:6);
  c_vec_sq = sum((c2 - c1)'.^2); c_vec = sqrt(c_vec_sq);
  scl_vec = (a_vec + b_vec)/4;
  bary = (a1 + a2 + b2)/3;

  th1_vec = -atan(b_vec./2./a_vec);
  th2_vec = pi - atan(2.*b_vec./a_vec);
  th3_vec = 3*pi/2 - atan(b_vec./a_vec);

  t1_vec = b_vec/3./sqrt(1 + b_vec.^2./a_vec.^2).*log(tan(.5*(th2_vec...
     + atan(b_vec./a_vec)))./tan(.5*(th1_vec + atan(b_vec./a_vec)))) );
  t2_vec = -a_vec/3.*log(tan(th3_vec./2 + pi/4.)./tan(th2_vec./2 +
pi/4.));
  t3_vec = -b_vec/3.*log(tan(th1_vec./2)./tan(th3_vec./2));

  L_self_vec = (t1_vec + t2_vec + t3_vec)./(2*pi*eps0.*a_vec.*b_vec);

  r_vec = [xi - bary(:,1), yi - bary(:,2), zi - bary(:,3)];
  r = sqrt(r_vec(:,1).^2 + r_vec(:,2).^2 + r_vec(:,3).^2);
  L = 1./(4*pi*eps0*r + 1./L_self_vec'./(1 + r./scl_vec'));

  V = sum(q.*L);

end
```

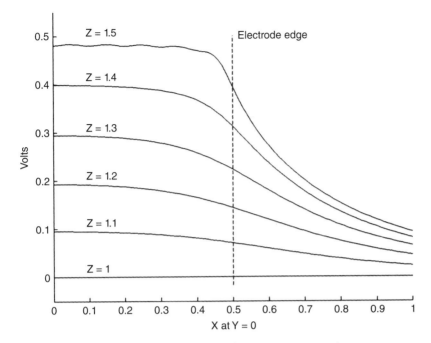

FIGURE 6.9 Voltage profiles of `mom_tri_2.m` example.

A call to `get_V` specifying the point of interest as a 1×3 vector $[x_i, y_i, z_i]$, along with the `rt_tris` and q arrays, returns the value of V at that point. Figure 6.9 shows the results of calculating voltage as a function of x at $y = 0$ for several values of z.

Almost everything in Figure 6.9 is what would be expected. The potential at $x = 0$ increases linearly with z from $v = 0$ at $z = 0.5$ to $v = 0.5$ at $z = 1.5$ The voltage equipotentials are parallel to the electrodes most of the way to the end of the electrodes, beginning to fall as they approach (being under the) electrode edge, and then falling toward 0 with increasing x. The farther the equipotential is from the electrode (decreasing z in this figure), the less control the electrode has on the shape of the equipotential as the edge is approached.

On closer inspection we can see that the $z = 1.5$ equipotential line isn't quite right. First, the equipotential is at ~ 0.48 V, not the 0.5 V set by the electrode voltage. Also, there is an undulation in the line itself. The problem here is that we are too close to the charged triangles and the λ approximation [equation (6.20), as used in `get_V.m`] is not that accurate. To do the job better, we would need a `get_V` function that actually integrates (numerically) over all the charged triangles. A significant weakness in the λ approximation is that it ignores the orientations of the charged triangles — only the barycenter locations are considered. While a function for calculating L that does the job properly would run very slowly, due to the size of the job, a function for calculating λ would be a reasonable undertaking; remember that, to calculate λ, we need (number of triangles) integrations, while to calculate L, we need the square of this amount.

6.8 CALCULATING THE ELECTRIC FIELD

The empirical relationships used in the L and λ approximations of the previous sections were easy to create. All we had to do was find a smooth function (with the approximately correct shape) that connected the two asymptotic regions, the simple $V \sim q/r$ relationship for large $r_{i,j}$ and the known V function (the L self term) at $r = 0$.

The L self term resulted from the integration of

$$\iint_j \frac{dx\,dy}{r} = \iint_j dr\,d\theta \tag{6.21}$$

where region j is the area of the triangle (or rectangle) under consideration and includes the point $r = 0$. While the details of the integration can be very involved, depending upon the shape of the region, this integral always exists.

On the other hand, finding the (magnitude of the) electric field at the barycenter of a uniformly charged planar region requires calculating the integral

$$\iint_j \frac{dx\,dy}{r^2} = \iint_j \frac{dr\,d\theta}{r} \tag{6.22}$$

where once again, $r = 0$ is in the integration region. Since

$$\int \frac{dr}{r} = \ln(r) \tag{6.23}$$

any integration with $r = 0$ as one of the limits does not exist.

In other words, there is no electric field equation analog to the self L term. The electric field goes to zero at the barycenter of a uniformly charged region. We must look elsewhere for a simple estimate of the electric field. We could, of course, perform the (numerical) integrations of the charged regions, but this approach carries the same caveats as those discussed in Section 6.7 for approximating V.

A simple but in most cases effective approximation is to use a centered numerical derivative, such as

$$E_z(x_i, y_i, z_i) = \frac{-V(x_i, y_i, z_i + \delta) - V(x_i, y_i, z_i - \delta)}{2\delta} \tag{6.24}$$

where δ is small enough to allow for the desired resolution, but large enough so as to avoid numerical roundoff errors in the calculations.

Setting $\delta = 0.01$, Figure 6.10 is a repeat of Figure 6.9 except that now the z component of the electric field (E_z) is plotted. The average electric field between two plates 1 m apart with 1 V difference between them is, of course, 1 V/m. For $z = 1.0$ through $z = 1.4$, we see this at $x = 0$, within a few percent. As we increase x, for $z = 1.0$ through $z = 1.2$ the field falls off as x nears the electrode edge, then falls toward 0 with increasing x. For $z = 1.3$ and particularly at $z = 1.4$, the electric field begins to peak near the electrode edge, a consequence of the charge density concentration at the electrode edges.

At $z = 1.5$ we see a very nonphysical situation. Figure 6.10 shows an essentially constant E_z of -0.2 V/m. How could the same calculation that gave such reasonable results for the other values of z perform so badly here?

The answer to this puzzle is that the program is calculating—as computer programs almost always do—exactly what we programmed it to calculate. Equation (6.24) prescribes a centered derivative (with respect to z) calculation about the chosen point. Since our electrode in this example has 0 thickness, the program is subtracting a voltage below the electrode from a voltage above the electrode.

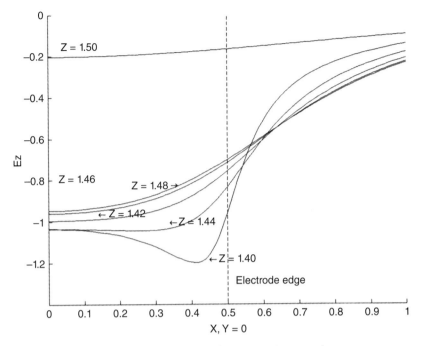

FIGURE 6.10 E_z profiles of `mom_tri_2.m` example.

Figure 6.11 is a closeup view of Figure 6.10, with z values between 1.4 and 1.5. There is a range of results going from the nonsensical value at $z = 1.50$ to the reasonable value at $z = 1.40$. Of interest is that in this model, $a = b = 0.1$ for all of the triangles. While not an actual calculation, the size of the triangles involved provides a good rule of thumb as to how useful calculations of V and E can be: If you are farther away from an electrode than the larger of a and b for one of the larger triangles in the region, then the values of V and E will be reasonable. If not, anything can happen.

6.9 THREE-DIMENSIONAL STRUCTURES

There are an infinite number of possible structures that can be described and analyzed. The limiting factors are the needs of the user and the sophistication of the meshing package being used. In this section we'll examine a few examples to show the possibilities and some of the issues that arise. The first example is shown in Figure 6.12.

This meshing is accomplished by changing just two lines in `mesh_gen_1.m`:

1 Change `h = 1.0;` to `h = 1.5;` for the top electrode.
2 On line 31, change $z = z + h$; to `z = exp(-(7*x).^2 - (7*y).^2);`.

Note that this change is transparent to the calling program `mom_tri_2.m`. As long as the data file is in the correct format, `mom_tri_2.m` will process the data, calculate the charge on each triangle's barycenter, and calculate the capacitance.

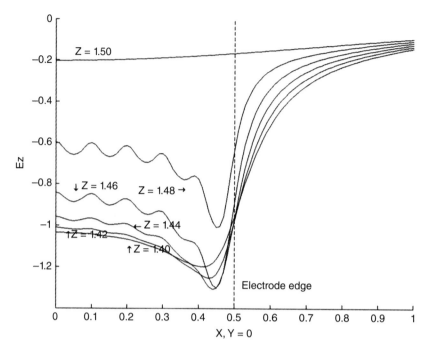

FIGURE 6.11 E_z profiles of mom_tri_2.m example, z near top electrode.

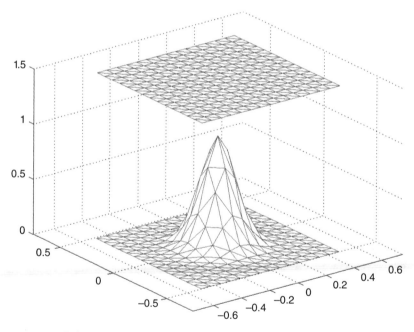

FIGURE 6.12   Parallel plate with a Gaussian "lump" on one plate.

The tip of the Gaussian curve in Figure 6.12 doesn't look particularly Gaussian; the resolution is too low in this region. There are numerous ways to improve this situation. Resolution everywhere could be improved, and the tip of the curve would benefit from this

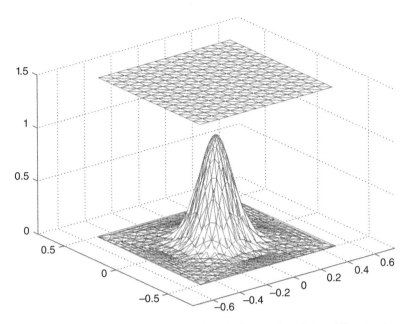

FIGURE 6.13 Structure shown in Figure 6.12 with increased resolution of Gaussian tip.

improvement. This solution is always possible but rarely desirable because the number of cells grows quickly, and the number of $L_{i,j}$ matrix elements to calculate grows with the square of the number of cells. Extra nodes could be added manually, but this is a time-consuming and usually haphazard process.

The two lines labeled "improved tip resolution" in mesh_gen_1.m handle this situation by recalculating the grid spacing of the lower conductor. Instead of a linear spacing (such as used in the upper conductor), all of the points are moved closer to $x = 0$, $y = 0$. This results in the greatly improved shape of the tip, shown in Figure 6.13.

The freely available distmesh package[4] creates meshings using MATLAB code that add greatly to our capabilities.[5] This package comes with extensive documentation and examples that won't be repeated here. We will, however, use two distmesh examples as mesh sources for our own examples. Consider the following program, distmesh_example.m—a distmesh code for a square plate with a circular hole:

```
% distmesh example code

fd = @ (p) ddiff (drectangle (p,-1,1,-1,1), dcircle (p,0,0,0.5));
fh = @ (p) 0.05 + 0.3*dcircle (p,0,0,0.5);
[p,t] = distmesh2d (fd,fh,0.05,[-1 -1; 1 1], [-1 -1; -1, 1; 1,-1; 1,1]);
```

The three lines of MATLAB code shown produce the sophisticated zoning of Figure 6.14.

This simple distmesh code has created (the meshing for) a rectangular electrode with a hole in the center. The distmesh code also produces output arrays that are easily translatable to the all_tris array format used by the examples in this book.

The easiest way to show how to convert distmesh output to the all_tris format is with an example. The MATLAB function mesh_gen_3.m combines the distmesh

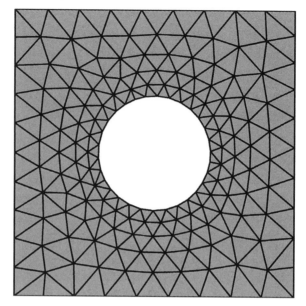

FIGURE 6.14   Results of the `distmesh` sample code `distmesh_example.m`.

example (displayed above) with the Gaussian post example shown previously and generates the graphics to display the result. Here is the program `mesh_gen_3.m`—a structure built by combining previous examples:

```
function all_tris = mesh_gen_3
% mesh_gen_3.m 3d figures generated w/distmesh and Matlab
%   this is the emitter region of a Spindt Tip structure

  figure(1)
  fd = @ (p) ddiff (drectangle (p,-1,1,-1,1), dcircle (p,0,0,0.5));
  fh = @ (p) 0.05 + 0.3*dcircle (p,0,0,0.5);
  [p,t] = distmesh2d (fd,fh,0.05, [-1 -1; 1 1], [-1 -1; -1, 1; 1,-1; 1,1]);

  figure(2)
  X = p (:,1);
  Y = p (:,2);
  Z = ones (length(X),1);
  trimesh (t,X,Y,Z);
  hold on
  C = [.5,.5,.5; .5,.5,.5; .5,.5,.5];
  colormap (C);

  all_tris = [];
  nr_tris = length(X);      %%
  for i = 1:nr_tris
    n1 = [X(t(i,1)), Y(t(i,1)), Z(t(i,1))];
    n2 = [X(t(i,2)), Y(t(i,2)), Z(t(i,2))];
    n3 = [X(t(i,3)), Y(t(i,3)), Z(t(i,3))];
    all_tris = [all_tris; [n1,n2,n3], .5];
  end
```

```
% ------------------------------------

X = p(:,1);
Y = p(:,2);
Z = .98*ones(length(X),1);
trimesh(t,X,Y,Z);
nr_tris = length(X);      %%
for i = 1:nr_tris
  n1 = [X(t(i,1)), Y(t(i,1)), Z(t(i,1))];
  n2 = [X(t(i,2)), Y(t(i,2)), Z(t(i,2))];
  n3 = [X(t(i,3)), Y(t(i,3)), Z(t(i,3))];
  all_tris = [all_tris;[n1,n2,n3], .5];
end


% --------------------------------------------------------

% Gaussian post, compress grid spacing near the center
test = [-1.2 : .14 : 1.2];
[x,y] = meshgrid(test);
rad = sqrt(x.^2 + y.^2);
x = x.*rad/2; y = y.*rad/2;
tri = delaunay(x,y);

z = exp(-(7*x).^2 - (7*y).^2);
tr = TriRep(tri, x(:), y(:), z(:));
trimesh(tr);

nodes = tr.X;
cons = tr.Triangulation;
[nr_tris,p] = size(cons);
for i = 1:nr_tris
    all_tris = [all_tris;[nodes(cons(i,1),:), nodes(cons(i,2),:), ...
          nodes(cons(i,3),:), -.5]];
end

axis ([-1.5 1.5 -1.5 1.5 0. 1.2])

end
```

There are actually two stacked identical distmesh structures in this example. The upper electrode has effectively been given some thickness. If an even thicker electrode were desired, in addition to separating the two identical layers, more sidewall layers would have to be added.

The simple script mom_tri_3.m calls mesh_gen_3 to generate the electrode meshes (see Figure 6.15), triangles_1 to find the charge on the electrodes, and then get_V to examine a voltage profile.

The program mom_tri_3.m—a calling script for mesh_gen_3 structures—is as follows:

```
% mom_tri_3.m script for handling triangular cell mom problems

close all; clear
```

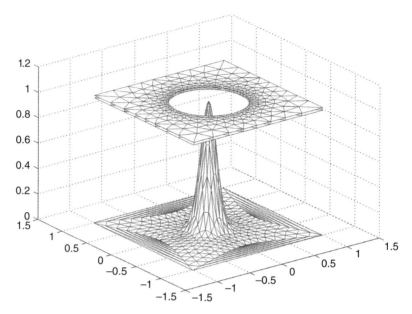

FIGURE 6.15 Combined electrode structure created by `mesh_gen_3`.

```
all_tris = mesh_gen_3;

[q, rt_tris] = triangles_1(all_tris);

q = q(1:end-1);
Cl = sum(q.*(q > 0))

% look at a voltage profile

figure(3)
x_plot = [-.8 : 0.002 : .8]; V = [];

for i = 1 : length(x_plot)
  pt = [0,x_plot(i), .99];
  V = [V, get_V(pt,rt_tris,q)];
end
plot(x_plot, V)
axis ([-.8 .8 -.6 .6])
xlabel ('X at Y = 0, Z = 0.99')
ylabel ('Voltage')
```

In `mesh_gen_3`, the two (identical) planar electrode layers are at $z = 1.00$ and $z = 0.98$, both at the same potential (0.5 V) (see Figure 6.16). The Gaussian post structure peaks at $z = 1.00$, it is at -0.5 V. The voltage profile is taken as a function of x at $y = 0, z = 0.99$. This profile starts and ends at the (x) edges of the hole in the upper layers, at a z value midway between them. It passes directly through the tip of the Gaussian post, 0.01 below the peak of the tip.

Figure 6.17 shows two concentric spheres. This structure was generated using `distmesh`, as shown in MATLAB function `mesh_gen_4.m`.

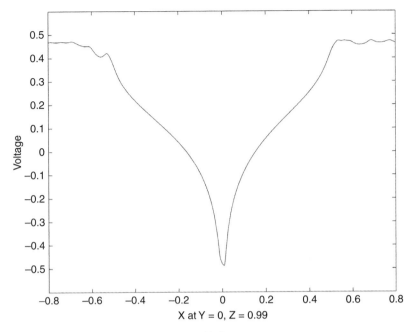

FIGURE 6.16 Voltage profile for mesh_gen_3 structure.

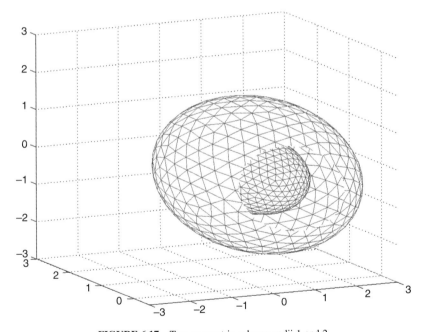

FIGURE 6.17 Two concentric spheres, radii 1 and 2.

Here is program mesh_gen_4.m—a function that can be used to generate concentric sphere meshing:

```
function all_tris = mesh_gen_4
```

```
% mesh_gen_4.m 3d figures generated w/distmesh
%  These are concentric spheres, r = 1 and r = 2

   figure(2)
   all_tris = [];
   C = [.5,.5,.5; .5,.5,.5; .5,.5,.5];
   colormap(C);

   % ------ outer sphere ----------------------

   fd = @ (p) dsphere(p,0, 0, 0, 2);
   [p,t] = distmeshsurface(fd, @huniform, 0.4, 1.1*[-2 -2 -2; 2 2 2]);

   X = p(:,1);
   Y = p(:,2);
   Z = p(:,3);
   trimesh(t,X,Y,Z);
   hold on
   nr_tris = length(t);
   for i = 1:nr_tris
     n1 = [X(t(i,1)), Y(t(i,1)), Z(t(i,1))];
     n2 = [X(t(i,2)), Y(t(i,2)), Z(t(i,2))];
     n3 = [X(t(i,3)), Y(t(i,3)), Z(t(i,3))];
     all_tris = [all_tris;[n1,n2,n3], .5];
   end

% -- inner sphere ------------------------------------

   figure(3)
   fd = @ (p) dsphere(p,0,0,0, 1);
   [p,t] = distmeshsurface(fd, @huniform, 0.2, 1.1*[-1 -1 -1; 1 1 1]);
   close 3

   X = p(:,1);
   Y = p(:,2);
   Z = p(:,3);
   figure(2)
   trimesh(t,X,Y,Z);

   nr_tris = length(t);
   for i = 1:nr_tris
     n1 = [X(t(i,1)), Y(t(i,1)), Z(t(i,1))];
     n2 = [X(t(i,2)), Y(t(i,2)), Z(t(i,2))];
     n3 = [X(t(i,3)), Y(t(i,3)), Z(t(i,3))];
     all_tris = [all_tris;[n1,n2,n3], -.5];
   end

   axis ([-3 3 -.5 3-3 3])
end
```

Figure 6.17 shows the outer sphere partially cut away so as to expose the inner sphere (see line 50 in mesh_gen_4).

MATLAB script mom_tri_4 calculates the capacitance of this structure and produces the curve of the voltage versus the radius between the spheres. Because the electric field in this structure is confined to the region between the spheres, an offset voltage is required to correct for the voltage at infinity—the outside world sees only the outer sphere. This voltage is automatically calculated as q (end) by mom_tri_4 and must be added to the voltage predicted (lines 8 and 22 in mom_tri_4) to get the correct results.

The mom_tri_4 function predicts a capacitance of 227.2 pF. From basic considerations, where a and b are the inner and outer radii, respectively, we obtain

$$V_{a,b} = \frac{q}{4\pi\varepsilon_0} \left(\frac{1}{a} - \frac{1}{b} \right) \tag{6.25}$$

From which

$$C = \frac{q}{V} = \frac{4\pi\varepsilon_0}{(1/a)-(1/b)} = 222.5 \,\text{pF} \tag{6.26}$$

The model's prediction is 2% above the exact answer.

The voltage between the spheres is only a function of r (symmetry demands this) and again, from basic considerations [a direct extension of equation (6.25)], we obtain

$$V(r) = \frac{q}{4\pi\varepsilon_0} \left(\frac{1}{r} - \frac{1}{b} \right) + V(a) \tag{6.27}$$

The program mom_tri_4.m—the script used to model concentric spheres—is as follows:

```
% mom_tri_4.m script for handling triangular cell mom problems

close all; clear

all_tris = mesh_gen_4;
[q, rt_tris] = triangles_1(all_tris);

V_off = q(end)
q = q(1:end-1);
C1 = sum(q.*(q>0))

% look at a voltage profile

figure(4)
x_plot = [1. : .05 : 2]; V = []; V_exact = [];

for i = 1 : length(x_plot)
  pt = [x_plot(i), 0, 0];
  V = [V, get_V(pt,rt_tris,q)];
  V_exact = [V_exact, -.5 - 2*(1/x_plot(i) - 1)];
end
plot(x_plot, V + V_off, 'kx', x_plot, V_exact, 'k')
xlabel ('X at Y = 0, Z = 0')
ylabel ('Voltage')
```

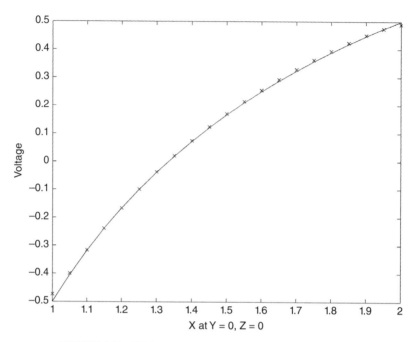

FIGURE 6.18   *V*(*r*) between the spheres: model and analytic results.

The `mom_tri_4.m` program plots the comparison of the model results and the analytic voltage between the spheres. The result is shown in Figure 6.18. For the most part the results of the model (the *x* values) agree very well with the analytic expression (the continuous line).

There is some small degree of error at both spheres (the left and right extremes of the figure), as there was in previous structure voltage scans. Again, the error arise when we try to "get too close" to an electrode.

6.10   Charge Profiles

In previous sections, when electrodes were rectangular and consisted of square cells, it was easy to examine a charge profile. In the more general case of arbitrarily sized and placed triangular cells, unfortunately, this is not the case.

MATLAB program `mom_tri_5.m` creates a worst-case situation to illustrate the issues:

- Program `mom_tri_5.m`—MATLAB script used to generate random node points in a square region:

```
% mom_tri_5.m script for looking at a charge profile

close all; clear

% note that mesh_get_5 uses a random number generator, results won't
repeat
```

```
all_tris = mesh_gen_5;
[q, rt_tris] = triangles_1(all_tris);

V_off = q(end)
q = q(1:end-1);
C1 = sum(q.*(q > 0))

% get some barycenters and areas for all the rt_tris
[barys, areas] = process(rt_tris);

tol = .02;
% find all triangles with |y| of barycenter < tol
n = length(q); x_used = []; rho_used = [];
for i = 1: n
  if abs(barys(i,2)) < tol
    if q(i) > 0
      x_used = [x_used, barys(i,1)];
      rho_used = [rho_used, q(i)/areas(i)];
    end
  end
end

figure(2)
plot(x_used, rho_used, 'x')
xlabel ('x at y = 0, top electrode')
ylabel ('charge density')
```

- Program mesh_gen_5.m:

```
function all_tris = mesh_gen_5
% mesh_gen_5.m
  %figure(1)
  %hold on
  %C = [.5,.5,.5; .5,.5,.5; .5,.5,.5];
  %colormap(C);

  all_tris = [];
  x = -.5 + rand(400,1);
  y = -.5 + rand(400,1);
  tri = delaunay(x,y);

  z = ones(size(x));
  tr = TriRep(tri, x(:), y(:), z(:));
  trimesh(tr);
  hold on
  C = [.5,.5,.5; .5,.5,.5; .5,.5,.5];
  colormap(C);

  nodes = tr.X;
  cons = tr.Triangulation;
  [nr_tris,p] = size(cons);
  for i = 1:nr_tris
```

```
        all_tris = [all_tris; [nodes(cons(i,1),:), nodes(cons
(i,2),:), ...
                nodes(cons(i,3),:), .5]];
    end


%  -------------------------------------------------------


    % flat electrode
    test = [-.5 : 0.05 : .5];
    [x,y] = meshgrid(test);

    tri = delaunay(x,y);

    z = zeros(size(x));
    tr = TriRep(tri, x(:), y(:), z(:));
    trimesh(tr);

    nodes = tr.X;
    cons = tr.Triangulation;
    [nr_tris,p] = size(cons);
    for i = 1:nr_tris
        all_tris = [all_tris; [nodes(cons(i,1),:), nodes(cons
(i,2),:), ...
                nodes(cons(i,3),:), -.5]];
    end
    axis([-.7 .7 -.7 .7 -.2 1.2])

end
```

- Program `process.m`:

```
function [barys, areas] = process(rt_tris)
%

    %n = max(size(used_tris)
    %fprintf('Number of used right triangles = %d \n', n)

    a2 = rt_tris(:,4:6); a1 = rt_tris(:,1:3);
    a_vec_sq = sum((a2 - a1)'.^2); a_vec = sqrt(a_vec_sq);
    b2 = rt_tris(:,7:9); b1 = rt_tris(:,1:3);
    b_vec_sq = sum((b2 - b1)'.^2); b_vec = sqrt(b_vec_sq);

    barys = (rt_tris(:,1:3) + rt_tris(:,4:6) + rt_tris(:,7:9))/3;
    areas = (a_vec.*b_vec/2)';

end
```

Then `mom_tri_5.m`, using the function `mesh_gen_5.m`, creates a new square electrode parallel plate capacitor (see Figure 6.19). One of the electrodes is identical to the previously generated simple square electrode with triangular cells. The other electrode,

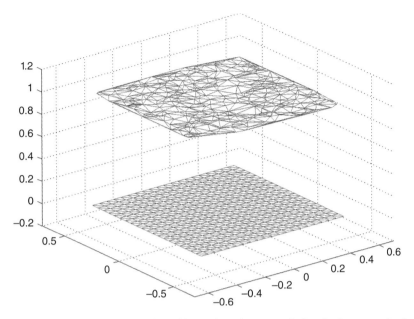

FIGURE 6.19 Parallel plate capacitor with one electrode composed of randomly generated cells.

however, is created using a set of points randomly scattered about the square electrode region.

Remember that this program will not repeat exactly; not only will your results not duplicate the results shown, they won't be the same twice.

The random cell electrode looks enough like a square electrode that the capacitance calculation repeats fairly well. Suppose, however, that we want to calculate a charge density profile. For this example, we choose the charge density as a function of X at $Y = 0$ along the upper electrode.

The issues in performing this calculation start surfacing immediately. Not only are all the triangles different from each other; their vertices, and therefore their barycenters, are also random distributions.

One way to proceed is to select the triangles whose barycenters are within some small distance of $y = 0$ and use these barycenters "as if" they are at $y = 0$. Figure 6.20 shows the result of this calculation.

When all the cells in an electrode were identical squares, the charge profile using appropriate q values gave us the correct distribution curve. Since the triangles' areas are all different in this situation, it is necessary to divide each value of q by the area of its triangle to obtain a charge density value.

Figure 6.20 is a reasonable but not excellent portrait of the charge distribution in question. It could be improved by fitting the data to a polynomial and discarding outliers. Alternatively, the charge density in several cells with barycenters close to each data point's x value could be fit to a polynomial and then the value at $(x, y = 0)$ determined. There is no ideal way to resolve this issue.

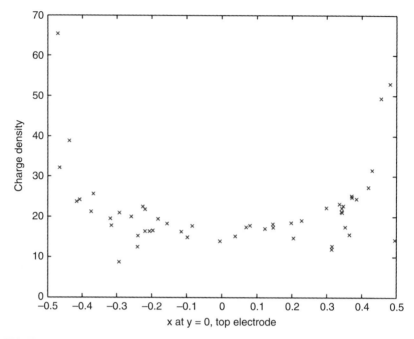

FIGURE 6.20   Charge density versus X at $Y = 0$ approximated by using triangles with $|Y| < 0.02$.

Voltage and electric field values in this example will be reasonable far away from the electrode(s). Values near the electrodes will be more erratic than in previous examples because the charge distribution itself is more erratic.

PROBLEMS

Two preliminary notes are useful here:

- There is almost never only one way to code an algorithm to achieve a particular task. For creation of node layouts for mesh generation, there is similarly almost never only one good solution. The listings shown below are examples that work; nothing more is claimed for them.
- In the example solutions to follow, the color mapping is left at the MATLAB default and the option to change it to black-and-white format is shown commented out. For more complicated structures (and for a data-file-driven package), a more inventive use of the available color mapping would be very valuable.

6.1   Create a version of `mesh_gen`, called by `mom_tri_2`, that creates a parallel-plate structure with circular electrodes.

6.2   Upgrade the results of Problem 6.1 to allow the creation of *washers*, namely, circular disks containing centered holes.

6.3 Upgrade the results of Problem 6.2 so that the washers interlock (in a chain-link configuration).

6.4 Create a `mesh_gen` function (and possibly supporting functions) to allow for the generation of meshed rectangles parallel to any axis. Use these functions to generate a box-within-a-box structure.

6.5 Create a `mesh_gen` function to generate spheres. Construct two concentric spheres with radii = 1, 2, and calculate the capacitance between them. Compare this result to the result using the `distmesh` package.

REFERENCES

1. `http://en.wikipedia.org/wiki/Center_of_mass`.

2. P. Lazic, H. Stefancic, and H. Abraham, The Robin Hood method—a novel numerical method for electrostatic problems based on a non-local charge transfer, `Physics.com-ph` (Nov. 20, 2004).

3. `http://en.wikipedia.org/wiki/Delaunay_triangulation`.

4. `http://persson.berkeley.edu/distmesh/`.

5. P.-O. Persson and G. Strang, A simple mesh generator in MATLAB, *SIAM Rev.* **46** (2): 329–345 (2004).

7

Summary and Overview

7.1 WHERE WE WERE, WHERE WE'RE GOING

This very short chapter marks a break, ending the discussion of the method of moments and beginning discussion of the finite difference (FD) and finite element (FE) methods. This is a good place to compare and contrast these methods.

All three methods are approaches to solving for the electrostatic variables charge and voltage in a given geometric structure of electrodes and possibly dielectric interfaces. They are all based on developing a set of linear algebraic equations that approximate continuous variables with approximate locally defined variables.

The method of moments (MoM) takes as its solution domain all of (three-dimensional) space. Charge is constrained to exist on conducting electrodes. These electrodes can have thickness (i.e., they themselves can be three-dimensional bodies), but since the laws of electrostatics guarantee that all the charge will move to the surface of three-dimensional (3d) electrodes, it doesn't matter whether if the electrodes have thickness or are simply arbitrarily shaped thin skins. This is very important mathematically, because even though we are dealing with 3d space, our variable — the charge — exists only on two-dimensional (2d) surfaces.

We divided the electrode surfaces into either square or triangular planar regions, or cells, each of which is assumed to have a uniform charge density. Since the cells each have a finite area (not necessarily all the same) and a linear equation will express the relationship between the (applied) voltage at each cell and the charge on all of the cells, the number of cells must be finite. This, in turn, implies that there must be a finite volume of conductors in our infinite space.

Introduction to Numerical Electrostatics Using MATLAB, First Edition. Lawrence N. Dworsky.
© 2014 John Wiley & Sons, Inc. Published 2014 by John Wiley & Sons, Inc.

The set of equations is written by demanding that the known, applied, voltage on each cell is identically the voltage due to the charge on all of the cells. The equation set is solved for these charges; the voltage, and then the electric field, anywhere in space may be found by summing the contributions to this voltage (and field) of all the charges.

The capacitance between two electrodes falls out easily as simply the sum of the charge on an electrode divided by the voltage difference between two electrodes.

Mathematically, there are as many equations as there are cells. Since the voltage on any one cell depends on contributions from all of the cells, the coefficient matrix of the equation set is fully populated. This means that even though only 2d surfaces are being considered, the number of matrix terms grows very rapidly with the number of cells.

While writing an expression for each coefficient matrix term is very easy, accurate evaluation of these terms is not so easy. Except for the diagonal terms, numerical integration and/or approximation is necessary. A 1000-cell structure, for example, has approximately 1,000,000/2 coefficients to evaluate, 499,000 of them numerically. In the previous chapters empirical approximations were developed for these coefficients that, although not the most accurate approximations possible, allowed us to solve many problems and treat many situations in a reasonable and efficient manner.

The method of moments (MoM) is one case of a much more general and powerful approach called the *boundary element method*.[1] Further discussion of the boundary element method is outside the scope of this book but is certainly a worthwhile endeavor for anyone wishing to pursue these types of approximate solutions.[2]

The finite difference (FD) and finite element (FE) methods take a very different approach. With only a few special exceptions, only a finite region of space is considered. One-, two- and three-dimensional formulations are possible; one-dimensional (1d) formulations are sometimes useful for teaching purposes but for the most part not of any value for practical problems. A 2d structure is actually a cross section of a uniform infinitely long 3d structure, so in general we're always talking about 3d structures even when the third dimension isn't seen explicitly in the figures or the equations.

A finite region of (2d) space is enclosed by a boundary. At this boundary we specify boundary conditions that are either a voltage or the normal derivative of the voltage everywhere (on the boundary). There might also be electrode regions inside this boundary that have voltage boundary conditions. We then break up the region within the boundary into discrete regions (again, cells). In both the FD or FE methods we write a set of equations relating these cells to one another.

The FD method is derived from numerical approximations of Laplace's equation, and only the voltages at the corners of rectangular cells are defined. The FE method is derived from formal integral approximation techniques, and the voltage is defined throughout the cell. In both cases each cell communicates only with cells that share nodes. Also, in the FE method cell shape is arbitrary (although practical considerations will narrow our choices somewhat).

This latter statement has mathematical implications in that while we need an equation for each corner of a rectangular cell or vertex of a triangular cell, each of these points (nodes) is shared by only several cells. Also, and more important, this means that each cell "communicates" directly with only several of the many cells involved in a structure and hence the coefficient matrix that is developed is very sparse. We will utilize sparse matrix storage in our computer programs rather than carry around thousands of zeros (and limit our problem size and solution speed).

In the MoM method, calculating the coefficients was not trivial; however, in the FE and FD methods calculating the coefficients is easy — in the FD method, it is actually trivial.

The FD method, using rectangular cells, is very limited in the structural sophistication possible. However, the interconnection of the cells and generation of the coefficient equations is so simple that it pays to spend time with the FD method to see what useful problems can be solved and what electrostatics we can learn. The ties between the geometry and the resulting equations are very clear and facilitate understanding of the situation tremendously. The FE method allows for very generalized structure descriptions (which probably accounts for its huge popularity), but the tradeoff here is that there is a sophisticated "assembly" procedure necessary to describe (to the system of equations) the interconnection and relative locations of all the cells.

As is the case when dealing with the MoM, generation of the mesh for rectangular structures is easy but limited, while setting up the mesh for triangular problems is difficult but very powerful. For the same reasons, therefore, we will limit FE discussions to relatively simple structures (no Golden Gate Bridge frames) and leave the sophisticated mesh generation problem to specialized packages.

The principal purposes of introducing the FD method are to (1) demonstrate the basics of this technique and write useful programs without first engaging in the more involved mathematics of the FEM method and (2) introduce basic programs for generating the coefficient matrix, calculating stored energy, present techniques for calculating upper and lower bounds on the approximations, calculate properties of multielectrode and open-boundary systems, and so on. There is a tremendous amount of electrostatics to be shown here, and it is advantageous to show it using the simplest algorithms possible.

Because the FE method is so much more general than the FD method in its ability to handle complex problems, at the cost of computational complexity, two programming styles are adopted:

1   For the FD method, the structure is hardcoded into the programs themselves. Parameters and shapes must be changed by changing source code. This is very limiting in terms of general application of the technique, but keeps with the strategy of keeping the programming as simple as possible to demonstrate the FD technique and the electrostatics to be learned from the solution.

2   For the FE method, the programs are data-file-driven. Although mesh generation and zoning techniques were not the desired themes for this book, it was necessary to choose a software package to do the job and then to teach enough of its use so that examples could be generated and "getting the mesh data into the FEM program" could be demonstrated. Fortunately, a very capable free mesh generation package is available, as will be discussed in later chapters. Since all of the geometric description (including cell meshing) data reside outside the programs, these programs are very general. Within the limits of the FEM calculations being presented in any program, these programs can handle an infinite number of different structures.

As has been mentioned before, all of the programs have been written with the primary goal of showing how the analyses become algorithms, which then become (MATLAB) code. In most cases when it was necessary to choose between clarifying these correlations and writing the most compact and/or most efficient code possible, the former option was chosen.

REFERENCES

1. W.-T. Ang, *A Beginner's Course in Boundary Element Methods*, Universal Publishers, Boca Raton, FL, 2007.

2. M. V. K. Chari and S. J. Salon, *Numerical Methods in Electromagnetism*, Academic Press, London, 2000.

8

The Finite Difference Method

8.1 INTRODUCTION AND A SIMPLE EXAMPLE

We begin this chapter by examining a two-dimensional (2d) rectangular structure. Extending the analysis to three-dimensional (3d) structures in rectangular, cylindrical, or spherical coordinates is straightforward and will be discussed later. Also, for now consider only prescribed voltage boundary conditions.

Laplace's equation in rectangular coordinates is

$$\nabla^2 V(x,y,z) = \frac{\partial^2}{\partial x^2} V(x,y,z) + \frac{\partial^2}{\partial y^2} V(x,y,z) + \frac{\partial^2}{\partial z^2} V(x,y,z) = 0 \tag{8.1}$$

For a uniform structure infinitely long in the Z direction, all partial derivatives with respect to Z vanish, leaving us with the 2d equation

$$\nabla^2 V(x,y) = \frac{\partial^2}{\partial x^2} V(x,y) + \frac{\partial^2}{\partial y^2} V(x,y) = 0 \tag{8.2}$$

Figure 8.1 shows a rectangular grid of points in 2d space.

The lines in the figure are just for our eyes to follow easily. Of interest are the intersections of these lines, which we'll call *nodes*. The X-axis separation of nodes is h_x [equation (8.3)]; the Y-axis separation of nodes is h_y [equation (8.4)]:

$$h_x = x_{i+1} - x_i \tag{8.3}$$

$$h_y = y_{j+1} - y_j \tag{8.4}$$

Introduction to Numerical Electrostatics Using MATLAB, First Edition. Lawrence N. Dworsky.
© 2014 John Wiley & Sons, Inc. Published 2014 by John Wiley & Sons, Inc.

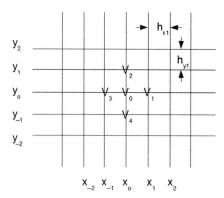

FIGURE 8.1   A rectangular grid of points.

For now, assume that $h_x = h_y = h$.

Voltages at the nodes are defined by the location of the node, for example, $V_{i,j}$. Five voltages of particular interest in the figure include:

$$V_0 \equiv V_{0,0} = V(x_0, y_0) \quad \text{and} \quad V_1 \equiv V_{1,0} = V(x_0 + h, y_0) \tag{8.5}$$

Expanding $V(x,y)$ in a Taylor series, we obtain

$$\begin{aligned} V_1 &= V(x_0 + h, y_0) \\ &= V(x_0, y_0) + h\frac{\partial}{\partial x}V(x_0, y_0) + \frac{h^2}{2}\frac{\partial^2}{\partial x^2}V(x_0, y_0) + \frac{h^3}{6}\frac{\partial^3}{\partial x^3}V(x_0, y_0) + \cdots \end{aligned} \tag{8.6}$$

and

$$\begin{aligned} V_3 &= V(x_0 - h, y_0) \\ &= V(x_0, y_0) - h\frac{\partial}{\partial x}V(x_0, y_0) + \frac{h^2}{2}\frac{\partial^2}{\partial x^2}V(x_0, y_0) - \frac{h^3}{6}\frac{\partial^3}{\partial x^3}V(x_0, y_0) + \cdots \end{aligned} \tag{8.7}$$

Adding these two equations together and updating the notation a bit, we get

$$V_1 + V_3 = 2V_0 + h^2\frac{\partial^2}{\partial x^2}V(x_0, y_0) + \frac{h_x^4}{12}\frac{\partial^4}{\partial x^4}V(x_0, y_0) + \cdots \tag{8.8}$$

Following the same procedure for V_2 and V_4, we obtain

$$V_2 + V_4 = 2V_0 + h^2\frac{\partial^2}{\partial y^2}V(x_0, y_0) + \frac{h^4}{12}\frac{\partial^4}{\partial y^4}V(x_0, y_0) + \cdots \tag{8.9}$$

All odd-order derivative terms cancel, and these approximations are accurate to the order of the fourth derivative term.

Combining equations (8.8) and (8.9), rearranging the terms, and dropping the higher-order terms, we obtain

$$\frac{\partial^2 V_0}{\partial x^2} + \frac{\partial^2 V_0}{\partial y^2} = \frac{V_1 + V_3 - 2V_0}{h^2} + \frac{V_2 + V_4 - 2V_0}{h^2} = 0 \tag{8.10}$$

We may therefore approximate Laplace's equation at (x_0, y_0) as

$$V_1 + V_2 + V_3 + V_4 - 4V_0 = 0 \tag{8.11}$$

or

$$V_0 = \frac{V_1 + V_2 + V_3 + V_4}{4} \tag{8.12}$$

Note that h has disappeared from these equations. This happens only when all four h values are equal.

Before considering how to do anything useful with these equation, we'll show an alternate derivation that does not give any insight into the size of the errors [equation (8.9)] but will prove easier to perform in different coordinate systems.

At (x_0, y_0), we may write

$$\frac{\partial}{\partial x} V(x,y) \simeq \frac{V_1 - V_0}{h} \simeq \frac{V_0 - V_3}{h} \tag{8.13}$$

The second derivative at the same point, therefore, is

$$\frac{\partial^2}{\partial x^2} V(x,y) \simeq \frac{[(V_1 - V_0)/h] - [(V_0 - V_3)/h]}{h} = \frac{V_1 + V_3 - 2V_0}{h^2} \tag{8.14}$$

The same relationship may be derived for the second partial derivative with respect to y and then by adding the two partial derivatives and setting this sum to zero (Laplace's equation); equation (8.12) follows immediately.

8.2 SETTING UP AND SOLVING A BASIC PROBLEM

Figure 8.2 shows the cross section of an asymmetric stripline transmission line. In this example $h_x = h_y = 1$; the width of the center conductor strip $= 1$, the box height is 4, and the width is 4. We should note, however, that

1 This is an extremely low-resolution example. Its purpose is to demonstrate the setup and solution of the relevant equations; the results will be physically reasonable but not accurate.

2 Units were not specified for the dimensions. This is because, in a 2d cross section of an infinitely long 3d problem, only the ratios of the dimensions are relevant, so all units cancel. Only the units used for ε in the final stored energy, charge, and capacitance calculations will matter.

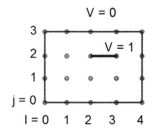

FIGURE 8.2 Stripline cross section.

As shown in Figure 8.2, the voltage along the outer nodes (the outer box of the stripline) is set to 0 V, and the voltage along the inner strip is set at 1 V. These voltages are now boundary conditions of the problem and will never change.

In this simple example there are more boundary condition nodes than free variable nodes. This is, of course, only because of the small size of the example. In general, the free variable nodes will greatly outnumber the boundary condition nodes.

The node locations (i,j) take on the values

$$0 \le i \le 4 = i_{\max} \text{ and } 0 \le j \le 3 = j_{\max} \tag{8.15}$$

In order to write a set of equations to solve, each node must be assigned a unique variable number. The simple MATLAB function `get_ijk.m` performs the useful functions of translating between the (i,j) node locations and the variable number:

```
function [ i_out, j_out, k ] = get_ijk( Imin, Imax, Jmin, i, j, k )
%   Function to convert from (i,j) coordinates to node numbers
%   and vice versa
%   Imin, Imax = left and right hand end values of i
%   Jmin = bottom end value of j
%   On input, if k == 0 then k is returned, i & j are not changed
%             if k > 0 then i and j are returned, k is not changed
%   Note that there is NO error checking, e.g. is i in the allowed
%     range? is k > number of nodes? etc.

  if k == 0                        % get k
    i_out = i; j_out = j;
    i = i - Imin; j = j - Jmin;
    k = i + 1 + j*(Imax-Imin + 1);
  else
    j = ceil(k/(Imax-Imin + 1)) - 1;    % get i and j
    i = k - 1 - j*(Imax-Imin + 1);
    j_out = j + Jmin;
    i_out = i + Imin;
  end

end
```

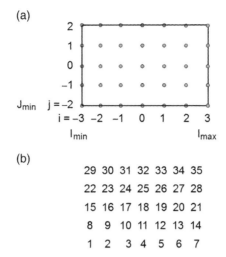

FIGURE 8.3 Example of node (a) and variable numbering (b) with (0,0) at the center.

This function has been generalized so that i and j do not have to start at 0 but may start at any specific value. This generalization will be useful for node layouts where (0,0) is not in the lower left hand corner.

Figure 8.3 shows an example of this latter node numbering choice and the accompanying variable numbering pattern.

Returning to the structure in Figure 8.2, our goal is to write equation (8.11) for every node in our system, 20 equations in all. V_0 in each of these equations is the voltage at each equation's "home" node.

[*Note*: Each node has both a unique node location (e.g., $V_{3,1}$) and a variable number (e.g., V_9). Since the node location has physical meaning while the variable number is just an arbitrary assignment, the node location will be used to describe a node when dealing with the physical system. Variable numbers will be used only when dealing with the equation set. Hopefully this will not cause confusion. Notationally, the presence or absence of a comma in the subscript should clearly indicate whether the node location or the variable number is being discussed.]

Equation (8.11) relates the voltage at a node to the voltages at the four surrounding nodes. This is clear at, say, $V_{3,1}$, but not at, say, $V_{3,0}$ since the latter node is on the outer boundary of the structure; there is one missing voltage point. At a corner node, say, $V_{4,0}$, there are two missing voltage points for the equation.

Fortunately, nature lets us resolve this issue simply. The outer boundary nodes are electrical boundary conditions. Equation (8.11) does not apply at these nodes. For convenience in writing general-purpose algorithms we'll keep these nodes in the equation set and freeze their voltages at prescribed the boundary condition values. Alternatively, we could reduce the size of the equation set by one for each of the boundary condition nodes (periphery and internal) by substituting the boundary condition voltage into the equation set and eliminating the variable.

Applying equation (8.11) and the boundary conditions to the structure in Figure 8.2, we get the following set of equations:

$$\begin{bmatrix} 1&0&0&0&0&0&0&0&0&0&0&0&0&0&0&0&0&0&0&0 \\ 0&1&0&0&0&0&0&0&0&0&0&0&0&0&0&0&0&0&0&0 \\ 0&0&1&0&0&0&0&0&0&0&0&0&0&0&0&0&0&0&0&0 \\ 0&0&0&1&0&0&0&0&0&0&0&0&0&0&0&0&0&0&0&0 \\ 0&0&0&0&1&0&0&0&0&0&0&0&0&0&0&0&0&0&0&0 \\ 0&0&0&0&0&1&0&0&0&0&0&0&0&0&0&0&0&0&0&0 \\ 0&1&0&0&0&1&-4&1&0&0&0&1&0&0&0&0&0&0&0&0 \\ 0&0&1&0&0&0&1&-4&1&0&0&0&1&0&0&0&0&0&0&0 \\ 0&0&0&1&0&0&0&1&-4&1&0&0&0&1&0&0&0&0&0&0 \\ 0&0&0&0&0&0&0&0&0&1&0&0&0&0&0&0&0&0&0&0 \\ 0&0&0&0&0&0&0&0&0&0&1&0&0&0&0&0&0&0&0&0 \\ 0&0&0&0&0&0&1&0&0&0&1&-4&1&0&0&0&1&0&0&0 \\ 0&0&0&0&0&0&0&0&0&0&0&0&1&0&0&0&0&0&0&0 \\ 0&0&0&0&0&0&0&0&0&0&0&0&0&1&0&0&0&0&0&0 \\ 0&0&0&0&0&0&0&0&0&0&0&0&0&0&1&0&0&0&0&0 \\ 0&0&0&0&0&0&0&0&0&0&0&0&0&0&0&1&0&0&0&0 \\ 0&0&0&0&0&0&0&0&0&0&0&0&0&0&0&0&1&0&0&0 \\ 0&0&0&0&0&0&0&0&0&0&0&0&0&0&0&0&0&1&0&0 \\ 0&0&0&0&0&0&0&0&0&0&0&0&0&0&0&0&0&0&1&0 \\ 0&0&0&0&0&0&0&0&0&0&0&0&0&0&0&0&0&0&0&1 \end{bmatrix} \begin{bmatrix} V_1 \\ V_2 \\ V_3 \\ V_4 \\ V_5 \\ V_6 \\ V_7 \\ V_8 \\ V_9 \\ V_{10} \\ V_{11} \\ V_{12} \\ V_{13} \\ V_{14} \\ V_{15} \\ V_{16} \\ V_{17} \\ V_{18} \\ V_{19} \\ V_{20} \end{bmatrix} = \begin{bmatrix} 0 \\ 0 \\ 0 \\ 0 \\ 0 \\ 0 \\ 0 \\ 0 \\ 0 \\ 0 \\ 0 \\ 0 \\ 1 \\ 1 \\ 0 \\ 0 \\ 0 \\ 0 \\ 0 \\ 0 \end{bmatrix} \qquad (8.16)$$

Two observations about this coefficient array are necessary here:

1. There are only four free variables out of the 20 total variables. As mentioned above, this occurs here only because of the very small size of this example; if there were more than 20 variables, it would have been impossible to show the entire a array. For practical problems, with hundreds to thousands of variables, the boundary condition variables will constitute a very small percentage of the total.

2. Most of the array is filled with zeros. We'll ignore this attribute for the first sample program, but address it immediately thereafter.

MATLAB script fd1.m sets up equation (8.16) starting with the geometry description of Figure 8.2 and performs finite difference (FD) analysis of that figure:

```
%fd1.m   first example finite difference program

Imin = 0; Imax = 4; Jmin = 0; Jmax = 3;     % definition of array
Ic1 = 2; Ic2 = 3; Jc = 2; % inner conductor
[i,j,Kmax] = get_ijk(Imin,Imax,Jmin, Imax,Jmax,0); % number of variables

a = sparse(Kmax,Kmax);    % zeros(Kmax);    % allocate the coefficient array
b = zeros(Kmax,1);   % allocate the bcs array

% initialize a as if every variable is free
for j = Jmin: Jmax
  for i = Imin : Imax
    [i,j,k0] = get_ijk(Imin,Imax,Jmin, i,j,0);
    a(k0,k0) = -4;
    if k0 < Kmax, a(k0,k0 + 1) = 1; end
    if k0 > 1,   a(k0,k0-1) = 1; end
```

```
    k1 = k0 + (Imax-Imin) + 1;
    if k1 < = Kmax, a(k0,k1) = 1; end
    k2 = k0 - (Imax-Imin) - 1;
    if k2 > 0, a(k0,k2) = 1; end
  end
end

% Perimeter (v = 0) bcs
% clean out the appropriate row, replace it w/ 1 on diagonal
for i = Imin : Imax
  [i,j,k] = get_ijk(Imin,Imax,Jmin, i,Jmin,0);
  a(k,:) = 0; a(k,k) = 1;
  [i,j,k] = get_ijk(Imin,Imax,Jmin, i,Jmax,0);
  a(k,:) = 0; a(k,k) = 1;
end
for j = Jmin + 1 : Jmax - 1
  [i,j,k] = get_ijk(Imin,Imax,Jmin, Imin,j,0);
  a(k,:) = 0; a(k,k) = 1;
  [i,j,k] = get_ijk(Imin,Imax,Jmin, Imax,j,0);
  a(k,:) = 0; a(k,k) = 1;
end

% inner strip (v = 1) bcs
% clean out the appropriate row, replace it w/ 1 on diagonal
% put 1 in b array
for i = Ic1 : Ic2
  [i,j,k] = get_ijk(Imin,Imax,Jmin, i,Jc,0);
  a(k,:) = 0;
  a(k,k) = 1;
  b(k) = 1;
end

% solve for voltages
v = a\b;

% list non-zero voltages by row, column, & variable nrs
fprintf ('  i   j   k   Volts \n \n')
for k = 1 : Kmax
  volts = v(k);
  if volts > 0
    [i,j,k] = get_ijk(Imin,Imax,Jmin, 0,0,k);
    fprintf ('%3d   %3d   %3d   %8.3f \n', i, j, k, volts)
  end
end
```

The program flow of fd.1 m is as follows:

1   Initialize every variable (row in the equations) of the coefficient matrix *a* as if the variable were a free variable. As explained above, the overwhelming majority of variables will generally be free variables, so that this is not an inefficient way to begin. Initialize the forcing function vector *b* to zero.

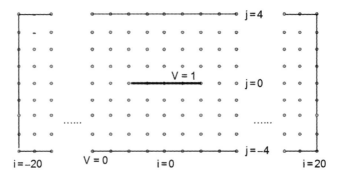

FIGURE 8.4 A practical example geometry for FD modeling.

2 Replace every variable (row) of a $V = 0$ boundary condition (in this example the outer perimeter) with zeros everywhere except the diagonal term. Set this term equal to one.

3 Repeat steps 1 and 2 for every V other than 0 boundary condition (in this case the inner strip); then set the corresponding term of the b vector equal to the prescribed voltage (in this case 1 Volt).

4 Solve the equation set.

One last optional calculation was added to the program just for some interesting information. Even in this very small example, only 8% of the coefficient matrix is not set to zero. Since there are (typically) five nonzero terms in each row of the coefficient matrix, this percentage will shrink drastically as problem sizes increase. This will be both a tremendous waste of computer time and memory when solving large problems. The situation is worse for 3d grids because while there will typically be seven coefficients per row, there will be many more rows.

Fortunately, MATLAB is prepared for this situation. In `fd1.m`, simply replace line 7, `a = zeros (Kmax);` with `a = sparse (Kmax, Kmax);` and MATLAB will understand what our situation is and how to handle it.

Figure 8.4 shows a more realistic geometry.

There are obvious symmetries to this structure that are not being exploited. This will be considered in later section.

Program `fd.1 m` is easily modified to handle this geometry by replacing lines 3 and 4, namely

```
Imin = 0; Imax = 4; Imin = 0; Imax = 3; %definition of array
Ic1 = 2; Ic2 = 3; Jc = 2; %inner conductor
```

with

```
ns = 1; %scaling factor for resolution studies
Imin = -20*ns; Imax = -Imin; Jmin = -4*ns; Jmax = -Jmin; %definition
of array
Ic1 = -2*ns; Ic2 = -IC1; Jc = 0; %inner conductor
```

The parameter `ns` lets us vary resolution by scaling all dimensions simultaneously. The implications of this, too, will be considered in a later section. For now, simply leaving $ns = 1$

creates the geometry shown in Figure 8.4. The `fd2.m` program is a simple stripline program with scalable geometry:

```
%fd2.m   A simple but practical FD example with adjustable resolution

ns = 10;     % scaling factor for resolution studies
Imin = -20*ns; Imax = -Imin ; Jmin = -4*ns; Jmax = -Jmin;   % definition
of array
Ic1 = -2*ns; Ic2 = -Ic1; Jc = 0; % inner conductor
 [i,j,Kmax] = get_ijk(Imin,Imax,Jmin, Imax,Jmax,0);   % number of
variables
a = sparse(Kmax,Kmax); % zeros(Kmax); % allocate the coefficient array
b = zeros(Kmax,1); % allocate the bcs array

% initialize a as if every variable is free
for j = Jmin: Jmax
  for i = Imin : Imax
    [i,j,k0] = get_ijk(Imin,Imax,Jmin, i,j,0);
    a(k0,k0) = -4;
    if k0 < Kmax,  a(k0,k0+1) = 1;  end
    if k0 > 1,   a(k0,k0-1) = 1;  end
    k1 = k0 + (Imax-Imin) + 1;
    if k1 < = Kmax, a(k0,k1) = 1;  end
    k2 = k0 - (Imax-Imin) - 1;
    if k2 > 0, a(k0,k2) = 1;  end
  end
end

%  Perimeter (v = 0) bcs
%  clean out the appropriate row, replace it w/ 1 on diagonal
for i = Imin : Imax
  [i,j,k] = get_ijk(Imin,Imax,Jmin, i,Jmin,0);
  a(k,:) = 0; a(k,k) = 1;
  [i,j,k] = get_ijk(Imin,Imax,Jmin, i,Jmax,0);
  a(k,:) = 0; a(k,k) = 1;
end
for j = Jmin + 1 : Jmax - 1
  [i,j,k] = get_ijk(Imin,Imax,Jmin, Imin,j,0);
  a(k,:) = 0; a(k,k) = 1;
  [i,j,k] = get_ijk(Imin,Imax,Jmin, Imax,j,0);
  a(k,:) = 0; a(k,k) = 1;
end

%  inner strip (v = 1) bcs
%  clean out the appropriate row, replace it w/ 1 on diagonal
%  put 1 in b array
for i = Ic1 : Ic2
  [i,j,k] = get_ijk(Imin,Imax,Jmin, i,Jc,0);
  a(k,:) = 0;
  a(k,k) = 1;
  b(k) = 1;
end
```

```
% solve for voltages
v = a\b;

% list non-zero voltages by row, column, & variable nrs
fprintf ('   i    j    k    Volts \n \n')
for k = 1 : Kmax
  volts = v(k);
  if volts > 0
    [i,j,k] = get_ijk(Imin,Imax,Jmin, 0,0,k);
    fprintf ('  %3d   %3d   %3d   %8.3f   \n', i,  j,  k,  volts)
  end
end

%  We'll need an xy voltage array for ongoing calcs
Sx = Imax - Imin + 1; Sy = Jmax - Jmin + 1;
Volts = zeros (Sx,Sy);
for ii = Imin:Imax
  i = ii - Imin + 1;
  for jj = Jmin:Jmax
    j = jj - Jmin + 1;
    [ii,jj,k] = get_ijk(Imin,Imax,Jmin, ii,jj,0);
    Volts(i,j) = v(k);
  end
end


C1 = C_Gauss_box(Volts)
C2 = C_Energy(Volts)
```

This program also introduces capacitance calculations, which will be discussed below.

8.3 THE GAUSS–SEIDEL (RELAXATION) SOLUTION TECHNIQUE

[*Note*: This section discusses an equation solution technique that historically has been tied very closely to FD models. It is not as efficient numerically as the technique(s) that MATLAB offers, but it links the mathematics of the solution so closely to the geometry of the structure that it is very interesting to examine. It is the most efficient technique in terms of computer memory utilization because only the voltages themselves (the desired solution) are stored. Although this used to be an important consideration, personal computer memory capacity has become so large that this is no longer true. This section may be skipped if so desired with no loss of continuity.]

Equation (8.16), ignoring the boundary condition entries for the moment, is merely a repeated application of equation (8.12), once per node. This latter equation has the property that the magnitude of the diagonal term of the coefficient matrix is equal to or greater than the sum of all the off-diagonal terms (in each row). The boundary condition row has the property that the diagonal term is the only nonzero term in its row of the coefficient matrix; therefore, it, too, is equal to or greater than the sum of all the off-diagonal terms in the row.

Linear equation sets that satisfy this property are guaranteed to converge to the correct solution when Gauss–Seidel (also called relaxation) iterations are applied.[1] Gauss–Seidel iterations consist of repeatedly applying equation (8.12) to all the nodes in the grid (ignoring the boundary condition nodes) until satisfactory convergence is obtained. The iterative procedure proceeds as follows:

1. Set all boundary condition nodes to their prescribed values.
2. Set all free variable nodes to an arbitrary value other than a boundary condition value; the closer to their actual (final value), the better for convergence efficiency.
3. Step through all the nodes: (a) if the node is a boundary condition, ignore it; (b) if the node is a free variable, apply equation (8.12), using the latest node voltages.
4. If convergence is acceptable, exit; otherwise go to step 2.

Returning to the example shown in Figure 8.2, MATLAB program `fd3.m` a—simple example of Gauss–Seidel iteration—implements this procedure:

```
%fd3.m  Relaxation grid solution technique example

% note that Imin and Jmin have been set to 1 just to keep this simple
Imax = 5; Jmax = 4;   % definition of array
Ic1 = 3; Ic2 = 4; Jc = 3; % inner conductor

v = .5*ones(Imax,Jmax);

% Perimeter (v = 0) bcs
v(:,1) = 0; v(:,Jmax) = 0; v(1,:) = 0; v(Imax,:) = 0;

% inner strip (v = 1) bcs
v(Ic1:Ic2,Jc) = 1;

fprintf ('  iter   v(2,3)   v(3,2)   v(4,2)   v(2,3) \n')
iter = 0;    %This is just for displaying the results
fprintf ('  %4d   %8.3f   %8.3f   %8.3f   %8.3f \n', iter,v(2,2),v(3,2), …
                  v(4,2), v(2,3))

% solve for voltages
max_iters = 7;
for iter = 1: max_iters
  for i = 2: Imax - 1
    for j = 2: Jmax - 1
      if v(i,j) ~ = 1
        v(i,j) = (v(i + 1,j) + v(i-1,j) + v(i,j-1) + v(i,j + 1))/4;
      end
    end
  end
  fprintf ('%4d  %8.3f  %8.3f  %8.3f  %8.3f \n', iter,v(2,2), v(3,2), …
                  v(4,2), v(2,3))
end
```

Since this demonstration program is not leading to complicated situations, some shortcuts were taken:

TABLE 8.1 Results of `fd3.m`.

Iterations	$V(1,1)$	$V(2,1)$	$V(3,1)$	$V(1,2)$
0	0.500	0.500	0.500	0.500
1	0.250	0.438	0.359	0.313
2	0.188	0.387	0.347	0.297
3	0.171	0.379	0.345	0.293
4	0.168	0.378	0.345	0.292
5	0.168	0.378	0.345	0.292
6	0.167	0.378	0.344	0.292
7	0.167	0.378	0.344	0.292

1 The node numbers were all increased by one as compared to the numbering shown in Figure 8.2. This facilitates easy usage of MATLAB array numbering, which always starts at one.

2 Since the outer boundary nodes are all prescribed voltage (boundary condition) nodes, they are simply not referenced in the iteration loop.

3 The only internal boundary condition nodes are (both) set to one volt, so a very simple check is made to ensure that these nodes are ignored.

The simplicity of programming a Gauss–Seidel iteration algorithm is apparent in `fd3.m`:

- There are no node numbers to be calculated from the (i,j) coordinates; the nodes are simply referred to by these coordinates.
- There is no coefficient (a) matrix, and there is no forcing function (b) vector. There are only the voltage values themselves.
- The entire calculation is performed in one simple equation (line 26 of `fd3.m`).
- A larger example (e.g., Figure 8.4) would not require more complicated coding.

Table 8.1 shows the results of `fd3.m`. For convenience, the node numbering has been converted back to agree with Figure 8.2.

Naturally, the results listed in Table 8.1 are identical to the results of `fd1.m`.

Since the Gauss–Seidel iteration technique is so easy to program, why do we not use it all the time? Why do we have to worry about sparse matrices, solving very large sets of equations, and continually converting back and forth between node locations and variable (index) numbers?

The answer lies in calculation efficiency. The number of relaxation iterations required for convergence increases (approximately) with the square of the number of variables. We need a definition of acceptable convergence (and there is no unique definition), and we have to monitor convergence — this complicates programming a bit and slows the iterations down a bit.

Convergence may be improved by a technique called *overrelaxation*.[2] We modify equation (8.12) to

$$V_{i,j,\text{new}} = (1-f)V_{i,j,\text{old}} + f\frac{V_{i+1,j} + V_{i-1,j} + V_{i,j+1} + V_{i,j-1}}{4} \tag{8.17}$$

where f is a number between 1 and 2.

When $f = 1$, equation (8.17) is identical to equation (8.12). As f approaches 2, the relaxation converges more rapidly. (*Note*: The expression $0 < f < 1$ is called *underrelaxation*; it is sometimes used to stabilize the convergence of interactive solutions of nonlinear equations.) Unfortunately, when dealing with dielectric interfaces and/or symmetry boundary conditions (two topics that haven't been addressed yet), overrelaxation is sometimes unstable until the iterations are somewhat converged (something that cannot be defined in general). This means that overrelaxation must be introduced slowly and convergence stability monitored.

The bottom line here is that the Gauss–Seidel iteration technique is easy and intuitive to program but is not efficient and in some cases is not stable.

8.4   CHARGE, GAUSS'S LAW, AND RESOLUTION

Gauss's law tells us that if we draw a continuous curve around any charge distribution and integrate the normal component of the electric field crossing that curve (multiplied by the permittivity of the medium), we will learn the total charge inside the curve.

A curve for performing such a calculation for the example of Figure 8.4 (modeled in program fd2.m) is shown in Figure 8.5, the dashed line just inside the outer boundary.

This curve is convenient because it passes between nodes in such a manner that only x- or y-directed field components need be considered, and near the outer box the electric field is essentially normal to the walls.

For example, E_y at (0,3) in Figure 8.5, is approximated by $(V_{0,2} - V_{0,4})/2$. We will be consistent in taking the inner voltage – the outer voltage as we move around the curve.

For convenience in this and some later calculations, at the end of fd2.m a short piece of code is added to convert from the v array (v as a function of the node number) to a volts array (volts as a function of (i,j)). Both i and j begin at 1, as they are offset by I_{min} and J_{min}, respectively. The volts array gives us no new information, but fits easily into the MATLAB scheme of matrix indexing (all array indices start at 1).

The MATLAB program C_Gauss_box.m, called at the end of fd2.m, produces a capacitance calculation based on integrating the charge around the Gaussian surface shown in Figure 8.5. A simple trapezoidal integration is used. The code is so simple that the trapezoidal integral code is written explicitly; there is no need for MATLAB numerical integration capability here.

Running this program (setting ns = 1), we get a result of $C = 36.54$ pF/m (assuming, of course, that our dimensions were in meters). At this point we have no measure of the accuracy of this result.

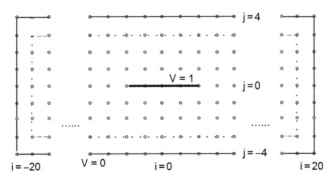

FIGURE 8.5   Repeat of Figure 8.4 with Gaussian surfaces for charge calculation shown.

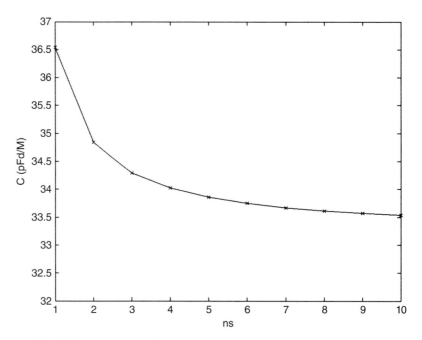

FIGURE 8.6   Capacitance C from Gauss's law as a function of resolution parameter ns.

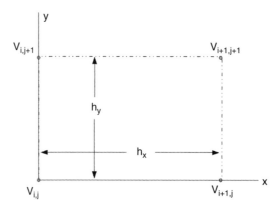

FIGURE 8.7   Definitions for the linear interpolation of voltage values.

Figure 8.6 shows C for values of ns from 1 to 10. As the resolution increases, the capacitance falls, apparently asymptotically approaching ~33.3 pF/m. Intuitively, it seems logical to assume that C will level off at (asymptoticallly approach) a value as ns grows. At this point, however, we have no way of validating or quantifying this assumption.

There are calculations available in the literature for this structure if I_{min} and I_{max} were so large as to have no effect on the results. Using ns = 5 as a reference calculation, we can vary I_{min} and I_{max} to examine this situation. For $-I_{min} = I_{max} = 10$, $C = 33.90$. For $-I_{min} = I_{max} = 15$, $C = 33.86$, and of course our original calculation $-I_{min} = I_{max} = 20$ gave us $C = 33.86$. It appears that $-I_{min} = I_{max} = 20$ is far enough away to let us compare results to the infinitely wide case.

There are many stripline calculators available online. The calculator used in this case[3] predicts a Z_0 (characteristic impedance) of 100.5 Ω for this geometry. Using the relationship presented in Chapter 2, we obtain

$$C = \frac{1}{v_0 z_0} = \frac{1}{(3 \times 10^8)(100.5)} = 33.2 \, \text{pF/m} \tag{8.18}$$

This agrees very well with the asymptotic approximation for the data shown in Figure 8.7. Another cross-check on our calculations is to look at the MoM stripline calculation results shown in Figure 5.7. For the wider line shown, $C = 52.0 \, \text{pF/m}$. Setting ns = 5 and setting $-I_{c1} = I_{c2} = 4$ so as to duplicate the geometry of Figure 5.7, fd2.m predicts $C = 51.7 \, \text{pF/m}$. This shows very good agreement between two totally different independent calculations, but we still do not know if they are accurate.

8.5 VOLTAGES AND FIELDS

The finite difference solution gives us the voltages $V_{i,j}$, at the nodes. We have said nothing, however, about the voltage at any point in space. We can create a simple approximation to give us this information using linear interpolation. Remember that the numerical FD solution is an approximation in the first place that was derived only for the node points. This additional approximation is, as advertised, a linear interpolation of the node voltages. It is an approximation based on an approximation, so we should not expect too much from it.

In a rectangular region with nodes at the four corners as shown in Figure 8.6, let

$$V(x,y) = V_{i,j} + \left(V_{i+1,j} - V_{i,j}\right)\frac{x}{h_x} + \left(V_{i,j+1} - V_{i,j}\right)\frac{y}{h_y} + \left(V_{i+1,j+1} - V_{i+1,j} - V_{i,j+1} + V_{i,j}\right)\frac{xy}{h_x h_y}$$
$$\tag{8.19}$$

The x–y axis shown is a local (i.e., referring only to this cell) axis, with $(0,0)$ at the lower left node point (i,j).

Since V is continuous and differentiable inside the cell, we can calculate the electric field in the cell:

$$E_x = -\frac{\partial V}{\partial x} = \frac{V_{i,j} - V_{i+1,j}}{h_x} + \left(V_{i+1,j} + V_{i,j+1} - V_{i+1,j+1} - V_{i,j}\right)\frac{y}{h_x h_y} \tag{8.20}$$

$$E_y = -\frac{\partial V}{\partial y} = \frac{V_{i,j} - V_{i,j+1}}{h_y} + \left(V_{i+1,j} + V_{i,j+1} - V_{i+1,j+1} - V_{i,j}\right)\frac{x}{h_x h_y} \tag{8.21}$$

Consider E_x and E_y along the left edge ($x = 0$) of the region shown in Figure 8.7, defined as $E_{x,l}$ and $E_{y,l}$, respectively:

$$E_{x,l} = \frac{V_{i,j} - V_{i+1,j}}{h_x} + \left(V_{i+1,j} + V_{i,j+1} - V_{i+1,j+1} - V_{i,j}\right)\frac{y}{h_x h_y} \tag{8.22}$$

$$E_{y,l} = \frac{V_{i,j} - V_{i,j+1}}{h_y} \qquad (8.23)$$

Now look at the electric field in the rectangular region just to the left of the region presented above. Remember that $V_{i,j}$ and related terms are now referring to different nodes than above. The electric field at the right edge ($x = h_x$) of this region is, in general, different than the electric field at the left edge of the region presented above. In other words, while our node point description of voltages and the interpolation function shown describe a voltage distribution that is continuous everywhere, the electric field is discontinuous across the rectangular region boundaries.

Equations (8.19), (8.20), and (8.21) let us describe the voltage and the electric field everywhere. The voltage function is smooth and continuous, albeit with discontinuous derivatives (aka the *electric field*) at cell boundaries. The electric field functions are poor; their best use is for averaging (i.e., integrating) over large regions as will be done in Section 8.6.

8.6 STORED ENERGY AND CAPACITANCE

Another approach to calculating the capacitance of a structure is to calculate the energy stored in the electric field U_E and then calculate the capacitance from the following circuit equation:

$$C = \tfrac{1}{2} U_E V^2 \qquad (8.24)$$

The energy stored in an electric field is given by

$$U_E = \frac{\varepsilon}{2} \iint \left(E_x^2 + E_y^2 \right) dx\, dy \qquad (8.25)$$

where the integration is over the (in this case 2d) region.

Referring to Figure 8.7, we know E_x and E_y in every cell in the structure. We therefore can calculate the stored electric field energy in each of these cells and then sum the results to get U_E. Substituting equations (8.20) and (8.21) into (8.25) and then (8.24), for each cell, we obtain

$$C_{i,j} = \frac{\varepsilon}{3}\left[\begin{array}{l} \frac{h_y}{h_x}\left[\left(V_{i,j}-V_{i+1,j}\right)^2 + \left(V_{i,j+1}-V_{i+1,j+1}\right)^2 + \left(V_{i,j}-V_{i+1,j}\right)\left(V_{i,j+1}-V_{i+1,j+1}\right)\right] \\ + \frac{h_x}{h_y}\left[\left(V_{i,j}-V_{i,j+1}\right)^2 + \left(V_{i+1,j}-V_{i+1,j+1}\right)^2 + \left(V_{i,j}-V_{i,j+1}\right)\left(V_{i+1,j}-V_{i+1,j+1}\right)\right] \end{array}\right]$$

$$(8.26)$$

and then the desired capacitance is the sum of all of these terms over the entire structure.

The MATLAB function C_Energy.m, called at the bottom of fd2.m, implements this calculation to determine the total energy capacitance:

```
function C = C_Energy(V)
% Calculate C in a rectangular box using Stored Energy
  [Sx,Sy] = size(V);
  eps0 = 8.854;
```

```
  Utot = 0;;
  for i = 1 : Sx - 1
    for j = 1 : Sy - 1
      a = V(i,j); b = V(i + 1,j); c = V(i + 1,j + 1); d = V(i,j + 1);
      Uxsq = 1./3.*((a - b)^2 + (d - c)^2 + (a - b)*(d - c));
      Uysq = 1./3.*((a - d)^2 + (b - c)^2 + (a - d)*(b - c));
      Utot = Utot + Uxsq + Uysq;
    end
  end

  C = Utot*eps0;
end
```

Figure 8.8 adds the results of this calculation to the graph of the results of the Gaussian surface calculation of capacitance (Figure 8.6) as a function of ns. The latter calculation (total stored energy) result is lower than the former calculation result for all values of ns shown, and both curves appear to be asymptotically approaching (approximately) the same value.

Both capacitances (C from charge and C from energy) fall as the number of iterations increases. The capacitance calculated from the stored energy, at least in this example, is always the lower of the two.

Is one of these calculations inherently more accurate, or are we looking at results that are just artifacts of this example?

The set of cell voltages [equation (8.18)] forms a piecewise continuous approximation to the actual voltage distribution. It satisfies the boundary conditions exactly. The stored energy in the electric field is exactly calculated as a function of this approximate voltage distribution. The exact voltage distribution, which, of course, is not known, is the voltage distribution that minimizes the stored energy.[4] The approximate distribution therefore leads to a stored energy that is

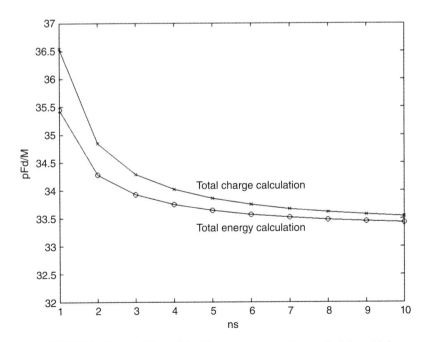

FIGURE 8.8   Repeat of Figure 8.6 with total energy capacitance calculation added.

larger than the exact stored energy. Since the predicted capacitance is directly proportional to this stored energy, it must be higher than the actual capacitance. Figure 8.8 shows that the predicted capacitance is falling as the resolution of the grid (ns) increases. We know, therefore, that the accuracy of the prediction is increasing as the grid resolution increases, with the predicted capacitance always greater than but falling toward the actual capacitance.

The stored energy calculation result is therefore bounded on one side (it is always too large). In this example it is always smaller than the total charge calculation result. The former is therefore (at least here) more accurate.

We will show in Chapter 9 that in many situations we can extend the total stored energy calculation, which produces an upper bound on the capacitance, to a second calculation that produces a lower bound on the capacitance. Once we have bounded capacitance estimates on both sides, we have a measure of the accuracy of our solution.

The equations for the electric field components [equations (8.20) and (8.21)] lead to fields that are discontinuous across cell boundaries. Physically, this can happen only if there is charge along these boundaries. This approximation artifact will cause the total charge calculation result to depend on the actual integration path chosen. In other words, whereas we can be comfortable in assuming that the total charge calculation accuracy is improving when the resolution improves, we don't know how to compare results when the chosen integration path changes. This can be a problem when, for example, we wish to tweak the geometry slightly and study the effects of this tweak on the capacitance.

Unless specifically stated otherwise, all capacitance calculations going forward will be stored energy calculations.

There are two final salient points regarding solution accuracy:

1   In principle, as in the examples shown, solution accuracy steadily increases as resolution increases. In practice, however, this is not always the case. Using Figure 8.8 as an example, if we continued to increase ns, we would find that the capacitance begins to increase. As the number of variables increases with the square of the resolution, problems arise because of the finite length (number of binary digits) of the computer representation of each number.

2   Figure 8.9 shows a piece of a structure that does not neatly conform to a rectangular grid and a simple choice of boundaries that approximate the structure and do conform to the rectangular grid. In this situation we are combining the inherent approximation of the FD technique with the fact that we are not exactly analyzing the actual structure. If we simply increase the resolution by decreasing gridpoint separation without

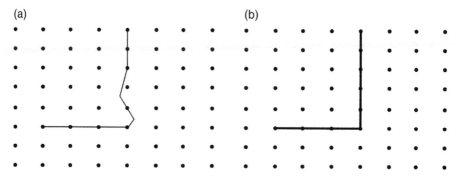

FIGURE 8.9   Example of an inherently inaccurate model: (a) actual structure; (b) modeled structure.

updating the chosen boundaries, then we will improve the solution accuracy for the chosen boundaries, but not necessarily for the actual structure. Chapter 9 discusses techniques for approximating *off-gridpoint* boundaries and improving the accuracy in this situation. In general, however, the finite element method should be considered to handle irregular boundaries.

PROBLEMS

8.1 Redo `fd2.m` for a square box of side 10 centered in a second square box of side 20. Set the inner box at 1 V and the outer box at 0 V. Calculate the capacitance for several values of `ns` and show the voltage distribution along the lines $j = 0$ and $j = J_{c2}$.

8.2 Redo `fd2.m` for a square box of side 20 with the octagon shown in Figure P8.2a centered in the box.

8.3 Repeat Problem 8.1 but rotate the entire structure 45° as shown in Figure P8.3a. Compare the results to those of Problem 8.1.

8.4 Figure P8.4a shows a structure with two unequal rectangles inside an enclosing, larger, rectangle. The inner rectangles are set at aribtrary voltages and the outer rectangle is not set to anything (is left "floating") with zero charge on it. Find the (mutual) capacitance between the inner rectangles and the voltage of the outer rectangle.

(*Hint*: As will be discussed in Chapter 9, a floating electrode's voltage will be that voltage that minimizes the stored energy in the system.)

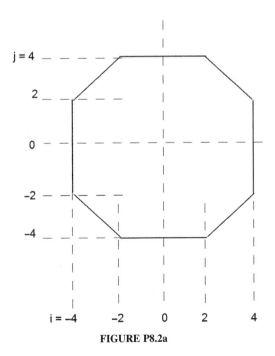

FIGURE P8.2a

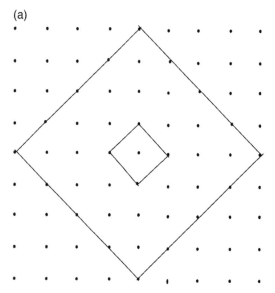

FIGURE P8.3 (a) Structure of Problem 8.1 rotated 45° on grid.

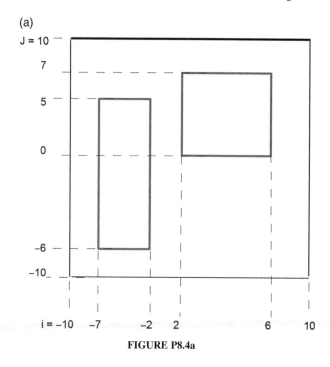

FIGURE P8.4a

REFERENCES

1. S. Frankel, *Multiconductor Transmission Line Analysis*, Artech House, 1977, pp. 364–366.

2. *Successive Over Relaxation* (available at www.en.wikipedia.org).

3. http://www.wheeler.com/technology/equations/index.html.

4. C. T. Carson, and G. K. Cambrell, Upper and lower bounds on the characteristic impedance of TEM mode transmission lines, *IEEE Trans. Microwave Theory Tech.* **MTT-14**: 497–498 (Oct. 1966).

9

Refining the Finite Difference Method

In this chapter we'll refine and expand the finite difference solution technique and its applications. We'll look at refined grids for better accuracy in high-field regions (regions with rapidly changing voltage), mixed dielectrics, other coordinate systems, multielectrode structures, and magnetic wall boundary conditions, including calculation of symmetric structures. We'll show how, in many cases, we can calculate both upper and lower bounds on our estimates of C (and for transmission line cross sections L and Z_0). We'll take a brief look at solving three-dimensional (3d) problems.

9.1 REFINED GRIDS

Regions such as outer corners of electrodes have very high electric fields as compared to other regions in a structure. Saying that the field magnitude is *high* is equivalent to saying that the voltage in that region changes rapidly with position. In such a situation we want to reduce the value of h to ensure that a significant portion of the voltage variation does not take place within one cell. The brute-force approach to this is to simply increase the value of h throughout by, say, a factor of 100 in both axes. This will, of course, increase the number of nodes by a factor of 10,000 (assuming a simple rectangular box geometry). If we had started out with a $50 \times 250 = 12,500$-node system, then increasing the number of nodes by a factor of 10,000 would not seem too practical. An alternative approach is to decrease h only where (we think) this increase is necessary. Before looking at examples of this, however, we need to recast the finite difference approximation to Laplace's equation in terms of different size values of h.

Introduction to Numerical Electrostatics Using MATLAB, First Edition. Lawrence N. Dworsky.
© 2014 John Wiley & Sons, Inc. Published 2014 by John Wiley & Sons, Inc.

Referring to Figure 8.1, let the h value separating nodes V_0 and V_1 be h_1, the value separating nodes V_0 and V_2 be h_2, and so on. The Taylor expansions from Chapter 8 are now

$$
\begin{aligned}
V_1 &= V(x_0 + h_1, y_0) \\
&= V(x_0, y_0) + h_1 \frac{\partial}{\partial x} V(x_0, y_0) + \frac{h_1^2}{2} \frac{\partial^2}{\partial x^2} V(x_0, y_0) + \frac{h_1^3}{6} \frac{\partial^3}{\partial x^3} V(x_0, y_0) + \cdots
\end{aligned}
\tag{9.1}
$$

$$
\begin{aligned}
V_3 &= V(x_0 - h_3, y_0) \\
&= V(x_0, y_0) - h_3 \frac{\partial}{\partial x} V(x_0, y_0) + \frac{h_3^2}{2} \frac{\partial^2}{\partial x^2} V(x_0, y_0) - \frac{h_3^3}{6} \frac{\partial^3}{\partial x^3} V(x_0, y_0) + \cdots
\end{aligned}
\tag{9.2}
$$

Multiplying the equation (9.1) by h_3, (9.2) by h_1, adding these two equations together and updating the notation a bit, we obtain

$$
h_3 V_1 + h_1 V_3 = (h_3 + h_1) V_0 + \frac{h_3^2 h_1 + h_1^2 h_3}{2} \frac{\partial^2}{\partial x^2} V(x_0, y_0) + \cdots
\tag{9.3}
$$

Following the same procedure for V_2 and V_4, we get

$$
h_4 V_2 + h_2 V_4 = (h_4 + h_2) V_0 + \frac{h_4^2 h_2 + h_2^2 h_4}{2} \frac{\partial^2}{\partial y^2} V(x_0, y_0) + \cdots
\tag{9.4}
$$

Solving for the second derivatives, adding the equations together, and setting the results to zero, we get

$$
\left(\frac{1}{h_1 h_3} + \frac{1}{h_2 h_4} \right) V_0 = \frac{V_1}{h_1 (h_1 + h_3)} + \frac{V_2}{h_2 (h_2 + h_4)} + \frac{V_3}{h_3 (h_1 + h_3)} + \frac{V_4}{h_4 (h_2 + h_4)}
\tag{9.5}
$$

Our first example of using equation (9.5) is to modify the stripline problem of Chapter 8, for two reasons: (1) to increase the resolution near the edge of the center strip electrode, as discussed above; and (2) to allow for a study of finite-thickness center strip electrode (the zero-thickness electrode was a good starting approximation, but we can do better).

Unfortunately, detailed structures take more involved descriptions than do simple structures and there is really no way around this annoyance. Figure 9.1 shows the detail of the new grid layout. Taking advantage of the symmetry of the structure, only the upper right quadrant of the layout is shown.

Figure 9.2 shows the grid layout used. Important properties of these layouts are as follows:

1   The sum of the h_i values is 40. This is the total width of the box.
2   The sum of the h_j values is 8. This is the total height of the box.
3   Node separation h_i is the distance between nodes (i, j) and $(i + 1, j)$ for all j. The graph of h_i in Figure 9.2 is therefore not perfectly centered — it is moved one node to the left; h_j has the same property.

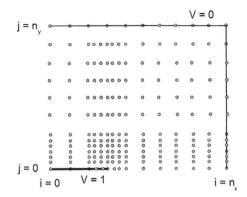

FIGURE 9.1 Stripline example with a refined grid layout.

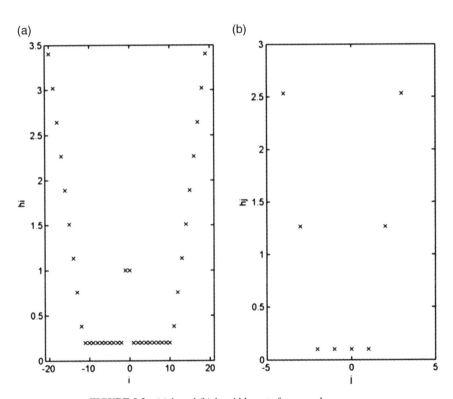

FIGURE 9.2 (a) h_i and (b) h_j grid layouts for example program.

4 Height h_i is laid out so that there is moderate resolution (1) near $x = 0$, high resolution (0.1) centered about the edges of the center conductor, and then steadily increasing resolution out to the right and left edges.

5 Height h_j is laid out so that there is high resolution (0.1) near $y = 0$ and then steadily increasing resolution out to the top and bottom edges.

6 The center conductor runs from $i = -2$ to $i = +2$ (this is not shown in the figure). This sets the center conductor width at 1.2.

Here are three programs related to this topic:

- [Program fd4.m] — modified fd2.m script using adjustable h_i and h_j:

```
%fd4.m  fd2 repeated with non-uniform hi and hj values
clear
ns = 5;          % scaling factor for resolution studies

% Make Sure Jc1 and Jc2 settings in get_grid.m are what is desired -
%    these setting do not scale with ns

[Imin, Imax, Jmin, Jmax, Ic1, Ic2, Jc1, Jc2, hi, hj] = get_grid(ns);

[i,j,Kmax] = get_ijk(Imin,Imax,Jmin, Imax,Jmax,0); % number of
variables

a = sparse(Kmax,Kmax);   % zeros(Kmax);   % allocate the
coefficient array
b = zeros(Kmax,1); % allocate the bcs array

% initialize a as if every variable is free
h1 = 0; h2 = 0; h3 = 0; h4 = 0;   % Terms using these will drop out
for j = Jmin: Jmax
  j1 = j - Jmin + 1;
  if j < Jmax, h2 = hj(j1); end;
  if j > Jmin, h4 = hj(j1-1); end;
  for i = Imin : Imax
    i1 = i - Imin + 1;
    if i < Imax, h1 = hi(i1); end;
    if i > Imin, h3 = hi(i1-1); end;
    [i,j,k0] = get_ijk(Imin,Imax,Jmin, i,j,0);
    a(k0,k0) = -1/h1/h3 - 1/h2/h4;
    if k0 < Kmax, a(k0,k0 + 1) = 1/h1/(h1 + h3); end
    if k0 > 1,    a(k0,k0-1) = 1/h3/(h1 + h3); end
    k1 = k0 + (Imax-Imin) + 1;
    if k1 < = Kmax, a(k0,k1) = 1/h2/(h2 + h4); end
    k2 = k0 - (Imax-Imin) - 1;
    if k2 > 0, a(k0,k2) = 1/h4/(h2 + h4); end
  end
end

% Perimeter (v = 0) bcs
for i = Imin : Imax
  [i,j,k] = get_ijk(Imin,Imax,Jmin, i,Jmin,0);
  a(k,:) = 0; a(k,k) = 1;
  [i,j,k] = get_ijk(Imin,Imax,Jmin, i,Jmax,0);
  a(k,:) = 0; a(k,k) = 1;
end
for j = Jmin + 1 : Jmax - 1
  [i,j,k] = get_ijk(Imin,Imax,Jmin, Imin,j,0);
  a(k,:) = 0; a(k,k) = 1;
  [i,j,k] = get_ijk(Imin,Imax,Jmin, Imax,j,0);
  a(k,:) = 0; a(k,k) = 1;
end
```

```
% inner strip (v = 1) bcs
for i = Ic1 : Ic2
  for j = Jc1 : Jc2
     [i,j,k] = get_ijk(Imin,Imax,Jmin, i,j,0);
     a(k,:) = 0;
     a(k,k) = 1;
     b(k) = 1;
   end
end

% solve for voltages
v = a\b;

% list non-zero voltages by row, column, & variable nrs
fprintf (' i   j   k   Volts  \n  \n')
for k = 1 : Kmax
  volts = v(k);
  if volts > 0
    [i,j,k] = get_ijk(Imin,Imax,Jmin, 0,0,k);
    fprintf ('%3d   %3d   %3d   %8.3f  \n',  i,  j,  k,  volts)
  end
end

% we'll need an xy voltage array for ongoing calcs

Sx = Imax - Imin + 1; Sy = Jmax - Jmin + 1;
Volts = zeros(Sx,Sy);
for ii = Imin:Imax
  i = ii - Imin + 1;
  for jj = Jmin:Jmax
    j = jj - Jmin + 1;
    [ii,jj,k] = get_ijk(Imin,Imax,Jmin, ii,jj,0);
    Volts(i,j) = v(k);
  end
end

     C = C_Energy2(Volts, hi, hj)
```

- [Program get_grid.m] — function for generating h_i and h_j lists:

```
function [Imin, Imax, Jmin, Jmax, Ic1, Ic2, Jc1, Jc2, hi, hj] =
get_grid(ns)

% get_grid.m   developing the non-uniform grid

nx = 20*ns; ny = 4*ns;
i_vec = [-nx : nx]; Ic1 = -2*ns; Ic2 = 2*ns;
j_vec = [-ny : ny];
Jc1 = 0; Jc2 = 0;

Imin = i_vec(1); Imax = i_vec(end); Jmin = j_vec(1); Jmax = j_vec(end);
```

```
hi = ones(1,ns);                    % inner x low resolution region (rhs only)
hi = [hi, .2*ones(1,10*ns)]; % x high resolution region
rem_len = nx - length(hi);
addon = [1:rem_len]; addon = addon*(nx-sum(hi))/sum(addon);
% tapered region
hi = [hi,addon];
hi = [fliplr(hi),hi];        % lhs


hj = [.1*ones(1,2*ns)];     % y high resolution region (top only)
rem_len = ny - length(hj);
addon = [1:rem_len]; addon = addon*(ny-sum(hj))/sum(addon);
% tapered region
hj = [hj,addon];
hj = [fliplr(hj),hj];        % bottom


end
```

- [Program C_Energy2.m] — modified C_energy.m function using adjustable h_i and h_j:

```
function C = C_Energy2(V, hi, hj)
%  C_energy routine modified to accept hi and hj arrays

   [Sx,Sy] = size(V);
   eps0 = 8.854;

   Utot = 0;
   for i = 1 : Sx - 1
     for j = 1 : Sy - 1
       a = V(i,j); b = V(i+1,j); c = V(i+1,j+1); d = V(i,j+1);
       Utot = Utot ...
             + hj(j)/hi(i)*((a-b)^2 + (d-c)^2 + (a-b)*(d-c)) ...
             + hi(i)/hj(j)*((a-d)^2 + (b-c)^2 + (a-d)*(b-c));
     end
   end

   C = Utot*eps0/3;
end
```

MATLAB script fd4.m is fd2.m upgraded to allow for user-defined grids. The function get_grid.m defines the grids shown in Figure 9.2. The function C_Energy2.m is C_energy.m upgraded to allow for user defined grids.

The results of this program are very similar to the results of fd2.m shown in Figure 8.8. The accuracy is improved at ns = 1 but as expected, the improvement decreases as ns is increased. The accuracies easily available from such a simple numerical analysis are good enough to surpass the tolerance and dielectric accuracies of all except laboratory-grade fabricated devices.

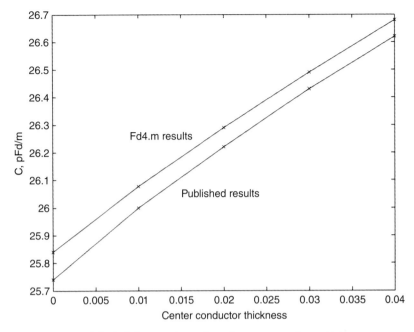

FIGURE 9.3 Stripline capacitance for various center conductor thicknesses.

The fd4.m program offers other useful capabilities; for $ns = 10$, h_j near $y = 0$ is 0.1, while the the sum of all h_j is 80. This means that scaling to a total thickness of 4, each (y direction) grid step near $y = 0$ is 0.005. Keeping the center conductor centered in the box, we can therefore vary the thickness of the center conductor in 0.01 increments.

Figure 9.3 shows the results of calculating the capacitance for several different center conductor thicknesses ($ns = 10$). Published values1 are shown for the same calculation. The excellent absolute error and even better relative error (incremental change in capacitance as conductor thickness is changed) is apparent.

The y-direction location of the center conductor can also be varied in small steps about $y = 0$ just as easily as the center conductor's thickness was varied. The data resulting from doing this are very useful for establishing tolerances on the location of the center conductor when the required tolerance values on the capacitance are known. In many cases the relative error is more important than the absolute error because the latter can be minimized experimentally. Then, knowing the variation to expect due to manufacturing tolerances is important information.

9.2 ARBITRARY CONDUCTOR SHAPES

Equation (9.5) is also useful when one is confronted with boundary conditions that don't lend themselves to falling along rectangular grids. Figure 9.4, for example, shows a circular (actually the cross section of a long cylinder) metal conductor. It does not matter whether

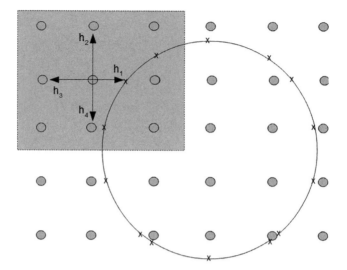

FIGURE 9.4   A boundary condition circle on a square grid.

this conductor is solid or hollow — electrostatically the entire region is at the same potential as the surface. The × tickmarks are points on the surface of the circle that lie along the grid connecting the voltage node points (grid not shown).

The shaded region of Figure 9.4 shows a central node adjacent to four other nodes, one of which lies on the circle boundary. The h values for calculating the voltage at this node are shown. The definitions shown refer to equation (9.5). This procedure is extended to all nodes for which a tickmark × is located between the V_0 node and a node interior to the circle.

As an example, consider a stripline with a circular center conductor centered (in y) between the strip electrodes. The ratio of the strip separation to the circle diameter is $20:1$. This number is the same as in the last example in Chapter 5.

The program fd5.m implements this technique, with the function get_grid_circle.m setting up the proper h values. In get_grid_circle.m, the radius of the circle is artificially increased by 0.001 to prevent the circle from passing through nodes (this is only a programming convenience; nothing in the analysis demands this maneuver as long as the boundary conditions and h values are correctly handled):

- [Program fd5.m here] — MATLAB script for a circle in a (wide) box:

```
%fd5.m   fd4.m with a circular center conductor
clear
ns = 1;    % scaling factor for resolution studies

[Imin, Imax, Jmin, Jmax, hi,hj, bc_nodes] = prob9_1(ns);

[i,j,Kmax] = get_ijk(Imin,Imax,Jmin, Imax,Jmax,0); % number of
variables

a = sparse(Kmax,Kmax);   % zeros(Kmax);   % allocate the
coefficient array
b = zeros(Kmax,1);   % allocate the bcs array
```

```
% initialize a as if every variable is free
h1 = 0; h2 = 0; h3 = 0; h4 = 0;   % Terms using these will drop out
for j = Jmin: Jmax
  j1 = j - Jmin + 1;
  if j < Jmax, h2 = hj(j1); end;
  if j > Jmin, h4 = hj(j1-1); end;
  for i = Imin : Imax
    i1 = i - Imin + 1;
    if i < Imax, h1 = hi(i1); end;
    if i > Imin, h3 = hi(i1-1); end;
    [i,j,k0] = get_ijk(Imin,Imax,Jmin, i,j,0);
    a(k0,k0) = -1/h1/h3 - 1/h2/h4;
    if k0 < Kmax,   a(k0,k0 + 1) = 1/h1/(h1 + h3); end
    if k0 > 1,   a(k0,k0-1) = 1/h3/(h1 + h3); end
    k1 = k0 + (Imax-Imin) + 1;
    if k1 < = Kmax, a(k0,k1) = 1/h2/(h2 + h4); end
    k2 = k0 - (Imax-Imin) - 1;
    if k2 > 0, a(k0,k2) = 1/h4/(h2 + h4); end
  end
end

% Perimeter (v = 0) bcs
for i = Imin : Imax
  [i,j,k] = get_ijk(Imin,Imax,Jmin, i,Jmin,0);
  a(k,:) = 0; a(k,k) = 1;
  [i,j,k] = get_ijk(Imin,Imax,Jmin, i,Jmax,0);
  a(k,:) = 0; a(k,k) = 1;
end
for j = Jmin + 1 : Jmax - 1
  [i,j,k] = get_ijk(Imin,Imax,Jmin, Imin,j,0);
  a(k,:) = 0; a(k,k) = 1;
  [i,j,k] = get_ijk(Imin,Imax,Jmin, Imax,j,0);
  a(k,:) = 0; a(k,k) = 1;
end

% circle (v = 1) bcs
for k = 1 : Kmax
  if bc_nodes(k) == 1
    a(k,:) = 0;
    a(k,k) = 1;
    b(k) = 1;
  end
end

% solve for voltages
v = a\b;

% list non-zero voltages by row, column, & variable nrs
fprintf (' i j k Volts \n \n')
for k = 1 : Kmax
  volts = v(k);
  if volts > 0
    [i,j,k] = get_ijk(Imin,Imax,Jmin, 0,0,k);
```

```
        fprintf ('%3d %3d %3d %8.3f \n', i, j, k, volts)
    end
end

% we'll need an xy voltage array for ongoing calcs

Sx = Imax - Imin + 1; Sy = Jmax - Jmin + 1;
Volts = zeros(Sx,Sy);
for ii = Imin:Imax
  i = ii - Imin + 1;
  for jj = Jmin:Jmax
    j = jj - Jmin + 1;
    [ii,jj,k] = get_ijk(Imin,Imax,Jmin, ii,jj,0);
    Volts(i,j) = v(k);
  end
end

C1 = C_Gauss_box(Volts)
C = C_Energy2(Volts, hi, hj)
```

- [Program get_grid_circle.m]:

```
function [Imin, Imax, Jmin, Jmax, hi, hj, bc_nodes] =
get_grid_circle(ns)

% developing the non-uniform grid
% This version generates the circle boundary conditions
% The circle radius is 2.   Note that nx and ny have been
% changed from previous programs
% This function returns hi and hj and a list of bc nodes

  nx = 100*ns;   ny = 20*ns;
  rad = 2*ns + .001; % Circle radius, circle centered at (i,j) = (0,0)

  Imin = -nx;   Imax = nx;   Jmin = -ny;   Jmax = ny;        % box parameters
  [i,j,Kmax] = get_ijk(Imin,Imax,Jmin, Imax,Jmax,0); % number of
variables

% Initialize arrays to their correct sizes
  hi = ones(1,2*nx); hj = ones(1,2*ny);
  bc_nodes = zeros(1,Kmax);

% Walk through the whole array except the outer box
  for i = Imin + 1 : Imax-2
    for j = Jmin + 1 : Jmax - 2
      r_pt_sq = i^2 + j^2;
      if (r_pt_sq < rad^2)
        [~,~,k] = get_ijk(Imin,Imax,Jmin, i,j,0);
        bc_nodes(k) = 1;
      end
    end
  end
```

```
% Below could have been kept inside previous nested loops but it should
%   be easier to follow this way
% We now know which nodes are inside the circle, we want h values
  for k = 2 : Kmax-1
    if bc_nodes(k) == 1
      [i,j,~] = get_ijk(Imin,Imax,Jmin, i,j,k);
      j1 = j-Jmin + 1; i1 = i-Imin + 1;
      if bc_nodes(k-1) == 0        % look to the left
        x = sqrt(rad^2 - j^2);
        hi(i1-1) = -x-(i-1);
      elseif bc_nodes(k + 1) == 0    % look to the right
        x = sqrt(rad^2 - j^2);
        hi(i1) = x-i;
      elseif bc_nodes(k + Imax-Imin + 1) == 0        %      look up
        y = sqrt(rad^2 - i^2);
        hj(j1) = y-j;
      elseif bc_nodes(k - (Imax-Imin + 1)) == 0        %      look down
        y = sqrt(rad^2 - i^2);
        hj(j1-1) = -y-(j-1);
      end
    end
  end

end
```

Figure 9.5 shows the results produced by this program and the result from the calculation in Chapter 5. The stored energy calculation, unlike calculations in previous examples, does

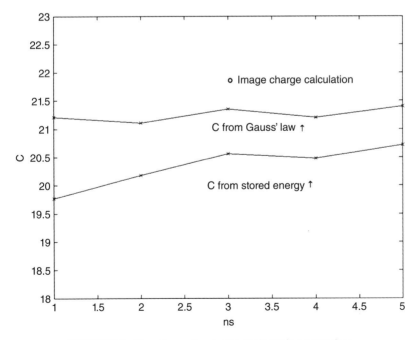

FIGURE 9.5   Capacitance of a circular conductor in a square box.

not fall with increasing resolution (ns). This is because, in this situation, the calculation is not correct. Referring to Figure 9.4, many of the cells near the circle are not complete rectangles; they are rectangles with a piece sliced off by the circle. This fact has not been considered in the stored energy calculation. The stored energy calculation could be rewritten to correct this problem, but writing a general-purpose program to do this is very difficult.

(*Note*: Automatically correctly handling this issue will be one of the advantages of the finite element technique.)

Returning to our example, the total charge calculation, using a Gaussian surface just inside the outer box, as was done in Chapter 8, is still correct. The results of this calculation do not change monotonically with ns. This is because changing ns changes not only the resolution, but also the number of points on the circle that become boundary conditions, and as a result the capacitance meanders a bit. Overall, the results of the calculation agree well with the result produced in Chapter 5.

9.3 MIXED DIELECTRIC REGIONS AND A NEW DERIVATION OF THE FINITE DIFFERENCE EQUATION

The results to be shown in this section are presented, for now, without derivation. The results are easy to apply to existing and upcoming examples, and including them at this point is convenient and reasonable. The derivation of these results follows very easily and logically from the energy-minimization-based derivation to be presented in Chapters 12, 15, and 16 (and the FD and FEM results for this topic are identical). The derivation will therefore be postponed until discussion of the finite element method.

Figure 9.6 shows a square grid with the voltage node references presented (in Chapter 8).

When solving Laplace's equation using the FD technique, the only places where the existence of the dielectric is relevant is at the dielectric interface (J_{int} in the figure).

The a matrix coefficient terms are modified to include the dielectric constant at each node; the dielectric constant at an interface node is the average of the dielectric constants. For the air dielectric interface shown in Figure 9.6, therefore, we obtain

$$2(1+k)V_0 = \frac{1+k}{2}V_1 + V_2 + \frac{1+k}{2}V_3 + kV_4 \tag{9.6}$$

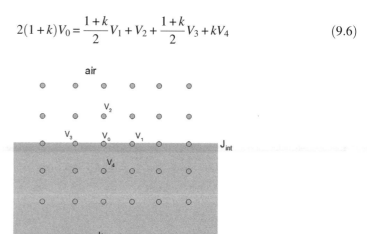

FIGURE 9.6 Square grid shows dielectric surface.

In calculating the total energy stored in the electric field, we sum over all of the rectangular "cells" in the structure. We refer to the cell by the node number of the lower left-hand node. Previously, since the structures being studied had a uniform dielectric constant ($k = 1$), we multiplied the entire expression by the permittivity of free space. This must now be modified so that each cell's contribution can be multiplied by its own permittivity.

9.4 EXAMPLE: STRUCTURE WITH A DIELECTRIC INTERFACE

Figure 9.7 shows the boxed stripline of Figure 8.4, now called a *microstripline* because the center conductor is sitting on a dielectric. Figure 9.7 shows only the right-hand side of the symmetric left-to-right structure.

The dielectric constant in this example is $K = 9.7$. (This 9.7 value is the dielectric constant of alumina, also termed *amorphous aluminum oxide*, a common microwave frequency dielectric substrate material.

The results of this calculation, for a zero-thickness conductor, is that the capacitance predicted by fd2.m increases by a factor of ~5.3. In other words, the effective relative dielectric constant of the structure is 5.35.

The modifications of fd2.m used to perform this calculation are

1 Define er = 9.7 at or near the top of the program.

2 At the bottom of the loops where the *a* values are initialized, add

```
if j == 0;
  a(k0,k0) = -2*(1+er_diel);
  a(k0,k0+1) = (er_diel+1)/2;
  a(k0,k0-1) = (er_diel+1)/2;
```

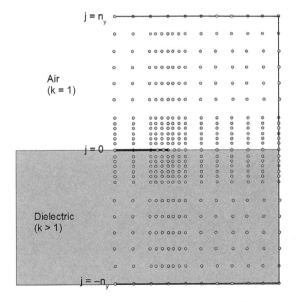

FIGURE 9.7 Boxed microstrip transmission line.

```
    a(k0,k2) = er_diel;
end
```

3 Pass the `er` value to the capacitance calculation function along with the voltage array.

4 Inside the capacitance calculation function, near the top, define the j interface value

```
j_interface = ceil(Sy/2);
```

5 Modify the calculation of the total energy to read

```
term = Uxsq + Uysq;
if j < j_interface, term = term*er_diel; end;
Utot = Utot + term;
```

If different center conductor thicknesses are to be considered, the center conductor typically is not symmetric in y — the practical situation is that the center conductor sits "on" the dielectric material.

An example of a nonuniform dielectric structure will be presented in Section 9.5.

9.5 AXISYMMETRIC CYLINDRICAL COORDINATES

In three-dimensional (3d) cylindrical coordinates, in order to emphasize that there are three terms — one for each spatial direction — Laplace's equation is typically written as follows:

$$\nabla^2 V = \frac{1}{r}\frac{\partial}{\partial r}\left(r\frac{\partial V}{\partial r}\right) + \frac{1}{r^2}\frac{\partial^2 V}{\partial \phi^2} + \frac{\partial^2 V}{\partial z^2} \qquad (9.7)$$

For our purposes, it is useful to expand the first term:

$$\nabla^2 V = \frac{\partial^2 V}{\partial r^2} + \frac{1}{r}\frac{\partial V}{\partial r} + \frac{1}{r^2}\frac{\partial^2 V}{\partial \phi^2} + \frac{\partial^2 V}{\partial z^2} \qquad (9.8)$$

We will consider only problems that are axisymmetric, that is, in which there is no ϕ dependence. The step sizes in i and j are h_i and h_j, respectively. When r is not 0, quoting earlier results, we obtain

$$\nabla^2 V \approx \frac{V_{i+1,j} + V_{i-1,j} - 2V_{i,j}}{h_i^2} + \frac{V_{i+1,j} - V_{i-1,j}}{2r_i h_i} + \frac{V_{i,j+1} + V_{i,j-1} - 2V_{i,j}}{h_j^2} = 0 \qquad (9.9)$$

where r_i is the conventional (cylindrical coordinate) radial distance from the axis $(0,j)$ to the point (i,j).

This formula becomes problematic at $r = 0$. However, since we are dealing only with situations that are axially symmetric, the derivative of V with respect to r at $r = 0$ must be 0, and L'Hospital's rule applies. In the limit as r goes to 0, we have

$$\lim \frac{\partial V/\partial r}{r} = \frac{\partial^2 V/\partial r^2}{1} = \frac{\partial^2 V}{\partial r^2} \tag{9.10}$$

and therefore for $r = 0$, we get

$$\nabla^2 V \approx 2 \frac{2V_{1,j} - 2V_{0,j}}{h_i^2} + \frac{V_{0,j+1} + V_{0,j-1} - 2V_{0,j}}{h_j^2} = 0 \tag{9.11}$$

The numerical equations to be solved are as follows: (1) for $r_i = 0$

$$2V_{0,j} \left[\frac{2}{h_i^2} + \frac{1}{h_j^2} \right] = \frac{4V_{1,j}}{h_i^2} + \frac{V_{0,j+1} + V_{0,j-1}}{h_j^2} \tag{9.12}$$

and (2) for $r_i \neq 0$

$$2V_{i,j} \left[\frac{1}{h_i^2} + \frac{1}{h_j^2} \right] = V_{i+1,j} \left[\frac{1}{h_i^2} + \frac{1}{2r_i h_i} \right] + V_{i-1,j} \left[\frac{1}{h_i^2} - \frac{1}{2r_i h_i} \right] + \frac{V_{i,j+1} + V_{i,j-1}}{h_j^2} \tag{9.13}$$

Figure 9.8 shows two parallel circular electrodes, 500 units apart. The bottom electrode has a conical metal tip sitting on it. The tip is centered at $r = 0$. If we set $h_r = h_z = 1$, the tip is 20 units wide at its base and 10 units high. Setting the tip dimensions to these numbers is convenient because the boundary condition of the tip surface passes through node points $(10,0)$, $(9,1)$, ..., $(0,10)$. The two electrodes are set to 1 V apart, with the tip at the potential of the bottom electrode.

The boundary condition shown at the perimeter of the circular electrodes has not been discussed thus far. The voltage is specified as a linear gradation with y between the voltages

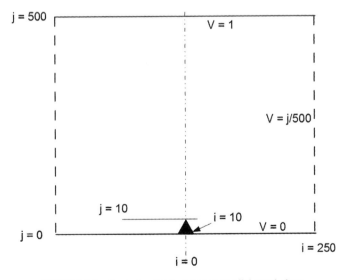

FIGURE 9.8 Conical metal tip between parallel metal plates.

at the two electrodes. If we had only two parallel plates with lateral dimensions much larger than their separation, we would expect to find this voltage distribution, equivalent to a uniform vertical electric field of value $\frac{1}{500} = 0.002$ (volts per length unit), near $r = 0$. Since there is a small tip sitting at $r = 0$ on the bottom electrode, here we should expect some change from this simple uniform voltage distribution in this tip in this region. However, moderately far from this region, with very large (even infinitely large) parallel plates, the distribution will be extremely close to the uniform distribution. In other words, this boundary condition is an approximation of infinite parallel plates.

MATLAB program fd7.m — which presents an example of cylindrical coordinates with a sharp tip between parallel plates — sets up and runs this simulation.

```
%fd7.m  Small tip between two plates, cylindrical coordinates
close; clear

hi = 2;   hj = 1;   % r and z step sizes
Itip = 20;
Jtip = Itip;   % tip dimensions
Imin = 1; Imax = 100; Jmin = 1; Jmax = 250;  % ultimately 250 and 500
[i,j,Kmax] = get_ijk(Imin,Imax,Jmin,  Imax,Jmax,0);  % number of
variables

a = sparse(Kmax,Kmax);   % zeros(Kmax);   % allocate the coefficient array
b = zeros(Kmax,1);   % allocate the bcs array

% initialize a as if every variable is free
for j = 1 : Jmax
  for i = 1 : Imax
    [i,j,k0] = get_ijk(Imin,Imax,Imin, i,j,0);
    if i == 1
      a(k0,k0) = -2*(2/hi^2 + 1/hj^2);
      a(k0,k0+1) = 4/hi^2;
      if j < Jmax, a(k0,k0 + Imax) = 1/hj^2; end;
      if j > 1, a(k0,k0-Imax) = 1/hj^2; end;
    else
      r = (i-1)*hi;
      a(k0,k0) = -2*(1/hi^2 + 1/hj^2);
      if k0 < Kmax, a(k0,k0 + 1) = 1/hi^2 + 2/r/hi; end;
      if k0 > 1, a(k0,k0-1) = 1/hi^2 - 2/r/hi; end;
      if j < Jmax, a(k0,k0 + Imax) = 1/hj^2; end;
      if j > 1, a(k0,k0-Imax) = 1/hj^2; end;
    end
  end
end

% bottom plate bc (v = 0), top plate bc (v = 1)
for i = 1 : Imax
  [i,j,k] = get_ijk(Imin,Imax,Jmin, i,1,0);
  a(k,:) = 0; a(k,k) = 1;
  [i,j,k] = get_ijk(Imin,Imax,Jmin, i,Jmax,0);
  a(k,:) = 0; a(k,k) = 1;
  b(k) = 1;
end
```

```
% tip bc (V = 0)
for i = 1 : Itip
  for j = 1 : Jtip - i +1
    [i,j,k] = get_ijk(Imin,Imax,Jmin, i,j,0);
    a(k,:) = 0; a(k,k) = 1;
  end
end

% perimeter bc (v = y/Jmax)

i = Imax
for j = 1 : Jmax
    [i,j,k] = get_ijk(Imin,Imax,Jmin, i,j,0);
    a(k,:) = 0; a(k,k) = 1;
    b(k) = (j-1)/(Jmax-1);
end

% solve for voltages
v = a\b;

% xy voltage array for ongoing calcs

Volts = zeros(Imax,Jmax);
for i = 1 : Imax
  for j = 1 : Jmax
    [i,j,k] = get_ijk(Imin,Imax,Jmin, i,j,0);
    Volts(i,j) = v(k);
  end
end

% Plot the region near the tip
Isize = Imax/2; Jsize = Isize;
Volts2 = Volts(1:Isize,1:Jsize);
[r,z] = meshgrid(1:Jsize, 1:Isize);
contour (z,r,Volts2,[.025:.025:.15],'k')
hold on
%set (H, 'color', [0,0,0])
x_tri = [1, Itip, 1]; y_tri = [1, 1, Jtip];
fill (x_tri, y_tri, [.8,.8,.8])
axis ([1 50 1 50])
text (35,9,'0.025 volts')
text (35,15,'0.050 volts')
text (35,21,'0.075 volts')
text (35,27,'0.100 volts')
xlabel ('r')
ylabel ('z')
text (5,7, 'tip')

% Some E field information
E0 = Volts(1,Jmax) - Volts(1,Jmax-1)
Etip = Volts(1,Jtip + 1)
Emult = Etip/E0
```

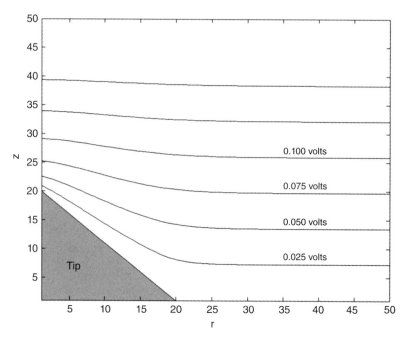

FIGURE 9.9 Equipotential surfaces near the sharp tip.

Figure 9.9 shows several equipotential surfaces near the "tip of the tip" (ToT). These surfaces are equally spaced in terms of voltage, 0.025 V apart. Away from the ToT, in any direction, they are also equally spaced geometrically, and the simple calculation $E_z = V(i,j+1) - V(i,j) = 0.004$ gives the expected result of the value of the applied voltage (1 V) divided by the number of gridpoints in y (250).

Near the ToT, however, the equipotential surfaces bunch together. This means that the electric field peaks near the ToT. The same simple E_z calculation as found above at the ToT yields 0.058 V, a factor of 14.5 greater than 0.004.

The effect on tip sharpness on peak electric field can easily be investigated by changing the sharpness of the tip. We could do this, in fd7.m, by changing either the I_{tip} or J_{tip} value, but this would be an inconvenient approach because then the surface of the tip structure would not pass through gridpoints and we'd have to upgrade the program to properly handle the boundary conditions. An easier solution would be to change h_i and/or h_j. Changing h_i to 0.5, for example, increases the peak field multiplication factor to 18.5; changing h_i to 2.0 decreases the factor to 7.7. Note that the axes of the contour plot generated by fd7.m (Figure 9.9) do not scale with these changes.

These results point out a basic property of electric fields — they peak at sharp points; the sharper the point, the higher the degree of peak of the field. This is very bad news when materials breakdown is a concern, but it is very useful news when you want a localized high electric field. At very a high electric field (of the correct polarity), electrons tunnel through the surface of a metal electrode and launch themselves (typically into a vacuum).[2] This is a useful way to get electrons into a vacuum for devices such as a scanning electron microscope, as discussed in Chapter 2.

Two typical methods for building a field emission device: (1) using the *Spindt tip* structure[3] and (2) placing a long needle protruding from the base electrode.

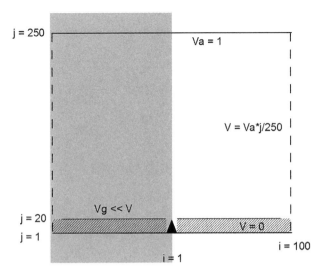

FIGURE 9.10 Spindt tip field emitter structure.

The important characteristics of the Spindt tip structure (method 1) are shown in Figure 9.10. A tip as described in the previous figure is "grown" on an electrode in a circular well in a dielectric layer, typically about 1 μm deep. A common dielectric material is silicondioxide with a relative dielectric constant of 4.0. An additional metal layer, the *gate*, is deposited onto the surface of the dielectric layer. Geometries and voltage are arranged so that the voltage on the upper layer (the anode) is not sufficient to create field emission. The local field due to the gate voltage, which is much lower than the anode voltage, is used to create the field emission and modulate the current flow.

Since this structure will be modeled using cylindrical coordinates, the region shown shaded in Figure 9.10 does not appear explicitly in the calculations.

In this structure, the only dielectric air interface is at the sides of the well. Therefore, the only change needed is that for $i = i_{\text{tip}}$ and $j < j_{\text{gate}} = j_{\text{tip}}$, equation (9.13) must become

$$(1+K)V_{i,j}\left[\frac{1}{h_i^2}+\frac{1}{h_j^2}\right] = KV_{i+1,j}\left[\frac{1}{h_i^2}+\frac{1}{2r_ih_i}\right] + V_{i-1,j}\left[\frac{1}{h_i^2}-\frac{1}{2r_ih_i}\right] + \frac{K+1}{2}\frac{V_{i,j+1}+V_{i,j-1}}{h_j^2}$$

$$(9.14)$$

MATLAB program fd8.m — describing the FD Spindt tip model — sets up and solves the cylindrical coordinate, dielectric interface, structure described above.

```
%fd8.m  fd7 modified to a Spindt Tip structure
close

hi = 1;  hj = 1;  % r and z step sizes
Itip = 20;
Jtip = Itip;  % tip dimensions
Imin = 1;  Imax = 100; Jmin = 1;  Jmax = 250;  % ultimately 250 and 500
[i,j,Kmax] = get_ijk(Imin,Imax,Jmin, Imax,Jmax,0);  % number of
variables
```

```
Jgate = Jtip;
Vanode = 1000; Vgate = 100; % Anode and gate voltages
Krel = 4.0;   % dielectric constant

a = sparse(Kmax,Kmax);   % zeros(Kmax);   % allocate the coefficient array
b = zeros(Kmax,1); % allocate the bcs array

% initialize a as if every variable is free
for j = 1 : Jmax
  for i = 1 : Imax
    [i,j,k0] = get_ijk(Imin,Imax,Imin, i,j,0);
    if i == 1
      a(k0,k0) = -2*(2/hi^2 + 1/hj^2);
      a(k0,k0 + 1) = 4/hi^2;
      if j < Jmax, a(k0,k0 + Imax) = 1/hj^2; end;
      if j > 1, a(k0,k0-Imax) = 1/hj^2; end;
    else
      r = (i-1)*hi;
      a(k0,k0) = -2*(1/hi^2 + 1/hj^2);
      if i < Imax, a(k0,k0 + 1) = 1/hi^2 + 2/r/hi; end;
      a(k0,k0-1) = 1/hi^2 - 2/r/hi;
      if j < Jmax, a(k0,k0 + Imax) = 1/hj^2; end;
      if j > 1, a(k0,k0-Imax) = 1/hj^2; end;

      if (i == Itip) & (j < Jtip) % dielectric interface
        a(k0,k0) = a(k0,k0)*(1 + Krel)/2;
        if i < Imax, a(k0,k0 + 1) = a(k0,k0 + 1)*Krel; end;
        a(k0,k0 + Imax) = a(k0,k0 + Imax)*(Krel + 1)/2;
        if j > 1, a(k0,k0-Imax) = a(k0,k0-Imax)*(Krel + 1)/2; end;
      end
    end
  end
end

% bottom plate bc (V = 0), top plate bc (V = 1)
for i = 1 : Imax
  [i,j,k] = get_ijk(Imin,Imax,Jmin, i,1,0);
  a(k,:) = 0; a(k,k) = 1;
  [i,j,k] = get_ijk(Imin,Imax,Jmin, i,Jmax,0);
  a(k,:) = 0; a(k,k) = 1;
  b(k) = Vanode;
end

% tip bc (V = 0)
for i = 1 : Itip
  for j = 1 : Jtip - i +1
    [i,j,k] = get_ijk(Imin,Imax,Jmin, i,j,0);
    a(k,:) = 0; a(k,k) = 1;
  end
end

% gate bc (V = Vgate)
j = Jtip;
```

```
for i = Itip : Imax*.8
  [i,j,k] = get_ijk(Imin,Imax,Jmin, i,j,0);
  a(k,:) = 0; a(k,k) = 1;
  b(k) = Vgate;
end

% perimeter bc (V = Vanode*y/Jmax)
i = Imax;
for j = 1 : Jmax
  [i,j,k] = get_ijk(Imin,Imax,Jmin, i,j,0);
  a(k,:) = 0; a(k,k) = 1;
  b(k) = (j-1)/(Jmax-1)*Vanode;
end

% solve for voltages
v = a\b;

% xy voltage array for ongoing calcs
Volts = zeros(Imax,Jmax);
for i = 1 : Imax
  for j = 1 : Jmax
    [i,j,k] = get_ijk(Imin,Imax,Jmin, i,j,0);
    Volts(i,j) = v(k);
  end
end

% Some E field information
Etip = Volts(1,Jtip+1:Jtip+21) - Volts(1,Jtip:Jtip+20)
plot([1:21],Etip)
%axis([1 21 0 .08])
```

Figure 9.11 shows (some of) the output of this program for $V_{anode} = 1$ V, $V_{gate} = 0.0$, and 0.1 V.

Figure 9.11 shows that the gate voltage has significant control over the electric field at the ToT. The peak field quickly falls back to the background field as z increases (away from) the ToT. Note that the background field is slightly higher when $V_g = 0.0$ than when $V_g = 0.1$ because this field is determined by the voltage difference $V_a - V_g$.

In typical applications, a Spindt tip structure's anode will be the ~10 mm away from the tip, while the tip itself is about 1 μm high and wide. Scaling V_a to keep the same background field as in this example, we quickly approach the situation where V_a can be thousands of volts while V_g is typically ~100 V. This latter voltage is determined by the diameter of the gate well and the physical properties of the gate material. There is tremendous circuital advantage to controlling electron emission by modulating 100 V as compared to modulating, say, 5000 V. Selection of the anode voltage level is based on the application — this is the energy that the electrons will have when they arrive at the anode.

Method 2, mentioned above, consists in building a field emission tip by simply placing a long (relative to width) needle sticking up from the base electrode. This structure has been implemented with fine wires and more recently with carbon nanotube;[4] it is typically used when only one emitter — rather than an array of emitters (e.g., in a scanning electron microscope) — is needed.

Figure 9.12 shows the same tip as used in the previous two examples, but now it is sitting on a circular pedestal of the same diameter as that of the base of the tip. The peak field at the

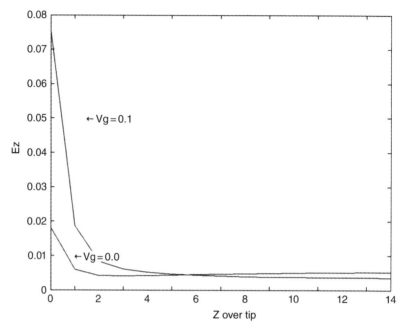

FIGURE 9.11 E_z near ToT of a Spindt tip structure.

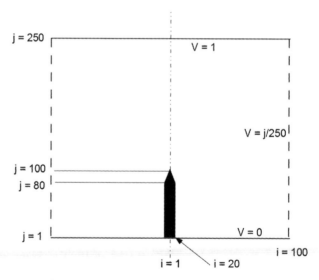

FIGURE 9.12   Sharp needle tip between two electrodes.

tip is now 0.132 (V/m). Wresting the ToT away from the base electrode is clearly an important factor in ensuring a high electric field.

Modifying fd8.m to model this structure is very simple. Remove all references to a dielectric interface; remove the gate boundary condition and replace it with a simple pedestal boundary condition.

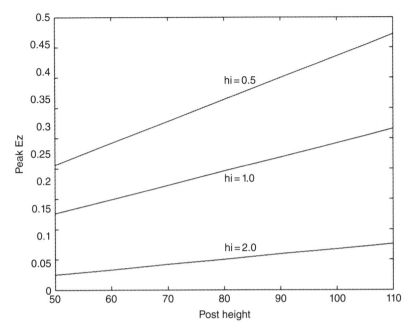

FIGURE 9.13   Peak field versus post height and post diameter for a sharp tip.

Figure 9.13 shows the relationship between post height, post diameter, and peak E_z.

The way to achieve a high peak field is clearly to use a higher, narrower, post. As Figure 9.13 shows, the field nearly doubles when the post height doubles. This cannot simply be attributed to the tip's closer proximity to the anode — the sharp tip is distorting the field near it and creating the high field condition.

In the tip examples in this section, the actual details of the ToT were glossed over and an ideal (0 curvature of radius) tip was assumed. This simplification becomes problematic on close proximity to the ToT (*close proximity* is a relative term here, difficult to define because it is usually measured in comparison to the radius of curvature).

9.6   SYMMETRY BOUNDARY CONDITION

Figure 9.14 shows a situation where existing geometric symmetry can be utilized to create a boundary condition that reduces the extent of the problem to be solved.

The dashed vertical line in the figure is clearly an axis of symmetry. By inspection, then, $V_3 = V_1$ and so forth. We can solve for all the node voltages by just finding the voltages on the left side of the symmetry line up to and including the voltages of the nodes on the symmetry line. For the node shown, we obtain

$$V_0 = \frac{V_2 + 2V_3 + V_4}{4} \tag{9.15}$$

Figure 9.15 shows a second example of a structure with a line of symmetry. In this case, however, the line of symmetry passes through two lines of nodes rather than a single line of nodes.

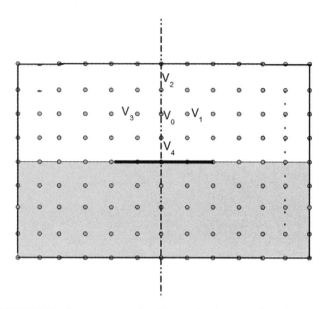

FIGURE 9.14   Symmetry example with row of nodes on the line of symmetry.

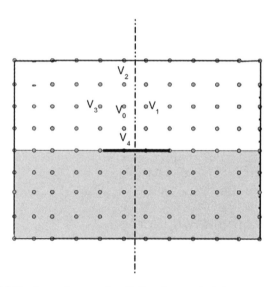

FIGURE 9.15   Symmetry example with line of symmetry between rows of nodes.

In this case, for the node shown the symmetry dictates that, $V_1 = V_0$, so that

$$V_0 = \frac{V_2 + V_3 + V_4}{3}$$ (9.16)

Examples of establishing symmetry boundary conditions in FD models are presented in Section 9.7.

9.7 DUALITY, AND UPPER AND LOWER BOUNDS TO SOLUTIONS FOR TRANSMISSION LINES

As explained in Chapter 9, the capacitance calculated using the stored energy calculation will always exceed the exact capacitance; that is, it is an upper bound to the estimate of the capacitance. This is why we typically see capacitance falling as we increase the resolution of a calculation — the higher resolution solution results in a more accurate description of $V(x,y)$.

In a transmission line structure, as shown in Chapter 2, the capacitance (per unit length) of a transmission line is related to the inductance (per unit length) of the same line by

$$C = \frac{1}{v_0^2 L} \tag{9.17}$$

where, for an air dielectric system, v_0 is the speed of light.

If we had a technique for calculating the inductance of the transmission line that produced an inductance estimate that is an upper bound to the exact inductance, then, using equation (9.17), we would obtain a capacitance that is a lower bound to the exact capacitance. We wouldn't know exactly where in between these two bounds the exact capacitance lies, but by averaging the two, we would get an estimate that is more accurate than either of the bounds.

Fortunately, such a technique exists for many practical geometries, and it is hardly more difficult to implement than was the original capacitance calculation.

Figure 9.16 is the cross section of a square transmission line, shown on a very low-resolution grid.

Figure 9.17a shows the upper right quarter of the same line. All four such quarters of the line are symmetric, and the dashed line edges shown are the symmetry boundary conditions

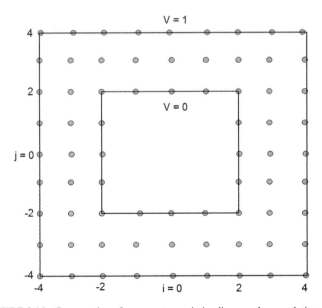

FIGURE 9.16   Cross section of a square transmission line on a low-resolution grid.

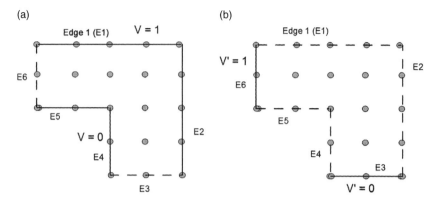

FIGURE 9.17   Upper right quarter of structure in Figure 9.16; (a) $\frac{1}{4}$ th of original line; (b) dual of (a).

discussed in Section 9.6. These edges are also called *magnetic walls*. This means that, if there were a current flowing through the line, the magnetic fields (which are normal to the electric fields) would be normal to these edges. This is directly analogous to the electric fields because the applied voltage is normal to the electric walls: the solid lines in Figure 9.17a.

The edges of the structure are labeled E1 – E6 in Figure 9.17a,b. The individual labels have no particular physical or mathematical significance; this is merely for convenience in matching edges to the computer code to be presented below.

The principle of duality in electrostatics states that if we take a structure such as the one shown in Figure 19.17a, interchange the electric and the magnetic walls, and drive the system with a current at the (new) electric walls, we will get a solution that will enable us to calculate the energy stored in the magnetic rather than the electric field of the original system.[5]

Interchanging the electric and magnetic walls (Figure 9.17b), we get only this dual structure If we apply V' to this structure as shown in the figure and solve for the voltage distribution v'_{ij}, we would find the total stored energy now located in the magnetic field, which is the dual of the electric field:

$$2U_m = \frac{4\sum \iint \left[(\partial V'/\partial x)^2 + (\partial V'/\partial y)^2 \right]}{16} \tag{9.18}$$

The summation is over all the cells, and the integral is over the area of each cell — this is identical to the formula for finding the energy stored in the electric field in Chapter 8. The multiplier 4 in the numerator to gives us the total energy (per unit length) of the line; the structure shown occupies $\frac{1}{4}$ th only the line. This multiplier is also placed in the original line energy equation if only $\frac{1}{4}$ th of the structure is in use or under study.

The 16 in the denominator of equation (9.18) is unique to the U_m equation, it doesn't show up in the U_e equation. When working with the original line, the applied voltage is identically applied to all four quarter sections and the energy calculated is correct. In the dual case, however, V' is really a current. Applying a current of 1 to a quarter of the structure means that we're applying a total current of 4 to the entire structure.

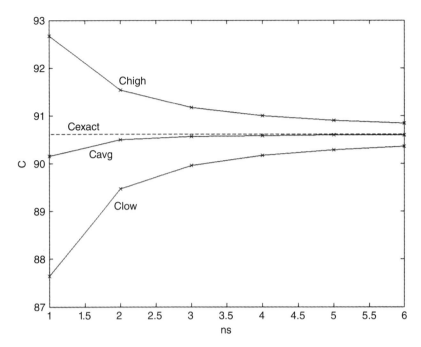

FIGURE 9.18   High and low boundary capacitances for the square transmission line.

Going back to the original line, we have

$$V^2 C_e \varepsilon_0 U_e \tag{9.19}$$

Since, for convenience, we have previously always set one electrode voltage at 0 and the other electrode voltage at 1, the voltage difference $V = 1$ never appeared explicitly in the calculations.

In the case of the magnetic field, however, $V' = 1$ in this example means a current of 4, and hence we must normalize the energy by placing the $4^2 = 16$ in the denominator of the RHS of equation (9.18).

Once we know U_m, we obtain

$$L' = \mu_0 U_m \tag{9.20}$$

$$C' = \frac{1}{v_0^2 L'} = \frac{\varepsilon_0 \mu_0}{\mu_0 U_m} = \frac{\varepsilon_0}{U_m} \tag{9.21}$$

We now have the bounded approximation for the exact value of C that we want:

$$C_{\text{low}} = \frac{\varepsilon_0}{U_m} < C_{\text{exact}} < \varepsilon_0 U_e = C_{\text{high}} \tag{9.22}$$

Performing the calculations using the nodes shown in Figure 19.17a,b, we get the results shown in Figure 9.18.

This problem can be solved analytically using a conformal transformation,[6] yielding 90.612 pF.[7] For ns $= 1$, while both the high and low capacitances are a few percent away from the exact answer, the average of these two capacitances is lightly less than 0.5% away from the exact answer. As ns is increased, both the high and low capacitances converge to the exact answer, and for ns ~>4, the averages are almost exact.

Program fd10.m — describing an example of high and low bounding capacitance calculation — performs the calculations used to generate Figure 9.18:

```
%fd10.m   A very simple FD example with adjustable resolution
%   showing upper and lower bounded capacitance solutions

eps0 = 8.854;
ns = 6;   % scaling factor for resolution studies
Imax = 4*ns + 1 ;   Jmax = 4*ns + 1;   % definition of array
Ic = 2*ns + 1;   Jc = 2*ns + 1;   % inner edges
[i,j,Kmax] = get_ijk(1,Imax,1, Imax,Jmax,0);     % number of variables

a = sparse(Kmax,Kmax);   % allocate the coefficient array
b = zeros(Kmax,1);       % allocate the bcs array

% First solve the physical structure -----------------------------

% initialize a as if every variable is free
for j = 1 : Jmax
  for i = 1 : Imax
    [i,j,k0] = get_ijk(1,Imax,1, i,j,0);
    a(k0,k0) = -4;
    if i < Imax, a(k0,k0 + 1) = 1; end
    if i > 1,    a(k0,k0-1) = 1; end
    k1 = k0 + Imax;
    if j < Jmax, a(k0,k1) = 1; end
    k2 = k0-Imax;
    if j > 1, a(k0,k2) = 1; end
  end
end

% bcs:
% If electrical, clean out the appropriate row, replace it w/ 1 on diagonal,
%   replace with 0 or 1 in b array as appropriate
% If magnetic, correct appropriate off-diagonal term.

%E1 - Electrical, V = 1
j = Jmax;
for i = 1 : Imax
  [i,j,k] = get_ijk(1,Imax,1, i,j,0);
  a(k,:) = 0; a(k,k) = 1;
  b(k) = 1;
end

%E2 - Electrical, V = 1
i = Imax;
for j = 1 : Jmax
  [i,j,k] = get_ijk(1,Imax,1, i,j,0);
  a(k,:) = 0; a(k,k) = 1;
```

```
  b(k) = 1;
end

%E3 - Magnetic, down facing
j = 1;
for i = Ic + 1 : Imax - 1
  [i,j,k] = get_ijk(1,Imax,1, i,j,0);
  a(k,k + Imax) = 2;
end

%E4 - Electric, V = 0
i = Ic;
for j = 1 : Jc
  [i,j,k] = get_ijk(1,Imax,1, i,j,0);
  a(k,:) = 0; a(k,k) = 1;
end

%E5 - Electric, V = 0
j = Jc;
for i = 1 : Ic
  [i,j,k] = get_ijk(1,Imax,1, i,j,0);
  a(k,:) = 0; a(k,k) = 1;
end

%6 - Magnetic, left facing
i = 1;
for j = Jc + 1 : Jmax - 1
  [i,j,k] = get_ijk(1,Imax,1, i,j,0);
  a(k,k + 1) = 2;
end

% solve for voltages
v = a\b;

% We'll need an xy voltage array for ongoing calcs

Volts = zeros(Imax,Jmax);
for i = 1:Imax
  for j = 1:Jmax
    [i,j,k] = get_ijk(1,Imax,1, i,j,0);
    Volts(i,j) = v(k);
  end
end

C_high = 4*C_Dual(Volts, Ic, Jc, eps0)
% Now solve the dual structure --------------------------------

a = sparse(Kmax,Kmax);   % allocate the coefficient array
b = zeros(Kmax,1);   % allocate the bcs array

% initialize a as if every variable is free
for j = 1 : Jmax
```

```
   for i = 1 : Imax
     [i,j,k0] = get_ijk(1,Imax,1, i,j,0);
     a(k0,k0) = -4;
     if i < Imax, a(k0,k0 + 1) = 1; end
     if i > 1,    a(k0,k0-1) = 1; end
     k1 = k0 + Imax;
     if j < Jmax, a(k0,k1) = 1; end
     k2 = k0-Imax;
     if j > 1, a(k0,k2) = 1; end
   end
end

%E1 - Magnetic, looking up
j = Jmax;
for i = 2 : Imax
  [i,j,k] = get_ijk(1,Imax,1, i,j,0);
  a(k,k-Imax) = 2;
end

%E2 - Magnetic, looking right
i = Imax;
for j = 2 : Jmax
  [i,j,k] = get_ijk(1,Imax,1, i,j,0);
  a(k,k-1) = 2;
end

%E3 - Electric, V = 0
j = 1;
for i = Ic : Imax
  [i,j,k] = get_ijk(1,Imax,1, i,j,0);
  a(k,:) = 0; a(k,k) = 1;
end

%E4 - Magnetic, looking left
i = Ic;
for j = 2 : Jc-1
  [i,j,k] = get_ijk(1,Imax,1, i,j,0);
  a(k,k + 1) = 2;
end

%E5 - Magnetic, looking down
j = Jc;
for i = 2 : Ic-1
  [i,j,k] = get_ijk(1,Imax,1, i,j,0);
  a(k,k + Imax) = 2;
end

%6 - Electric, V = 1
i = 1;
for j = Jc : Jmax
  [i,j,k] = get_ijk(1,Imax,1, i,j,0);
  a(k,:) = 0; a(k,k) = 1;
```

```
  b(k) = 1;
end

% zero out the cut away region - not efficient, but easy this way
for i = 1 : Ic-1
  for j = 1 : Jc - 1
    [i,j,k] = get_ijk(1,Imax,1, i,j,0);
    a(k,:) = 0; a(:,k) = 0; a(k,k) = 1;
  end
end
% solve for voltages
v = a\b;

% We'll need an xy voltage array for ongoing calcs

Volts = zeros(Imax,Jmax);
for i = 1:Imax
  for j = 1:Jmax
    [i,j,k] = get_ijk(1,Imax,1, i,j,0);
    Volts(i,j) = v(k);
  end
end

C_low = eps0/(4*C_Dual(Volts, Ic, Jc, eps0)/16/eps0)
C_avg = (C_high + C_low)/2
```

- Program C_dual.m is also useful here:

```
function C = C_Dual(V, Ic, Jc, eps0)
% Calculate C in a rectangular box using Stored Energy
  [Sx,Sy] = size(V);
  eps0 = 8.854;

  Utot = 0;
  for i = 1 : Sx - 1
    for j = 1 : Sy - 1
      if (i >= Ic) | (j >= Jc)
        a = V(i,j); b = V(i+1,j); c = V(i+1,j+1); d = V(i,j+1);
        Uxsq = 1./3.*((a - b)^2 + (d - c)^2 + (a - b)*(d - c));
        Uysq = 1./3.*((a - d)^2 + (b - c)^2 + (a - d)*(b - c));
        Utot = Utot + Uxsq + Uysq;
%         fprintf ('%d %d %f %f %f %f\n', i, j, a,b,c,d)
      end
    end
  end

  C = Utot*eps0;
end
```

Program fd10.m is written to explicitly show how each boundary condition is handled for both the physical structure and its dual. Consequently there are very many lines of code

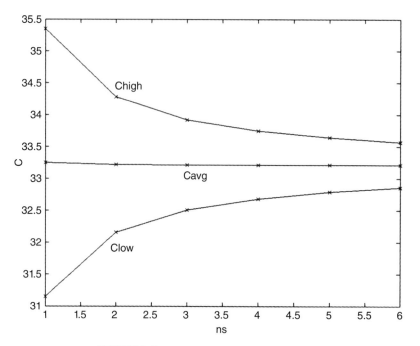

FIGURE 9.19   Repeat of example from Chapter 8.

for a relatively simple job. Note also that the capacitance calculation C_dual.m had to be modified from its previous version to avoid acquisition of spurious numbers from the hole in the center of the line. Also, the integrals used to calculate the stored energy are shown without scaling factors ε or μ. As such, they are not dimensionally electric or magnetic field energies. The program performs all the proper scalings.

MATLAB program fd10.m can be used to repeat the example of Chapter 8. In fd10.m simply set, we obtain Imax = 20*ns + 1; in line 6 and Jc = 1; in line 7.

The resulting program, as a result of the latter change, is now carrying some baggage that it doesn't actually need, but the calculations are correct.

Figure 9.19 shows the result of this exercise. The high capacitance results are, of course, a repeat of the results in Chapter 8. They were calculated much more quickly this time because the program took advantage of the symmetries of the structure. The average of the high and low capacitances in this example barely change with increase in ns, demonstrating the capability of this technique to predict structural capacitances accurately at very modest grid resolutions.

Daly presents a number of similar examples, showing in detail how to set up the dual boundary conditions.[7]

9.8 EXTRAPOLATION

Calculation of capacitance (and transmission line properties) using the dual network technique works so well that we'd like to use it all the time. Unfortunately, in certain situations it doesn't work. For transmission line cross sections, we must be able to specify magnetic walls. When there are no axes of symmetry, this cannot be done easily, if at all. Figure 9.20 illustrates such a situation.

FIGURE 9.20 Transmission line cross section with no symmetry axes.

Another situation that won't work is described in the example presented in Section 9.9: a three-dimensional (3d) problem with an electrode structure positioned somewhere in a box. We cannot draw a dual structure because we cannot describe a current path. Stating this another way, while we can define a capacitance between the inner conductor(s) and the box, we cannot define an inductance.

Working only with the electrical (physical) circuit capacitance, all we know is that we get asymptotically closer to the exact answer as we increase the grid size.

MATLAB program fd12.m performs a fit to the function

$$C = c_1 + \frac{c_2}{ns^{c_3}} \tag{9.23}$$

This program and a related one are shown here:

- Program myfit.m:

```
function [ fit ] = myfit( params, Input, Actual)
% Trial fitting function for extrapolating capacitance data

    c1 = params(1); c2 = params(2); c3 = params(3);
    Fit_Curve = c1 + c2./Input.^c3;
    Err_Vec = Fit_Curve - Actual;
    fit = sum(Err_Vec.^2);

end
```

- Program fd12.m—curve fit program for extrapolating capacitance value:

```
% Extrapolation curve fitting example
close

% data from stripline program
%ns = [1 : 6];
%C = [35.35 34.28 33.92 33.75 33.64 33.57];
```

```
ns = [1 : 3];
C = [69.73 64.43 62.76]

Starting = [60 5 1];
%Starting = [33, 3, 1];   % Guess at a fit

options = [];   % Accept the default optins

Fits = fminsearch(@myfit, Starting, options, ns, C)

plot (ns, C, 'kx') % Data
hold on

x = [1 : .1 : 6];
plot (x, Fits(1) + Fits(2)./x.^Fits(3), 'k') %Fit
xlabel ('ns')
ylabel ('C')
H = line ([5 6], [Fits(1), Fits(1)])
set (H, 'color', 'k')
text (3.5, Fits(1), 'Extrapolated Value')

fit_errs = (C - (Fits(1) + Fits(2)./ns.^Fits(3)))./C
```

Program `fd12.m` uses the MATLAB function `fminsearch.m` to perform a downhill simplex fit.[8] For the dataset shown in Figure 9.19 (C_{high} data), the program produces

$$C = 33.22 + \frac{2.128}{ns^{1.01}} \tag{9.24}$$

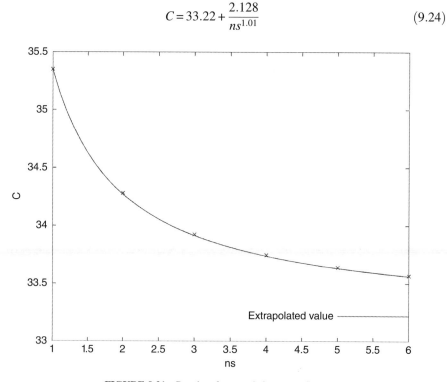

FIGURE 9.21   Results of extrapolation curve fit.

This predicted curve and the original data are shown in Figure 9.21.

The resulting fit agrees with the (six) data points to four significant figures. It extrapolates (letting n go to infinity) a capacitance of 33.22 pF/m. This is approximately 0.2% different from the mean value predicted by the upper-and-lower bound calculations (at ns = 6).

Unfortunately, there's no formal way to determine the best curve fitting function or to know whether the resolutions chosen are optimum; there's some art involved here. Three rules of thumb are that (1) there must be enough data points, (2) the highest resolution for these points is high enough for the capacitance–resolution curve to plateau, and (3) the fit to the existing data should be very good. One empirical caveat from experience — don't include very low-resolution data in the curve fit. Use at least three values that (and again, this is an experience judgment call) describe the geometry adequately.

9.9 THREE-DIMENSIONAL GRIDS

Figure 9.22 shows a square plate (1×1 m) centered 0.5 m above the bottom electrode of a cubic box. The box is $20 \times 20 \times 20$ m high. The intent here is to ensure that the sidewalls and topwalls are far enough away from the square plate that they do not contribute meaningfully to the plate-to-box capacitance. This assumption will be checked.

Writing Laplace's equation on a 3d grid is a simple extension of writing it on a 2d grid:

$$V_{i,j,k} = \frac{V_{i+1,j,k} + V_{i-1,j,k} + V_{i,j+1,k} + V_{i,j-1,k} + V_{i,j,k+1} + V_{i,j,k-1}}{6} \qquad (9.25)$$

In three dimensions, the number of gridpoints can be very large and it is worthwhile to look for symmetries. This example has symmetry about the X and Y axes. Working in the right left quadrant of the X–Y plane, there are symmetry planes at $x = 0$ and $y = 0$.

In two dimensions we didn't have to worry about scaling. If we wanted to change the resolution of our problem, we would simply multiply all dimensions by an integer. For example, a 5-unit-wide strip sitting in a 10×20 box would become a 20-unit-wide strip sitting in a 40×80 box. Since only the $y : x$ ratio determined capacitance, the results would not change. The voltage distribution would adjust itself properly and electric fields calculated based on the (1×1) voltage cell would be correct.

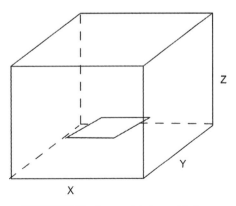

FIGURE 9.22 Square plate in a cubic box.

In three dimensions this automatic scaling does not work. Consider, for example, a parallel plate capacitor and its ideal parallel plate capacitance. If the electrode dimensions are 1×1 m with a 1 m separation, then

$$C = \varepsilon_0 \frac{1 \times 1}{1} = \varepsilon_0 \qquad (9.26)$$

Now, suppose that we scale all dimensions by n:

$$C = \varepsilon_0 \frac{n \times n}{n} = \varepsilon_0 n \qquad (9.27)$$

When scaling a geometry from its original dimensions, we must divide the new calculated capacitance by the scaling factor, assuming that the grid node separations are kept constant. This simple derivation can be repeated formally using the stored electric field integral and the electric field calculation; the conclusion will, of course, be the same.

The stored energy (in a cell) expression is very lengthy. It is derived by rewriting the voltage interpolation function in Chapter 8 for three dimensions and then following the same procedure that was shown in that chapter. The reader who needs this expression can copy it directly from the source code listing for this example. Here are three lated programs:

- Program `fd13.m` — plate in a 3d box:

```
%fd13.m   3d example, taking advantage of symmetry
clc

ns = 3;  % scaling factor for resolution studies
Imin = 0;   Imax = 10*ns ;   Jmin = 0; Jmax = 10*ns;   % definition of array
Mmin = -2*ns; Mmax = 10*ns;
Ic1 = 0;   Ic2 = 2*ns;   Jc1 = 0; Jc2 = 2*ns;   % inner conductor
Mc = 0;

getk = build_ketk(Imin, Imax, Jmin, Jmax, Mmin, Mmax);

[i,j,m,Kmax] = get_ijkm(Imin,Imax,Jmin,Jmax,Mmin, Imax,Jmax,
Mmax,0); % number of variables

a = spalloc(Kmax, Kmax, 6*Kmax);  % allocate the coefficient array
b = zeros(Kmax,1);  % allocate the bcs array

tic
% initialize a as if every variable is free
for m = Mmin : Mmax
  for j = Jmin : Jmax
    for i = Imin : Imax
      [i,j,m,k0] = get_ijkm(Imin,Imax,Jmin,Jmax,Mmin, i,j,m,0);
      a(k0,k0) = -6;
      if i < Imax, a(k0,k0+1) = 1; end
      if i > Imin, a(k0,k0-1) = 1; end
      if j < Jmax
        k1 = k0 + (Imax-Imin) + 1;
```

```
        a(k0,k1) = 1;
      end
      if j > Imin
        k2 = k0 - (Imax-Imin) - 1;
        a(k0,k2) = 1;
      end
      if m < Mmax
        k3 = k0 + (Imax-Imin + 1)*(Jmax-Jmin + 1);
        a(k0,k3) = 1;
      end
      if m > Mmin
        k4 = k0 - (Imax-Imin + 1)*(Jmax-Jmin + 1);
        a(k0,k4) = 1;
      end
    end
  end
end
toc

% bcs
tic

i = Imax; % right wall
for j = Jmin : Jmax
  for m = Mmin: Mmax
    [i,j,m,k] = get_ijkm(Imin,Imax,Jmin,Jmax,Mmin, i,j,m,0);
    a(k,:) = 0; a(k,k) = 1;
  end
end

j = Jmax; % back wall
for i = Imin: Imax
  for m = Mmin : Mmax
    [i,j,m,k] = get_ijkm(Imin,Imax,Jmin,Jmax,Mmin, i,j,m,0);
    a(k,:) = 0; a(k,k) = 1;
  end
end

% top and bottom walls
for i = Imin : Imax
  for j = Jmin : Jmax
    [i,j,m,k] = get_ijkm(Imin,Imax,Jmin,Jmax,Mmin, i,j,Mmin,0);
    a(k,:) = 0; a(k,k) = 1;
    [i,j,m,k] = get_ijkm(Imin,Imax,Jmin,Jmax,Mmin, i,j,Mmax,0);
    a(k,:) = 0; a(k,k) = 1;
  end
end

% i image
i = Imin;
  for j = Jmin : Jmax-1;
    for m = Mmin + 1 : Mmax-1;
```

```
      [i,j,m,k] = get_ijkm(Imin,Imax,Jmin,Jmax,Mmin, i,j,m,0);
      a(k,k+1) = 2;
   end
end

% j image
j = Jmin;
for i = Imin : Imax-1
  for m = Mmin + 1 : Mmax-1
    [i,j,m,k] = get_ijkm(Imin,Imax,Jmin,Jmax,Mmin, i,j,m,0);
    a(k, k + Imax-Imin + 1) = 2;
  end
end

m = Mc; % center conductor
for i = Ic1 : Ic2
  for j = Jc1 : Jc2
    [i,j,m,k] = get_ijkm(Imin,Imax,Jmin,Jmax,Mmin, i,j,m,0);
    a(k,:) = 0; a(k,k) = 1; b(k) = 1;
  end
end
toc
% solve for voltages
v = a\b;

% we'll need an xy voltage array for ongoing calcs

Sx = Imax - Imin + 1; Sy = Jmax - Jmin + 1;
Sz = Jmax - Jmin + 1;
Volts = zeros(Sx,Sy,Sz);
for i = Imin : Imax;
  for j = Jmin : Jmax
    for m = Mmin : Mmax
      [i,j,m,k] = get_ijkm(Imin,Imax,Jmin,Jmax,Mmin, i,j,m,0);
      Volts(i-Imin + 1,j-Jmin + 1,m-Mmin + 1) = v(k);
    end
  end
end

C = C_Energy_3d(Volts, ns)
```

• Program get_ijkm.m:

```
function [ i_out, j_out, m_out, k ] = get_ijkm( Imin, Imax, Jmin, ...
    Jmax, Mmin, i, j, m, k )
% This is a 3-dimensional version of get_ijk. For consistency of
notation,
%   k is kept as the node number, the position indices are (i,j,m)

  if k == 0                % get k
    i_out = i; j_out = j; m_out = m;
```

```
   i = i-Imin; j = j-Jmin; m = m-Mmin;
   k = i + 1 + j*(Imax-Imin + 1) + m*(Imax-Imin + 1)*(Jmax-Jmin + 1);
 else
   m_out = ceil(k/(Imax-Imin + 1)/(Jmax-Jmin + 1)) + Mmin - 1; %
+ Mmin;
   k_mid = k - (m_out-Mmin)*(Imax-Imin + 1)*(Jmax-Jmin + 1);
   j = ceil(k_mid/(Imax-Imin + 1)) - 1;   % get i, j and m
   i = k_mid - 1 - j*(Imax-Imin + 1);
   j_out = j + Jmin;
   i_out = i + Imin;
 end

end
```

- Program C_energy_3d.m:

```
function C = C_Energy_3d(V, ns)
%  Calculate C in a rectangular box using Stored Energy
   [Sx,Sy,Sz] = size(V);
   eps0 = 8.854;

   Utot = 0;
   for i = 1 : Sx - 1
    for j = 1 : Sy - 1
      for m = 1: Sz - 1
        a = V(i,j,m); b = V(i + 1,j,m); c = V(i + 1,j + 1,m); d = V(i,j + 1,m);
        e = V(i,j,m + 1); f = V(i + 1,j,m + 1); g = V(i + 1,j + 1,m + 1); h = V
(i,j + 1,m + 1);

        term = (2*a^2 + 2.*b^2 + 2*c^2 + 2.*d^2 - c*e - d*e ...
          + 2*e^2 - c*f - d*f + 2*f^2 - d*g - e*g + 2*g^2 ...
          - c*h - f*h + 2*h^2 - b*(d + e + g + h) - a*(c + f + g + h))/6;
        Utot = Utot + term;

%          fprintf ('%d %d %f %f %f %f\n', i, j, a,b,c,d)
      end
    end
   end

   C = Utot*eps0/ns;
end
```

These programs are a straightforward extension of previous programs. The functions get_ijkm.m and C_energy_3d.m are 3d versions of their 2d counterparts. The predicted capacitance values are 70.06, 64.72, 62.99, and 62.14 pF/m for ns = 1, 2, 3, and 4, respectively. Using fd12.m again, this capacitance extrapolates to 59.73 pF.

If the sidewalls and topwalls of the box are far enough away from each other so as to not influence the capacitance meaningfully, then the structure is effectively modeling a square plate over an infinite ground plane. The problem was run again at ns = 3, and the sidewalls and topwalls were slowly receded. The change from, say, $L_{max} = 12$ and $L_{max} = 13$, is < 0.1%. Rerunning several values of ns at $L_{max} = 13$, for instance, we find that the resultant curve fit predicts a capacitance of 59.8 pF.

(a)   (b)

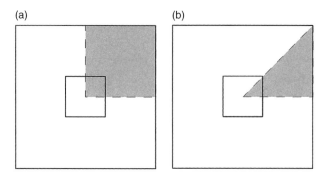

FIGURE 9.23   Symmetry choice used (a) and a possible better choice (b).

If we replace the (effectively) infinite ground plane with a second square electrode 2 units below the ground plane's location, we have created a parallel plate capacitor whose capacitance is half of the value calculated above: 29.9 pF. This agrees well with the MoM calculation in Chapter 4. The fact that a structure's capacitance is the same when calculated using two different methods is, of course, reasonable and expected, and is also reassuring.

Some final notes:

1   The problem could have been further simplified (and execution time significantly shortened) by further utilization of the symmetry of the structure (see Figure 9.23).

2   In `fd13.m`, the sparse matrix allocation statement was replaced by a `spalloc` statement (line 14). This improves the execution time noticeably. This same change could have been made in all of the FD examples, but since execution time in those programs was not an issue, the difference hardly matters. If further use of this type of program is planned, one should consider vectorizing and otherwise improving the program, particularly in the boundary condition establishment. The `tic-toc` statements included in the listing point out the opportunity areas for improvement.

3   Applications of a geometry such as that used in this example for many practical situations (misaligning the electrodes, using two electrodes of slightly different sizes, adding *feeder* lines to each of the electrodes, etc.) are clearly more easily implemented using the MoM technique than using a grid-based approach.

4   The FD approach is easy to implement, computer programs execute efficiently, and the accuracy of the results can be excellent. Why should we look for a better approach? We should do this for same reason that prompted us to go from square to triangular cells in the MoM technique: the ability to accommodate nonrectangular structures and boundary conditions.

Figure 9.24 illustrates an issue that limits the FD technique ability to handle boundary conditions.

Suppose that our actual boundary condition is the meandering (dashed) line shown. There is no way of differentiating this boundary condition from that of the straight line passing through the meandering line. We could make things better by increasing the resolution, but this approach becomes impractical very quickly because there is no way to localize a high-resolution grid in both x and y (and z if we have a 3d structure).

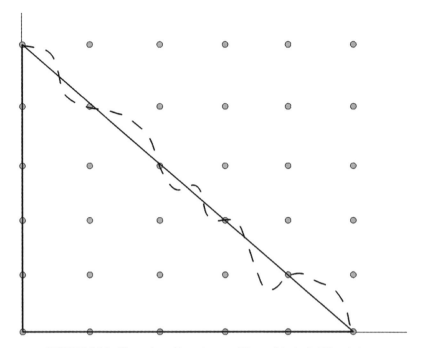

FIGURE 9.24   Illustration of boundary condition ambiguity in FD technique.

PROBLEMS

9.1   Set up a coaxial cable (concentric cylinders) using a rectangular grid and approximate boundary conditions and solve for the voltage grid. Use outer and inner radii of 10 and 5, respectively. Approximate the capacitance and compare it to the known exact solution. Note that this is not a good approach to this problem; it is merely an exercise in setting up the boundary conditions.

9.2   Figure P9.1 shows a structure consisting of two parallel plates in a box and a rectangular dielectric slab. The two parallel plates are set to two (different) voltages, and the rectangular box is floating. The dielectric slab may be moved in the X direction; its (X) center is I_c. Calculate the stored energy in the system as a function of I_c, and comment on the electrostatic forces on the dielectric slab.

9.3   Consider the example of Figure 9.10 and calculations of `fd8.m`. The (peak) field at the tip may be written as

$$E_{peak} = K_1 V_g + K_2 V_a \qquad (9.28)$$

where V_g and V_a are the gate and anode voltages, respectively, and K_1 and K_2 are constants dependent on the geometry. Using `fd8.m`, find these two constants for the geometry of the structure in `fd8.m`. This structure is a form of an electron vacuum tube triode. The results given above should show that, about a bias point of $V_g = 100\,\text{V}$ and $V_a = 1000\,\text{V}$, the peak field is principally determined by the gate

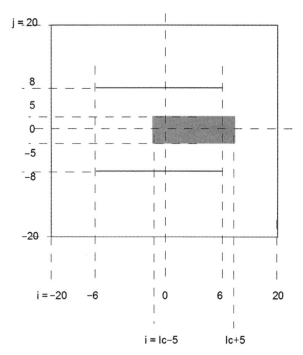

FIGURE P9.1 Two parallel plates inside a box and a rectangular dielectric slab.

voltage. The current emitted by the tip may be calculated using the Fowler–Nordheim equation:

$$I = aE^2 \exp\frac{-b}{E} \qquad (9.29)$$

Where, a and b are functions of the geometry and the tip material. Reasonable values for the structure shown with a molybdenum tip are $a = 1 \times 10^{-16}$ and $b = 5 \times 10^{10}$ with units such that E is in V/μm and I is in μA. Calculate I for the gate and anode voltages shown above, and then, using these values as a *bias point*, calculate the transconductance of the structure for small variations about this bias point:

$$\text{Transconductance} \equiv g_m \equiv \left[\frac{\partial I}{V_g}\right]_{V_a} \qquad (9.30)$$

The units of g_m are μA/V = μ℧. Can you describe how this device could function as a voltage-controlled current source and then be used as a linear power amplifier?

9.4 This problem is an exercise in demonstrating the importance of the ability to calculate upper and lower bounds on a transmission line capacitance. Consider the example of Section 9.7, shown in Figures 9.16 and 9.17. Repeat the calculations of this example using the simplest approximate voltage functions that you can imagine (they must meet all boundary conditions for all electric and magnetic walls). Calculate the high- and low-capacitance estimates, average them, and compare the results to those of the

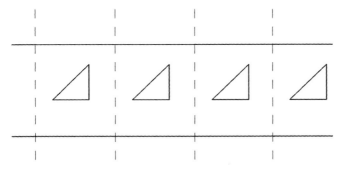

FIGURE P9.3   A periodic structure.

example in Section 9.7 No computer is necessary here — this is a paper-and-pencil numerical analysis exercise.

9.5 Figure P9.3 shows a sequence of identical center conductors evenly spaced between two plates. This is a section of an infinite structure or a region near the center of a "very large" structure. The boundary conditions on one of the sections are called *periodic boundary conditions* – they repeat periodically.

The (triangular) shape of the center conductor has no particular significance; it was chosen for this problem to emphasize that we are not talking about symmetric (image) structures here. Note that a periodic structure can have symmetric section boundary conditions, but this is not a necessary condition and again, is not the case in this problem. Now, modify `fd2.m` to describe this problem and find the voltage distribution. Show the voltage distribution along $j = 0$.

REFERENCES

1. www.wheeler.com/technology/equations/index.html.

2. G. Fursey, *Field Emission in Vacuum Microelectronics*, Springer, 2011.

3. C. A. Spindt, A thin-file field-emission cathode, *J. App. Phys.* 39(7):3504–3505 (1968).

4. W. A. de Heer, A. Chatelain, and D. Ugarte, A carbon nanotube field-emission electron source, *Science* **270**(5239):1179–1180 (Nov. 17, 1995).

5. C. T. Carson and G. K. Cambrell, Upper and lower bounds on the characteristic impedance of TEM mode transmission lines, *IEEE Trans. Microwave Theory Tech.* **MTT-14**: 497 (Oct. 1966).

6. L. N. Dworsky, *Modern Transmission Line Theory and Applications*, Wiley, New York, 1979.

7. P. Daly, Dual potential problems in transmission lines with limited or no symmetry, *IEE Proc.* (Pt. H) **132**(6): (Oct. 1985).

8. W. H. Press, S. A. Teukolsky, W. T. Vettering and B. P. Flannery, *Numerical Recipes in C*, Cambridge University Press, Section 10.5 [there are many editions of this excellent book, and several versions (e.g., *Numerical Recipes in Fortran*); some of the earlier versions are available online].

10

Multielectrode Systems

10.1 MULTIELECTRODE STRUCTURES

(*Note*: This chapter discusses applications, mostly circuital, of the structural models developed in Chapters 8 and 9. There is virtually nothing new in the computer code. Consequently, computer source listings are not included; only unique programming issues (if any) will be discussed. All capacitance values presented will be calculated as described in Chapter 9 — specifically, calculating the capacitance for several resolutions and then extrapolating the results to a hypothetical infinite resolution limit. There is no need for numerical analyses in this chapter's materials, so, formally, this chapter is outside the scope of this book. The materials of this chapter, however, are such a logical extension and typical application of the results of the previous chapters that including them seems reasonable and justified.

Figure 10.1a shows the cross section of a three-conductor transmission line. The electrodes are labeled 1 and 2 — the inner conductors, and 3 — the outer (enclosing) box.

Figure 10.1b shows an equivalent circuit of this structure. A circuit notation that has not been used previously is introduced in this figure. In this example, electrode 3 (the enclosing box) will always be fixed at zero volts. This is indicated by the "ground" symbol at the bottom of the figure. Once this zero-voltage, or ground, reference is established, every voltage not specifically referenced otherwise is assumed to be referenced to ground. C_1 is the capacitance from electrode 1 to ground; C_2 is the capacitance from electrode 2 to ground, and C_{12} is the capacitance from electrode 1 to electrode 2. These capacitances can be calculated using the same modeling techniques used in previous chapters, but some additional planning is now necessary. The three capacitances will be found by applying three different sets of boundary conditions:

Introduction to Numerical Electrostatics Using MATLAB, First Edition. Lawrence N. Dworsky.
© 2014 John Wiley & Sons, Inc. Published 2014 by John Wiley & Sons, Inc.

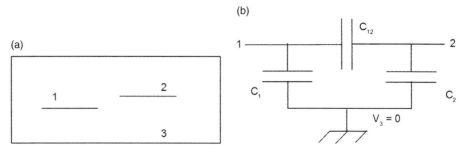

FIGURE 10.1   Three-conductor structure (a) and its circuital equivalent (b).

A   Apply 1 Volt to both terminals 1 and 2. Find the voltage distribution, the stored energy, and the capacitance C_A:

$$C_A = C_1 + C_2 \tag{10.1}$$

B   Apply 1 V to terminal 1, tie terminal 2 to ground (0 V), and find C_B:

$$C_B = C_1 + C_{12} \tag{10.2}$$

C   Apply 1 V to terminal 2, tie terminal 2 to ground, and find C_C:

$$C_C = C_2 + C_{12} \tag{10.3}$$

Inverting these three equations, we obtain

$$C_1 = \frac{C_B + C_A - C_C}{2} \tag{10.4}$$

$$C_2 = \frac{C_C + C_A - C_B}{2} \tag{10.5}$$

$$C_{12} = \frac{C_B + C_C - C_A}{2} \tag{10.6}$$

From a programming perspective, the three sets of boundary conditions could be set up one at a time, followed by application of equations (10.4)–(10.6). Or we could write the program to cycle through all three sets of boundary conditions and then finish the job [applying the conditions in equations (10.4–10.6)] automatically. This is a choice of convenience; the results will, of course, be the same.

The structure shown in Figure 10.1 could be a transmission line structure. This structure, using long but finite length lines, could be used as a frequency-selective filter or a directional coupler. Alternatively, it could represent two wires on a circuitboard (in practice, some dielectric would be present). In this case C_{12} would represent, in a circuit model, the capacitance coupling the two wires; this would almost always be an unwanted *parasitic* capacitance that would cause signals on wire 1 to show up on wire 2, and vice versa. In the case of analog signals, this is called *crosstalk*. In the case of digital signals, this will limit the speed and/or increase the error rate of the system.

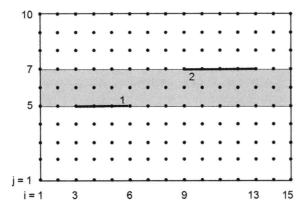

FIGURE 10.2   Two lines on either side of a dielectric board.

As a concrete example, consider Figure 10.2. Electrodes 1 and 2 are attached to a dielectric substrate of (relative) permittivity 2.0. This is approximately the permittivity of many of the Teflon-based printed circuitboards commonly used. Remember that terminal 3 is ground (reference voltage) by prior assumption.

The capacitances (in picofarads per meter) found for this structure are

$$C_a = 77.80 \quad C_b = 41.97 \quad C_c = 48.57$$
$$C_1 = 35.6 \quad C_2 = 42.2 \quad C_3 = 6.4$$

10.2   UTILIZING SUPERPOSITION

For detailed studies of subtle parameter variations in structures such as the last example, a high-resolution grid (possibly throughout the grid, possibly only near the electrodes) is needed. Detailed studies (using combinations of different electrode sizes and positions, dielectric constant tolerance, etc.) involving high-resolution grids are time consuming; having to solve three sets of equations for each case takes even more time. Fortunately we can reduce this problem to solving only two sets of equations very easily. We take advantage of the property of linear systems, called *superposition*.[1]

Superposition, in our situation, means that the response of a system to two electrodes, each driven at 1 V, is identically the sum of the (two) responses obtained when each electrode is driven at 1 V while the other electrode is set to 0 V. Putting this into practice, we have the following procedure:

1   Drive the system with condition B [equation (10.2)] as described above. Calculate C_B as above. Save the array containing the voltage distribution.

2   Drive the system with condition C as described above. Calculate C_C as above. Save the array containing the voltage distribution.

3   Add the two saved voltage distributions together. This new array contains identically the same voltage distribution that was found by driving the system with condition A.

4 Proceed to find C_A from this calculated voltage distribution exactly as if the voltage distribution had been found from condition A (as was done above.)

5 Proceed to find C_1, C_2 and C_{12} as above.

There is no need to quote the results of these calculations; they are identical to those given above.

10.3 UTILIZING SYMMETRY

Figure 10.3 shows two strip electrodes in a boxed microstrip structure.

This structure is symmetric about $i = 8$. Staying with the notation of Figures 10.1 and 10.2, in this case $C_b = C_c$. This means that the voltage distribution for condition c is the mirror image (about $i = 8$) of the voltage distribution for condition b. We can therefore simplify the preceding list of steps (in Section 10.2) by rewriting step 2:

2 (*Symmetric.*) Using the saved voltage distribution from step 1, create a new voltage distribution by mirroring the saved distribution in x [e.g., for all j, interchange $V(1,j)$ with $V(i_{max},j)$; $V(2,j)$, with $V(i_{max} - 1,j)$, etc.] Obtain C_c as above. Save this new array.

The complete characterization of the structure has now been reduced to solving only one set of equations. For the geometry shown in Figure 10.3, with $k = 2$, this yields

$$C_1 = C_2 = 65.5 \, \text{pF/m} \qquad C_{12} = 5.0 \, \text{pF/m}$$

10.4 CIRCUITAL RELATIONS AND A CAVEAT

Figure 10.4 shows a simple two ideal-capacitor circuit.

Three voltage nodes are labeled. Following the ground node convention mentioned in Section 10.3, label V_0 as the ground node. Then set V_1 as the voltage across capacitor C_1 and $V_2 - V_1$ as the voltage across capacitor C_2. The capacitors are connected in series so that V_2 is the sum of the voltages across the two capacitors.

To apply voltages to the system, we connect voltages sources between ground and the appropriate nodes. Since the voltage sources have internal resistances, there are time

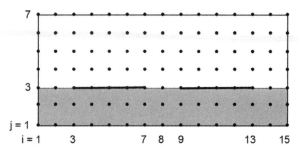

FIGURE 10.3 Two-electrode symmetric microstrip structure.

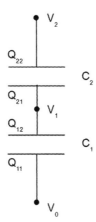

FIGURE 10.4 Two ideal-capacitor series circuit.

constants involved as the capacitors charge to the appropriate voltages. We will assume that the voltage sources are attached for a sufficiently long time (many time constants) that the capacitors are fully charged, that is, that all current flows have (effectively) decayed to zero. Removing the voltage sources after this charging time leaves all voltages the same as they were just before the voltage sources were removed. This is because, when nothing is connected to the capacitors, there is no way for charge to leave (or get to) any of the nodes.

In Figure 10.4, the charges on the capacitor plates are labeled. The charge at node 1, $Q_{21} + Q_{12}$, is free to rearrange itself on the two connected capacitor plates, but the total charge at this node cannot change unless some connection (current path) between this node and one of the other nodes is provided.

Now, suppose that you are told that $V_2 = 100$ V and are asked what V_1 is. The correct answer to the question as posed is "What would you like it to be?"

The reason for this somewhat odd and unsatisfying answer is that the question as posed is incomplete. We know that the total charge at node 1 cannot change because nothing has been connected to node 1 to provide a current path, but we do not know what this total charge is. A correctly stated question might be "We begin with no charge in the system. We connect a 10-V source to node 1, wait for many time constants, and then disconnect this voltage source. We then connect a 100-V source to node 2, again wait for many time constants and then disconnect this source. What are the node voltages?"

We have never considered this issue before — is everything that has been presented thus far incomplete? Fortunately, everything is all right. The use of Laplace's equation assumes that we're dealing with charge-free space; and equivalently, circuit models usually assume that we're starting out (our *initial conditions*) with no charge anywhere in the system. In other words, even though we haven't stated it explicitly, we have assumed that all nodes that do not constitute a boundary condition are charge-free. Since there is no way to change the total charge at a node that is not a boundary condition, these nodes forever remain charge-free. For the circuit shown in Figure 10.4, therefore

$$Q_{21} + Q_{12} = 0 \qquad (10.7)$$

Now we are in a position to finish the job. For the two capacitors, we obtain

$$Q_{22} = -Q_{21} = (V_2 - V_1)C_2 \tag{10.8}$$

$$Q_{12} = -Q_{11} = C_1 V_1 \tag{10.9}$$

Combining equations (10.7) and (10.8) and then setting the result $= (10.9)$, we get

$$Q_{12} = (V_2 - V_1)C_2 = C_1 V_1 \tag{10.10}$$

from which

$$V_1 = \frac{C_2 V_2}{C_1 + V_1} \tag{10.11}$$

and we have answered the question stated above.

   If the initial charge at node 1 is not zero, the RHS of equation (10.7) must be set to the value of this charge, and then updating equation (10.11) will give us the correct answer.

   In Chapter 8 it was stated that minimizing the total stored energy in the system results in satisfying Laplace's equation. Taking a look at this from the circuital perspective, the total stored energy in this circuit is

$$2U_E = C_2(V_2 - V_1)^2 + C_1 V_1^2 \tag{10.12}$$

Requiring that V_1 adjust itself to minimize this energy, we obtain

$$\frac{\partial 2U_E}{V_1} = 0 = C_2(V_2 - V_1) + C_1 V_1 \tag{10.13}$$

which is identical to equation (10.10).

   In other words, the finite difference scheme that we've been using, the numerical approximation to Laplace's equation (on a grid), is also telling us that there is no charge at any of the internal nodes (other than boundary condition nodes). This seems to be a shaggy-dog story (a long joke with virtually no punchline); we've gone quite a way just to get back to where we started: worrying about charge neutrality at internal gridpoints is of no value when these points (nodes) are just mathematical (grid) points in space. It is, however, a necessary albeit rarely stated initial condition when we are dealing with unconnected, or *floating*, electrodes in our system.

10.5   FLOATING ELECTRODES

Consider again Figure 10.1b. Consider a symmetric structure: $C_1 = C_2$. (This is not a necessary restriction, but it's convenient right now.) Set terminal 1 to 1 V and let terminal 2 float. From basic circuit theory, the input capacitance measured at terminal 1 is

$$C_{in} = C_1 + \frac{C_{12}C_1}{C_{12} + C_1} \tag{10.14}$$

and from equation (10.11), the voltage at terminal 2 (the floating terminal) is

$$V_2 = \frac{C_{12}}{C_{12} + C_1} \tag{10.15}$$

The units in equation (10.15) appear to be mismatched; remember, however, that there is an implied (1-V) voltage in the numerator of the RHS of this equation.

For the structure in Figure 10.3, we found the capacitance values to be

$$C_1 = C_2 = 65.5 \text{ pF/m} \qquad\qquad C_{12} = 5.0 \text{ pF/m}$$

and now we can calculate

$$C_{in} = 70.1 \text{ pF/m} \qquad\qquad V_2 = 0.071 \text{ V}$$

These values can also be calculated directly without first calculating C_1 and C_{12}. Still referring to the structure in Figure 10.3, we want to find the voltage distribution when electrode 1 is set to 1 V and electrode 2 is left floating. What hasn't been discussed yet is how to handle a floating electrode in a relaxation solution.

Figure 10.5 shows an example of a floating electrode in a grid. Only the nodes immediately surrounding this floating electrode are shown. Nodes (i,j) and $(i+1,j)$ are tied together by a floating electrode and are at voltage V_0. The notation here can be misleading — remember that V_0 is **not** an applied voltage boundary condition. It is the voltage that this electrode settles to (floats at) when 1 V is applied to the other electrode.

Writing the equations for these two nodes, we obtain

$$V_0 = \frac{V_0 + V_{i-1,j} + V_{i,j-1} + V_{i,j+1}}{4} \tag{10.16}$$

$$V_0 = \frac{V_0 + V_{i+2,j} + V_{i+1,j+1} + V_{i+1,j-1}}{4} \tag{10.17}$$

Adding these equations and then solving for V_0, we get

$$V_0 = \frac{V_{i-1,j} + V_{i+2,j} + V_{i,j+1} + V_{i+1,j+1} + V_{i,j-1} + V_{i+1,j-1}}{6} \tag{10.18}$$

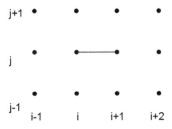

FIGURE 10.5   Example of a floating electrode in a grid.

The result is simple — the voltage on the floating electrode is the average of the voltages at all the nodes surrounding the electrode. This result is easily extended to electrodes that are more than two nodes wide and/or more than one node thick.

When an electrode sits on a dielectric interface, equation (10.18) must be modified to include the effects of this interface. For the structure in Figure 10.5, if the electrode is sitting on a dielectric or relative permittivity K, we let

$$<K> = \frac{K+1}{2} \tag{10.19}$$

and then

$$V_0 = \frac{<K>\left[V_{i-1,j} + V_{i+2,j}\right] + V_{i,j+1} + V_{i+1,j+1} + K\left[V_{i,j-1} + V_{i+1,j-1}\right]}{2\left(<K> + 1 + K\right)} \tag{10.20}$$

Returning to the structure in Figure 10.3, the results of this calculation are, of course, the same as the previous results, $C_{in} = 70.1$ and $V_2 = 0.071$.

It is also possible to calculate C_1 and C_{12} directly from these results by inverting equations (10.14) and (10.15):

$$C_1 = \frac{C_{in}}{1 - V_0} \tag{10.21}$$

$$C_{12} = \frac{C_{in} V_0}{1 - V_0^2} \tag{10.22}$$

When writing the full set of equations for finding all the node voltages, the voltage at a floating electrode is only one of the node voltages, with it equation (row) in the matrix specified by equation (10.20) and then all the nodes surrounding the floating electrode properly referring to the floating-electrode voltage.

PROBLEMS

10.1 Consider the circuit shown in Figure P10.1. The three capacitors are initially uncharged. Their capacitances are $C_1 = 1$, $C_2 = 5$, and $C_3 = 10$. Omission of units for these numbers is not an oversight. At this point only the ratios of the capacitances matter.

(a) Voltages $V_2 = 100$ and $V_3 = 50$ are applied; terminal V_1 is left floating. After waiting for a sufficient time, for the capacitors to charge, all charging wires are disconnected from the circuit. Calculate the charges on the three capacitors.

(b) What is V_1 at this time?

10.2 Voltage $V_1 = 200$ is now applied. After sufficient charging time has passed, this charging wire is disconnected. What are the three charges, and the three voltages?

10.3 Return to Problem 10.1. Assume that the capacitor values are expressed in microfarads. The charging voltage sources both have a Thevenin resistance of $100\,\Omega$. Describe the voltages as a function of time as the capacitors charge. [*Hint*: This

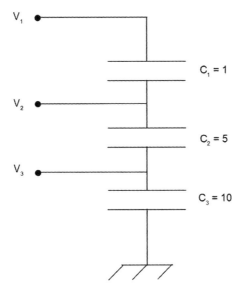

FIGURE P10.1 Three-capacitor circuit.

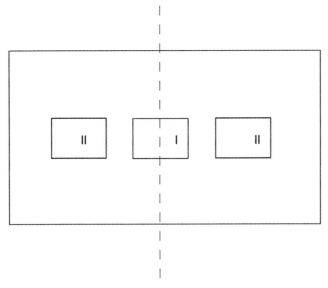

FIGURE P10.5 Symmetric three-electrode structure.

problem is easily solved using the *RC* charging formula shown in Chapter 2 and the principle of superposition (this chapter).]

10.4 Figure P10.5 shows a structure with three identical rectangular electrodes in a box. The structure is symmetric about the center in *X*, as shown. The electrodes are labeled I, II, and II; because of the symmetry, the labels of the two outer electrodes are identical.

Draw a circuit diagram showing all the capacitances in this structure, and then outline the simplest analysis procedure (under different voltage boundary conditions) needed to calculate the three capacitances.

REFERENCE

1. S. Ramo, J. R. Whinnery, and T. Van Duzer, *Fields and Waves in Communications Electronics*, Wiley, New York, 1994.

11

Probabilistic Potential Theory

11.1 RANDOM WALKS AND THE DIFFUSION EQUATION

Consider a grid with nodes (i, j) spaced one unit apart in both axes. These nodes, at this point, have nothing to do with electrostatics — there are no voltages, charges, or fields present. Instead, there is a particle undergoing a *random walk*. Brownian motion,[1] for example, is a random walk of particles. Random walks have various defining characteristics; the random walk discussed here is defined as follows:

1 There is a particle sitting at one of the nodes. At time $(m - 1)$ the probability of the particle being at node (i, j) is $P_{m-1}(i, j)$.

2 Every second the particle will quickly jump to an adjacent node. *Adjacent* means to the left $(-x)$, right $(+x)$, up $(+y)$ or down $(-y)$. Diagonal jumps and multinode jumps are not allowed.

3 All of the jumps have equal probability; that is, the probability of any given jump is $\frac{1}{4}$.

Four examples of such a walk, reminiscent of a drunkard wandering through a parking lot, are shown in Figure 11.1. In each of these examples the particle started at (0,0).

At a given time, a particle can arrive at node (i, j) only if it started from one of the (four) adjacent nodes. The probability of a particle arriving at node (i, j) is therefore

$$\tfrac{1}{4}[P_m(i-1, j) + P_m(i+1, j) + P_m(i, j-1) + P_m(i, j+1)] \tag{11.1}$$

If the particle had been at node (i, j) at time $m - 1$, since it had to jump to somewhere, there is a 100% probability (i.e., it is certain) that it is no longer at node (i, j) at time m.

Introduction to Numerical Electrostatics Using MATLAB, First Edition. Lawrence N. Dworsky.
© 2014 John Wiley & Sons, Inc. Published 2014 by John Wiley & Sons, Inc.

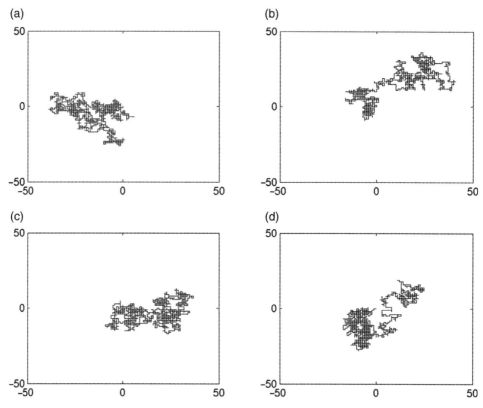

FIGURE 11.1   Four fixed-step two-dimensional random walks.

The net change in the probability of the particle being in cell (i,j) at time m as compared to the probability that it was at node (i,j) at time $m-1$ is therefore

$$\tfrac{1}{4}[P_m(i-1,j)+P_m(i+1,j)+P_m(i,j-1)+P_m(i,j+1)]-P_m(i,j) \qquad (11.2)$$

If we look over many time intervals and require that the net change be zero, that is, look for the steady-state situation, equation (11.2) is set equal to zero, and we have

$$P_m(i,j)=\tfrac{1}{4}[P_m(i-1,j)+P_m(i+1,j)+P_m(i,j-1)+P_m(i,j+1)] \qquad (11.3)$$

This equation should look familiar.

As an aside, a formal derivation, taking the proper limits and including proper physical constants, leads to the diffusion equation.[2] In rectangular coordinates and two dimensions, this is

$$\frac{\partial f}{\partial t}=D\nabla^2 f=D\left(\frac{\partial^2 f}{\partial x^2}+\frac{\partial^2 f}{\partial y^2}\right) \qquad (11.4)$$

In the steady state the left-hand side (LHS) of this equation is zero, leading to

$$\frac{\partial^2 f}{\partial x^2} + \frac{\partial^2 f}{\partial y^2} = 0 \qquad (11.5)$$

This is identically Laplace's equation, and then equation (11.3) is the finite difference numerical approximation to Laplace's equation that we've been dealing with all along.

One last point — the diffusion equation looks a lot like the wave equation:

$$\frac{\partial^2 f}{\partial t^2} = \frac{1}{v^2}\left[\frac{\partial^2 f}{\partial x^2} + \frac{\partial^2 f}{\partial y^2}\right] \qquad (11.6)$$

The wave equation and the diffusion equation describe different physical phenomena and are identical only when derivatives with respect to time are zero.

The equivalence of the steady-state wave and diffusion equations leads us to a solution technique for electrostatics problems, sometimes called *probabilistic potential theory*[3] (PPT), which is distinctly different from anything we've looked at so far. This technique relies on a *random-walk game* with a set of rules for play and scoring, while physical terms such as *electric field, charge*, or *energy* never appear.

11.2   VOLTAGE AT A POINT FROM RANDOM WALKS

Consider the two-dimensional (2d) cross section shown in Figure 11.2. The figure shows an inner rectangular box asymmetrically located inside an outer rectangular box. Not shown is a 1×1 grid of points that these boxes sit on. The outer box is set at 0 V; the inner box is set at 1 V. The point (0,12) — arbitrarily chosen — is the point at which we want to learn the voltage. There are several approaches to treating this as a random-walk probabilistic problem rather than an electrostatics problem.

The first approach is a random-walk game. Assume that a "random walker" is placed at point (0,12). He then takes his random walk, as described above and exemplified in

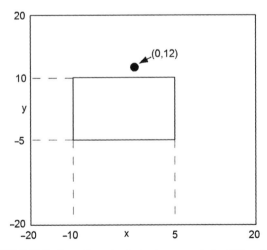

FIGURE 11.2   Two-dimensional structure example for PPT analysis.

Figure 11.1. When he hits either the inner or the outer box (boundary condition electrode), his walk is over; he is extracted from the game and the identity of the box that he hit is recorded.

This procedure is repeated a number of times (how many times is necessary will be discussed below). If n_1 is the number of extractions at the inner box and n_0 the number of extractions at the outer box, then the probability of starting at point (0,12), taking a random walk as described above and being extracted at the inner box, is

$$P \approx \frac{n_1}{n_1 + n_0} \tag{11.7}$$

Since both the random walk and the finite-difference Laplace equations are the same, this probability is identically the voltage at point (0,12) when the inner and outer boxes are electrodes at 1 and 0 volts, respectively. Equation (11.7) is an approximation to the probability. We expect this approximation to improve as the number of random walkers ($n_0 + n_1$) take their walk, but some discussion of "improve" is necessary — this equation does not converge to the correct answer in the same manner as a Gauss–Seidel relaxation converges to the correct answer. Also, of course, this solution technique can only be as good (or as bad) as the resolution of the grid. In this respect it is identical to the finite difference solutions discussed in previous chapters.

When ($n_0 + n_1$) random walkers take their walks, the probability of getting a certain number of these to be extracted at the inner conductor, when the probability of any one of them being extracted at the inner conductor is n_1, is calculated using the *binomial probability* formula. The binomial formula is standard probability-and-statistics text material.[4,5] Only relevant results and their interpretation will be presented here.

For a large number of samples (in statistics jargon, the results of each random walk discussed above is a sample), the binomial probability distribution is indistinguishable from the normal, or Gaussian, distribution (a bell curve).

Relating the two distributions, for $n = (n_0 + n_1)$, the number of samples and p the probability for each of the samples, the mean of the normal distribution is

$$\mu = np \tag{11.8}$$

and the standard deviation is

$$\sigma = \sqrt{np(1-p)} \tag{11.9}$$

Figure 11.3 shows a normal distribution. The curve is symmetric about the mean μ, which in this example is 5. The standard deviation of this curve σ is 1. The two vertical lines are at $x = \mu + 1.96\sigma$ and $\mu - 1.96\sigma$ (1.96 is often simply rounded to 2 in this situation). The area under the curve between the two lines occupies 95% of the total area under the curve. This is interpreted to mean that, for a very large number of samples, 95% of the results will fall between these two lines. This is referred to as the *95% confidence interval*. Had we chosen three sigma 3σ above and below the mean, we would find that ~99% of the area under the curve and this wider region would be our *99% confidence interval*.

Relating all of this back to our example, if the probability of n_1 extractions at the center conductors out of n total starts is

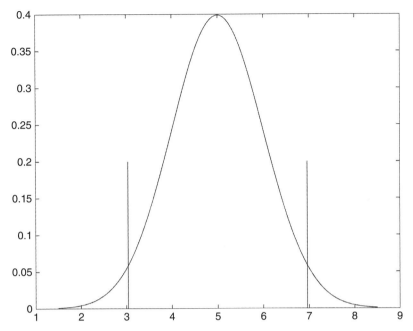

FIGURE 11.3   Graph of a normal distribution; $\mu = 5$, $\sigma = 1$.

$$p = \frac{n_1}{n} = \frac{n_1}{n_0 + n_1} \tag{11.10}$$

we expect the distribution of our n results to resemble a normal distribution (for n sufficiently large) that obeys equations (11.8) and (11.9).

Suppose that we want a 95% confidence interval that is 5% (of the mean) wide. Then

$$u + 1.96\sigma = 1.025\mu \tag{11.11}$$

Using equations (11.8) and (11.9) and solving for n, we find that

$$n = \frac{1.96^2}{0.025} \frac{1-p}{p} \tag{11.12}$$

where n is the number of random walk starts we need to give us a 5% wide confidence interval about p, where p is the voltage at the starting point — our desired result. At this point we seem to be chasing our collective tail. We want to know the minimum n that will let us find p with some measure of confidence, but we first need to know p in order to find n.

This dilemma is resolved by an iterative procedure, as will be shown in MATLAB program `ppt1.m` below. We take (what we hope to be) a representative number of random-walk starts and then calculate n using equation (11.12). If our number of starts is less than n, we repeat the process. Eventually the predicted value of n becomes relatively constant and then at some point we have made n starts.

Figure 11.4 shows the results of 10 runs of the program.

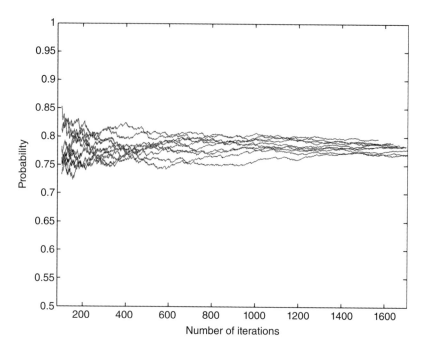

FIGURE 11.4 Repeated runs of program `ppt1.m` showing convergence of results.

The figure shows that in all of these runs the probability approaches approximately 0.78, but no two runs are identical. If we were to repeat this procedure hundreds of times, we'd find that approximately 95% of the probabilities were within 5% of the average of these probabilities, and if we repeated the procedure thousands of times to accumulate enough data to plot a smooth curve, we'd find a normal distribution of probabilities with the correct characteristics.

Here is MATLAB program `ppt1.m` — example of a 2d PPT calculation:

```
% ppt1   Simple 2d random walk ppt example

tic        % turn on timer function

% outer box is a square at x = + - max, y = + - max
max = 20;

% inner box is a rectangle, dimensions as shown
x1 = -10;   x2 = 5;   y1 = -5;   y2 = 10;

% set atarting point
x_start = 0; y_start = 12;

% Intialize counters
nr_iters = 500;              % starting max number of iterations
nr_ones = 0;                 % number of hits to 1 volt electrode
iter = 0;                    % random walker counter
nr_sigs = 1.96;              % number of sigmas for confidence factor
```

```
conf_width = .025;              % half-width fraction about mean for nr_sigs
confidence
iter_plot = [];  p_plot = [];  % arrays to store p(n) for a plot

while iter < nr_iters      % Random walks

  iter = iter + 1 ;
  hit_boundary = 0;               % wandering point flag
  x = x_start; y = y_start;       % starting point
  while hit_boundary < = 0
    p = rand;
    if p < 1/4
      x = x -1;
    elseif p < 2/4
      x = x + 1;
    elseif p < 3/4
      y = y - 1;
    else
      y = y + 1;
    end

    if abs(x) == max | abs(y) == max
      hit_boundary = 1 ; % outer box
    end

    if x > = x1 & x < = x2 & y > = y1 & y < = y2
      hit_boundary = 2;        % inner brick top or bottom
    end
  end

  if hit_boundary == 2      % Inner box hit counter
    nr_ones = nr_ones + 1;
  end

  if iter > 100              % Error predictor
      p = nr_ones/iter;      % working intermediate probability
      nr_iters = ceil((nr_sigs/conf_width)^2*(1 - p)/p);
      iter_plot = [iter_plot, iter];
      p_plot = [p_plot, p];
  end

end

volts = nr_ones/nr_iters;

fprintf ('Number of iterations = %d \n', nr_iters)
fprintf ('Voltage at starting point = %6.3f \n', volts)
toc

plot(iter_plot, p_plot, 'k')
axis ([80, nr_iters + 20, .5, 1.]);
xlabel ('Number of Iterations')
ylabel ('Probability')
hold on
```

An interesting attribute of this calculation is that the number of iterations required for a given confidence level of the result is a function of that result. If p (which in the example above is the voltage at the random walk's starting point) is close to one, the scaling factor $(1-p)/p$ in equation (11.12) is much less than one. On the other hand, if p is very small (e.g., for a point near the 0 V outer box), then this scaling factor begins to exceed 1 significantly. In other words, many more random walks are needed to calculate V at a point if its voltage is close to zero than to one.

This calculation property occurs because of the way in which the acceptable confidence level was specified in equation (11.11). A fixed percentage change of a small number is much smaller than the same fixed percentage change of a large number. Asking for a 5% window centered on (say) 0.01 V is asking much more than asking for a 5% window centered on (say) 0.99 V. From a conventional statistics perspective, we are saying that it's much easier, in terms of the number of data points needed, to be confident about the statistics of a fairly certain event than it is to be confident about the statistics of a fairly rare event.

Interchanging the electrode voltage settings doesn't reduce this problem. Suppose that we set the outer box in Figure 11.1 to 1 V and the inner box to 0 V, and we find the voltage at some point near the outer box to be 0.85 V. Then all we have to do to get the voltage at that point in the original configuration is to subtract this number from 1: specifically $1.00-0.85 = 0.15$ V. The problem is that if we had found the 0.85 V with a 95% confidence interval of (say) 0.02 V above and below 0.85, then we would know the 0.15 V answer to 0.02 V only above and below 0.15. This latter confidence interval is, on a percentage basis, much higher than the confidence interval at 0.85 V.

For high probability points we have another problem. The calculated necessary number of starting walkers gets so low that we have to make sure that we are getting meaningful numerical results. For example, if $p = 0.98$, then, on average, only 2% of the starting walkers will be extracted at the outer boundary. Starting 50 walkers at this point will produce numerical nonsense. We need to set a minimum number of iterations (about 1000 in this example) to ensure reasonable statistics.

Variations of the relaxation equation due to dielectric interfaces and/or unequal grid steps can be translated directly into probabilistic interpretations and the random-walk probabilities adjusted accordingly. For example, using the results in Chapter 9, at a dielectric interface along the y axis, we get

$$V(i,j) = \frac{kV(i,j-1) + V(i,j+1) + [(k+1)/2]V(i+1,j) + V(i-1,j)}{2(k+1)} \quad (11.13)$$

and the probability of a step in the $-y$ direction is

$$\frac{k}{2(k+1)} \quad (11.14)$$

and so on.

Uneven node separations, as treated in Chapter 9, are handled the same way. Similarly, a three-dimensional (3d) finite difference grid becomes a 3d random walk. For a simple uniform grid, the probability of any one step is $\frac{1}{6}$.

If the voltage (Figure 11.1) on the inner and outer boxes had been, say, 870 and 0 V, respectively, then the random-walk problem would be the same as above, but when p is

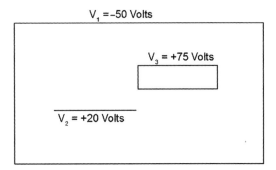

FIGURE 11.5 Structure with several electrodes and electrode voltages.

found, it must be scaled by a factor of 870. Combinations of multiple electrodes at arbitrary different voltages are handled in a similar manner.

Figure 11.5 shows a three-electrode structure with the electrodes set at voltages as indicated in the figure:

$$V_1 = -50\,\text{V} \qquad V_2 = 20\,\text{V} \qquad V_3 = 75\,\text{V}$$

The procedure for setting up the scaling and offsets is as follows:

1 Offset all the voltages so that the most negative of them is zero. In this case, subtract −50 from all the values:

$$V_1 = 0\,\text{V} \qquad V_2 = 70\,\text{V} \qquad V_3 = 125\,\text{V}$$

2 Scale all the voltages so that the highest voltage is 1.0. In this case, divide all the voltages by 125.

$$V_1 = 0\,\text{V} \qquad V_2 = +0.560\,\text{V} \qquad V_3 = 1.00\,\text{V}$$

3 Run the random-walk program with the following rules:
(a) The random walk starts at the desired gridpoint, as previously
(b) The walker is extracted on reaching any of the electrodes
(c) The number of extractions at each electrode is recorded (n_1 = number of extractions at electrode number 1, etc.). Then calculate the probability using the weighted sum

$$p = \frac{V_2 n_2 + V_3 n_3}{n_1 + n_2 + n_3} \tag{11.15}$$

4 Now suppose that we find that $p = 0.5$. First, reverse the scaling: $0.5(125) = 62.5$. Then reverse the offset:

$$V = 62.5 + V_1 = 62.5 - 50 = 12.5\,\text{V}.$$

The random-walk procedure is an interesting but at this point unsatisfying way to solve the finite difference Laplace equation. Two criticisms are that substantial effort has gone into finding the voltage at only one point rather than over the entire grid, and that the accuracy is a *statistical bet* — a 95% confidence interval doesn't really guarantee anything. For that matter, if this procedure were used over many points on a grid to determine the voltages at all these points, some of the voltages (about 5% of them) will definitely fall outside the confidence interval.

The procedure in Section 11.3 will be more satisfying in this latter respect because the diffusion equation will be solved without resorting to random walks, or for that matter to random numbers at all.

11.3 DIFFUSION

The random-walk solution technique as used above repeatedly started one random walker at the point of interest, *stayed with him* until he was extracted at an electrode, and then started another random walker. The same calculation could have been accomplished by starting a large number of random walkers simultaneously and tracking them all as they meandered about the grid. This is actually our intuitive picture of diffusion. If you put a small drop of cream into a cup of coffee, you can watch as molecules of cream take their random walks and spread out, or diffuse, until the cream is uniformly distributed in the coffee. If the walls of the coffee cup were chemical filters of some sort that absorbed the cream molecules rather than reflecting them back into the coffee, this coffee cup would be a good chemical analogy of our problem solving technique.

If the number of random walkers is very large, then we don't need a random-number generator to tell us which way each of them goes. We don't really care which walker goes which way; just knowing that (in two dimensions) 25% of the walkers at every gridpoint will go one step in each direction on the grid is all the information we need.

We can start with some large number of walkers at the starting point, but this becomes awkward when we get to a point that has, say, three walkers — how do we assign their position at the next step? Instead, start with 1.0 walker and send 0.25 each way. Physically we're following the same procedure; we're just sidestepping some numerical nonsense. The discussion below will still refer to *walker*, with the understanding that these are all numbers less than 1.0 rather than integers representing individuals.

The diffusion program proceeds in the following manner:

1 Put 1.0 walkers at the starting point.

2 At each iteration, at every point on the grid transfer $\frac{1}{4}$ of the walkers from that gridpoint to each of the (4) adjacent gridpoints. Extract walkers at the electrodes and keep score as in the random walker program.

3 When there is only some predetermined number of walkers left on grid, terminate the process and calculate the final probability. If we set the termination number, for example, for 10^{-6} walkers left on the grid, then we have a fixed voltage tolerance. On the other hand, if we update the (no longer) predetermined number to a fraction of the probability as we learn it, then we have a tolerance that is a percentage about the final probability.

MATLAB program ppt2.m—a repeat of the—previous problems, solved using diffusion performs this calculation:

```
%  ppt2 repeat of ppt1 example, solved by diffusion

%  All dimensions are re-referenced for Matlab indices
%  i.e. max has been added to every x & y value so that
%  array indices always start at 1

max = 20;   % These are the definitions from ppt1
%  inner box is a rectangle, dimensions as shown

x1 = 11;   x2 = 26;   y1 = 16;   y2 = 31;

%  set starting point
x_start = 21;   y_start = 33;

s = zeros(2*max + 1); % array showing current status
s(x_start, y_start) = 1;   % initialization

done = 0.;      % extraction monitor
prob = 0.;      % center conductor hit totaller

% Numbers will be moved from s to s2 so as to keep them
% separate, then after every point has been diffused, s2
% will be transferred back to s for the next iteration

x_plot = []; prob_plot = []; % Arrays for observing convergence

tic
while done < .999   % This sets .001 as the "done" level

    s2 = zeros(2*max + 1);      % clear s2
    for i = 2:2*max             % diffuse
      for j = 2:2*max
        s2(i-1,j ) = s2(i-1,j) + s(i,j)/4.;
        s2(i + 1,j) = s2(i + 1,j) + s(i,j)/4.;
        s2(i,j-1) = s2(i,j-1) + s(i,j)/4.;
        s2(i,j + 1) = s2(i,j + 1) + s(i,j)/4.;
      end
    end

    s2(1,:) = 0 ;       % extract at outer box
    s2(2*max + 1,:) = 0;
    s2(:,1) = 0 ;
    s2(:,2*max + 1) = 0;

    for i = x1:x2       % score then extract at inner box
      for j = y1:y2
        prob = prob + s2(i,j);
        s2(i,j) = 0;
      end
```

```
end

  done = 1 - sum(sum(s2));        % monitor extraction
  s = s2;                         % reestablish s

  prob_plot = [prob_plot,prob];

end
toc

fprintf ('Final probability = %6.3f \n', prob)

plot(prob_plot, 'k')
axis ([0, length(prob_plot) + 20, 0., 1.])
xlabel('Number of Iterations')
ylabel('Probability')

save s.txt s -ascii
```

Using a final remaining number of 0.001 as the termination condition, the program produces a voltage at the starting point of 0.782. Figure 11.6 shows the convergence of the probability with iterations.

Comparing Figures 11.6 and 11.4, it is clear how eliminating the random-number generator greatly improved the *smoothness* of convergence. Also, there is no variation — this program will produce the same results every time it is run.

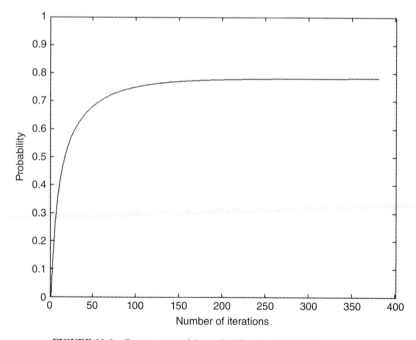

FIGURE 11.6 Convergence of the probability as calculated by ppt2.m.

11.4 VARIABLE-STEP-SIZE RANDOM WALKS

The two previous sections showed two techniques for performing a finite difference calculation by taking advantage of the equivalence between the numerical Laplace equation and the steadystate numerical diffusion equation. These calculations, although interesting, brought nothing to the party in terms of practical problem solution. Possibly, if information about the voltage is needed at only very few points, these solution techniques are more efficient than simply solving the entire set of equations.

Another approach to a random-walk solution is to eliminate the grid entirely.

Figure 11.7 shows the same example as shown in Figure 11.2. It has been reconfigured a bit for clarity, so it doesn't look exactly the same as Figure 11.2.

In Figure 11.7, a circle has been drawn about the starting point. The radius of the circle is chosen such that it is tangent to the boundary line (electrode) closest to the starting point. A temporary square grid is established as shown by the arrows, with the starting point and the tangent point of the circle establishing the gridpoint separations.

The random walker now takes a step from the starting point to one of the four gridpoints. She has a $\frac{1}{4}$ probability of landing on an electrode and being extracted.

Figure 11.8a shows one of the other possibilities, the next circle, and the next temporary grid. In this case the gridpoints do not lie along the x and y axes — and there is no need for them to be along these axes. Figure 11.8b shows a possible third circle and set of gridpoints. There is always a 25% probability that the walker will be extracted, this time at the outer electrode.

After three steps, the probability of having been extracted is greater than 50%:

$$p_3 = 1 - (0.75)^3 = 0.578 \tag{11.16}$$

This probability climbs rapidly with the number of steps taken. After six steps, it is

$$p_6 = 1 - (0.75)^6 = 0.955 \tag{11.17}$$

The average walk doesn't take long. Computationally, computer time needed for these shorter walks is offset by the complexity of finding the closest wall and calculating the

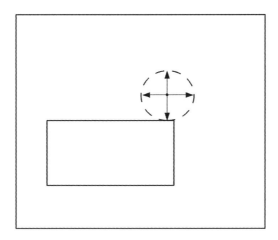

FIGURE 11.7 Two-dimensional potential problem with no grid.

(a)

(b)

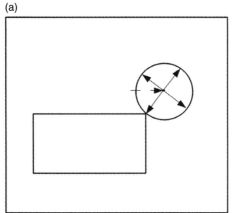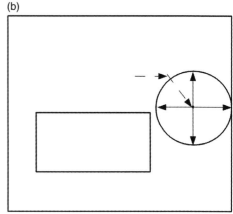

FIGURE 11.8 Further possible random steps, following Figure 11.7.

temporary grid. If there is a dielectric interface, then this interface should be considered a wall insofar as calculating a temporary grid, and we can use equation (11.13) directly to calculate the probabilities of steps. Consider the following programs:

- Program `ppt3.m` — repeat of previous example using nongridded PPT:

```
% ppt3   ppt1 example by non-gridded solution

tic            % turn on timer function
max = 20;      % This is the definition from ppt1

% inner box is a rectangle, dimensions as shown
x1 = -10;   x2 = 5;   y1 = -5;   y2 = 10;

% set starting point
x_start = 0; y_start = 12;

% The first task is to categorize all lines (electrode edges)
% The array al has a row for each line. Columns are:
% 1.   Line type. 1 = horizontal, 2 = vertical
% 2.   Location. y value for horizontal, x value for vertical
% 3,4.   Start & stop. x1, x2 for horizontal, y1, y2 for vertical
% 5.   Voltage

al = zeros(8,5);       % Block out the al (all_lines) array

% This is brute force - a better program would have a subroutine
doing this
% or it could be stored information in a data file

al(1,1) = 1;      % top outer box horizontal line
al(1,2) = max; al(1,3) = -max; al(1,4) = max; al(1,5) = 0.;

al(2,1) = 1;      % bottom outer box horizontal line
```

```
al(2,2) = -max; al(2,3) = -max; al(2,4) = max; al(2,5) = 0.;

al(3,1) = 2;  % left outer box vertical line
al(3,2) = -max; al(3,3) = -max; al(3,4) = max; al(3,5) = 0;

al(4,1) = 2;  % right outer box vertical line
al(4,2) = max; al(4,3) = -max; al(4,4) = max; al(4,5) = 0;

al(5,1) = 1;  % top inner box horizontal line
al(5,2) = y2; al(5,3) = x1; al(5,4) = x2; al(5,5) = 1.;

al(6,1) = 1;  % bottom inner box horizontal line
al(6,2) = y1; al(6,3) = x1; al(6,4) = x2; al(6,5) = 1.;

al(7,1) = 2;  % left inner box vertical line
al(7,2) = x1; al(7,3) = y1; al(7,4) = y2; al(7,5) = 1;

al(8,1) = 2;  % right inner box vertical line
al(8,2) = x2; al(8,3) = y1; al(8,4) = y2; al(8,5) = 1;

nr_walls = length(al);

% Intialize counters

nr_iters = 500;      % starting max number of iterations
nr_ones = 0;         % number of hits to 1 volt electrode
iter = 0;            % random walker counter
nr_sigs = 1.96;      % number of sigmas for confidence factor
conf_width = .025;   % half-width fraction about mean for nr_sigs
confidence
iter_plot = []; p_plot = [];  % arrays to store p(n) for a plot
nr_steps = [];                % monitor number of steps per walk

% Random walks

while iter < nr_iters

   iter = iter + 1 ;
   hit_boundary = 0;            % wandering point flag
   x = x_start; y = y_start;    % start at starting point
   steps = 0;                   % count nr steps per walk - not necessary

   while hit_boundary < = 0       % this is the particle wandering loop
   steps = steps + 1;

% Find the closest wall

     dists = [];
     for i = 1:nr_walls
       if al(i,1) == 1
         dist = hor_line_dist(al(i,2:4),[x,y]);
         dists = [dists;dist];
```

```
     else
       dist = vert_line_dist(al(i,2:4),[x,y]);
       dists = [dists;dist];
     end
end

[min_dist, index] = min(dists(:,1)); % min dist and col nr

p1 = dists(index,2:3);        % This is the point on a boundary
rad = norm([p1(1) - x,p1(2) - y]);   % size of step
theta = atan2(p1(2) -y,p1(1) -x);
p2 = [x - rad*sin(theta), y + rad*cos(theta)];
p3 = [2*x - p1(1), 2*y - p1(2)];
p4 = [2*x - p2(1), 2*y - p2(2)];

p = rand;
if p < 1/4
  x = p1(1);
  y = p1(2);
elseif p < 2/4
  x = p2(1);
  y = p2(2);
elseif p < 3/4
  x = p3(1);
  y = p3(2);
else
  x = p4(1);
  y = p4(2);
end

if abs(x) > = max || abs(y) > = max
  hit_boundary = 1; % outer box
end

if x > = x1 && x < = x2 && y > = y1 && y < = y2
hit_boundary = 2;      % inner brick
  end

end

nr_steps = [nr_steps, steps];

if hit_boundary == 2      % Inner box hit counter
  nr_ones = nr_ones + 1;
end

if iter > 100      % Error predictor
  p = nr_ones/iter;  % working intermediate probability
  nr_iters = ceil((nr_sigs/conf_width)^2*(1 - p)/p);
  iter_plot = [iter_plot, iter];
  p_plot = [p_plot, p];
```

```
end

end

volts = nr_ones/nr_iters;

t = toc;       % timer results
fprintf ('Run time = %6.3f \n', t)
fprintf ('Number of iterations = %d \n', nr_iters)
fprintf ('Voltage at starting point = %6.3f \n', volts)

plot(iter_plot, p_plot, 'k')
axis ([80, nr_iters + 20, .5, 1.]);
xlabel ('Number of Iterations')
ylabel ('Probability')
hold on

avg_nr_steps = sum(nr_steps)/length(nr_steps);
fprintf ('Average number of steps = %6.3f \n', avg_nr_steps)
```

- Program vert_line_dist.m — support function for ppt3.m:

```
function [stuff] = vert_line_dist(vert_line, pt)
% finds the shortest distance from the vert line at x
% which extends from yb to yt to the point pt,
% returns the distance and the point on the line xp,yp

x = vert_line(1); yd = vert_line(2); yu = vert_line(3);
pt_x = pt(1); pt_y = pt(2);

stuff = zeros(1,3);
if pt_y >= yd & pt_y <= yu       % we're broadside
  stuff(1) = abs(x – pt_x);
  stuff(3) = pt_y;
elseif pt_y < yd                 % point is below
  stuff(1) = sqrt((x – pt_x)^2 + (yd – pt_y)^2);
  stuff(3) = yd;
else                             % point is above
  stuff(1) = sqrt((x – pt_x)^2 + (yu – pt_y)^2);
  stuff(3) = yu;
end
stuff(2) = x;

end
```

- Program horiz_line_dist.m — support function for ppt3.m:

```
function [stuff] = hor_line_dist(hor_line, pt)
% finds the shortest distance from the hor line at y
% which extends from xl to xr to the point pt,
% returns the distance and the point on the line xp,yp
```

```
y = hor_line(1); xl = hor_line(2); xr = hor_line(3);
pt_x = pt(1); pt_y = pt(2);

stuff = zeros(1,3);
if pt_x >= xl & pt_x <= xr          % we're broadside
   stuff(1) = abs(y - pt_y);
   stuff(2) = pt_x;
elseif pt_x < xl                         % point is off to the left
   stuff(1) = sqrt((xl - pt_x)^2 + (y - pt_y)^2);
   stuff(2) = xl;
else                                     % point is off to the right
   %dist = sqrt((xr - pt_x)^2 + (y - pt_y)^2);
   stuff(1) = sqrt((xr - pt_x)^2 + (y - pt_y)^2);
   stuff(2) = xr;
end
stuff(3) = y;

end
```

MATLAB program ppt3.m calculates the voltage at any point in Figure 11.2 (the same as Figure 11.7). Using the same starting point as in the previous examples, the result is naturally the same as was calculated by the program ppt1.m. In ppt3.m, the algorithm for storing the electrode line information and for finding the closest electrode are not handled elegantly; the intent is to make the procedure as clear as possible to the reader.

Figure 11.9 shows the results of a typical run.

MATLAB program ppt3.m is a complicated program for the job that it does. Its potential advantage over ppt1.m and ppt2.m, however, is not apparent from the example

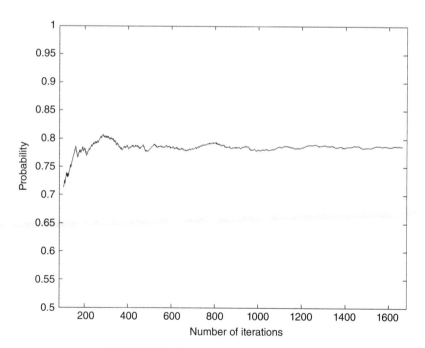

FIGURE 11.9 Typical result of running ppt3.m.

chosen. However, `ppt4.m` performs a task that is very difficult to do with grids. `ppt4` adds the capability of using a circle (actually a slice of a cylinder) as an electrode. In both `ppt3.m` and `ppt4.m` the geometry description is entered line by line. In `ppt4.m` the geometry description is entered into a data file that the program reads.

The following three PPT programs are useful for this example:

- Program `ppt4.m` here — general-purpose 2d gridless PPT model program:

```
function voltage = ppt4(x_start, y_start)
%   ppt4 data file driven no grid ppt 2-d solver

al = get_ppt_data();
%nr_walls = length(al);
[nr_walls,~] = size(al);

% Intializations
voltage = [];

% Random walks
for y_nr = 1 : length(y_start)   % walk through y data points from input
  tic % start the timer
  y_nr
  nr_iters = 50000;        % number of iterations
  % iter = 0;              % iteration counter
  nr_ones = 0;             % number of hits to 1 volt electrode

  for iter = 1 : nr_iters

    hit_boundary = 0;       % wandering point flag
    x = x_start; y = y_start(y_nr);   % start at starting point

    while hit_boundary < = 0       % this is the particle wandering loop

% Find the closest wall

    dists = zeros(nr_walls,4);
    for i = 1:nr_walls
      if al(i,1) == 1        % horizontal line
        dist = hor_line_dist(al(i,2:4), [x,y]);
      elseif al(i,1) == 2  % vertical line
        dist = vert_line_dist(al(i,2:4), [x,y]);
      else                   % circle
      dist = circle_dist(al(i,2:4), [x,y]);
      end
    dists(i,1:3) = dist; dists(i,4) = al(i,5);
                                  % the voltages should travel along

end

        [~ , index] = min(dists(:,1));   % col nr
```

```
        p1 = dists(index,2:3);       % This is the point on a boundary
%       rad_step = norm([p1(1) - x,p1(2) - y]);  % size of step
%       theta = atan2(p1(2)-y,p1(1)-x);
%       p2 = [x - rad_step*sin(theta), y + rad_step*cos(theta)];
%       p3 = [2*x - p1(1), 2*y - p1(2)];
%       p4 = [2*x - p2(1), 2*y - p2(2)];
        rad_step = dists(index,1);
        phi = atan2(p1(2)-y , p1(1)-x);
        p2 = [x + rad_step*cos(phi + pi/2) , y + rad_step*sin(phi +
pi/2)];
        p3 = [x + rad_step*cos(phi + pi) , y + rad_step*sin(phi + pi)];
        p4 = [x + rad_step*cos(phi + 3*pi/2) , y + rad_step*sin(phi +
3*pi/2)];
        volts = dists(index,4);

        p = rand;
        if p < 1/4
           x = p1(1);
           y = p1(2);
        elseif p < 2/4
           x = p2(1);
           y = p2(2);
        elseif p < 3/4
           x = p3(1);
           y = p3(2);
        else
           x = p4(1);
           y = p4(2);
        end

    % If p < 1/4 we have an extraction
       if p < 1/4
          if volts == 0
            hit_boundary = 1;
          else
            hit_boundary = 2;
          end
       end

    end

    if hit_boundary == 2 % 1 volt electrode hit counter
       nr_ones = nr_ones + 1;
    end

end

voltage = [voltage, nr_ones/nr_iters]
toc               % timer results
nr_iters

end
```

- Program `get_ppt_data.m`:

```
function a = get_ppt_data ()

    filename = uigetfile('*.txt');          % look at available
data files
    fid = fopen(filename);                    % open the file
    all_data = (fscanf(fid, '%g'))';          % read the file
    nr_lines = length(all_data)/5;
    closeresult = fclose(fid)

    a = [];      % parse the data
    for i = 1:nr_lines
        j = 5*(i-1) + 1;
        a = [a; all_data(j:j + 4)];
    end
end
```

- Program `circle_dist.m` — support function for `ppt4.m`:

```
function [stuff] = circle_dist(input, pt)
% finds the shortest distance from the circle
% centered at input(1:2) = center = (x0,y0), input(2) = radius = rad
% returns the distance and the point on the circle xp,yp

rad = input(1); x0 = input(2); y0 = input(3);
pt_x = pt(1); pt_y = pt(2);

stuff = zeros(1,3);

stuff(1) = abs(sqrt((pt_x - x0)^2 + (pt_y - y0)^2) - rad);
theta = atan2(pt_y - y0, pt_x - x0);
stuff(2) = rad*cos(theta) + x0;
stuff(3) = rad*sin(theta) + y0;

end
```

As set up, `ppt4` will accept either a single (X,Y) point or a single X value and a vector of Y points. In the former case a single voltage is returned; in the latter case a vector of voltages is returned. Note that `ppt4` is written as a function rather than a script so that it may be conveniently called from plotting and/or curve fitting programs.

MATLAB program `ppt4` can process arbitrary lines in both the x and y axes, and arbitrary circles. As set up, the voltages on the lines and circles may be only 1 or 0. The data file processing routine, `get_ppt_data.m`, is set up to only "see" `.txt` files. This, of course, is very easy to change.

The data file format is exemplified by the following dataset

1	40	−20	20	0
1	0	−20	20	0
2	−20	0	40	0
2	20	0	40	0
1	10	−10	10	1
1	5	−10	10	1
2	−10	5	10	1
2	10	5	10	1
2	−0.25	10	15	1
2	.25	10	15	1
3	0.25	0	15.	1

Figure 11.10 shows the geometry analyzed by ppt4.m using the data set data1.txt. While the program describes a circular electrode, burying half of the circle between the vertical line electrodes effectively creates a semicircle sitting on top of a rectangle.

Each line in the data file describes a particular electrode geometry. The entry in the first column of each line selects the geometry of that line. Currently allowed values of this entry are 1 — a horizontal line, 2 — a vertical line, and 3 — a circle. Each line has a total of five entries. The table below summarizes the available geometries:

Column1	Column2	Column 3	Column 4	Column 5
Horizontal line	Y location	Leftmost X	Rightmost X	Voltage
Vertical line	X location	Lower Y	Upper Y	Voltage
Circle	Radius	Center X	Center Y	Voltage

Figure 11.11 shows the results of a ppt4.m run using the data file data1.txt. The electric field (E_y above the post) was calculated from the voltage data using the simple finite difference derivative approximation. Superimposed on the data is the result of a curve fit to these data using the function

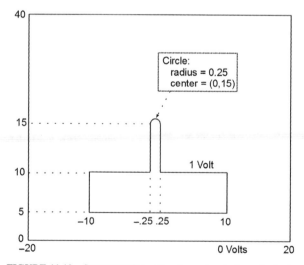

FIGURE 11.10   Structure using rectangles and a circle as electrodes.

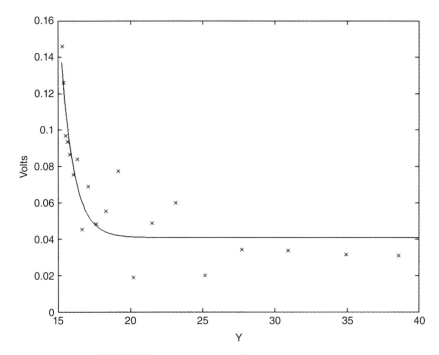

FIGURE 11.11   ppt4.m results for data1.txt dataset.

$$E_y = a + be^{-cy} \tag{11.18}$$

Several observations about this program run and the results are described here:

1 The convergence criterion [equation (11.12)] does not apply in this program because of the continual readjustment of the step size. For the results shown, a fixed number of iterations (50,000) was used at each starting point.

2 The curve fit is, of course, not unique. The results "look pretty good" but this is hard to quantify.

3 The results are very noisy. We can use these results to learn about the electrostatic nature of the system, but much more refined runs are necessary if quantitative results are desired.

4 The results (and the curve fit) demonstrate the peaking of an electric field at the tip of a post, and the falling back of the electric field to background level as we move away from the post.

The PPT method allows for an interesting interpretation of the peaking of electric fields near sharp objects. Consider the three cases shown in Figure 11.12.

In all of the three cases depicted in Figure 11.12, the black dot represents a point where we are calculating the voltage. Along the line normal to the surface and passing through the point, our best estimate of the electric field is

$$E_y\left(\frac{h}{2}\right) \approx \frac{V(0) - V(h)}{h} \tag{11.19}$$

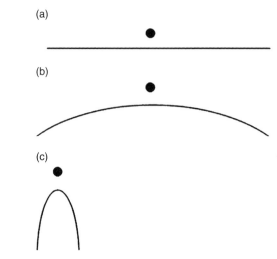

FIGURE 11.12   Examples of various degrees of sharpness near a field point.

where $V(0)$ is the known applied voltage (boundary condition) on the conductor and $V(h)$ is the voltage predicted by the calculation.

In the case of Figure 11.12a, particles beginning to wander randomly from the point have a high probability of reaching the nearby electrode surface (a plane in this case). This implies that $V(h)$ would be close to $V(0)$ and E_y would be small.

In the case shown in Figure 11.12b, the wandering particles would be less likely to hit the electrode surface before wandering off and possibly hitting another surface. Assuming that the other surface is at a different potential than $V(0)$ (e.g., the outside box in our previous example), the field would be somewhat higher than it was in Figure 11.12a.

In the case in Figure 11.12c, the particles will most likely miss the nearby electrode and wander off. In this case the voltage would be closer to the outside box voltage (in our previous example) than to the nearby electrode voltage, and the field would be very high. The sharper the tip and the higher it sits on a post and base electrode of the same voltage, the less likely that the particle will find this compound electrode structure, and consequently the higher the electric field will be.

11.5   THREE-DIMENSIONAL STRUCTURES

Extending the gridless approach to three dimensions is straightforward — the probability of a step is now $\frac{1}{6}$ for the six orthogonal choices. One issue that must be contended with is that finding the closest point on some surface and choosing one step as the step to this point does not uniquely define all the steps as it did in two dimensions.

In Figure 11.13, (x_0, y_0) is the current point and p_1 is the closest point to the current point on one of the surfaces in the geometry. Whereas P_2 is now uniquely defined, the other four points are not uniquely defined. They are orthogonal to each other and to the lines from the current point to p_1 and p_2, and all six distances from the current point are the same. The four lines (from the current point to p_2–p_6) may be regarded as a rigid cross that can be spun about the line from p_1 to p_2. In this case a second random number should be obtained to determine

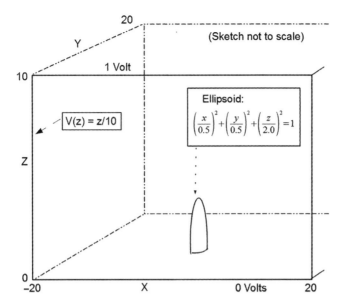

FIGURE 11.13   Example of three-dimensional gridless random-walk step generation.

the angle between, say, the line from the current point to p_2 and some fixed reference in the structure. Once this is chosen, all six points are specified.

PROBLEMS

11.1   Suppose that we want to use the gridded random-walk procedure (ppt1.m) to find all the node voltages in a structure. As the program stands, this is simply a job of rerunning the program with every free (nonboundary) node as the starting point (assuming, of course, that there are no symmetries in the structure to employ). Modify ppt1.m to do this and to store the results in a matrix V. In order to avoid excessive computer runtime, fix the number of iterations to 1000 per starting point.

11.2   Add the boundary conditions into the V matrix of Problem 11.1 and view the results using MATLAB'S mesh(x,y,V) function.

11.3   Calculate the 95% confidence interval at each matrix point voltage generated in Problem 11.1. View these results using the mesh function.

11.4   One way of improving (reducing) the computer runtime of Problem 11.1 is, as each node voltage is determined, to convert that node to a boundary condition fixed at this (now known) voltage. Modify the program written for Problem 11.1 (or simply create a new program) to do this. Compare the runtimes and the results of the two programs. Are there any new issues here?

11.5   Using ppt4.m, generate a dataset for two concentric cylinders (two circles in this 2d calculation), run the program, and compare the results to the exact answer.

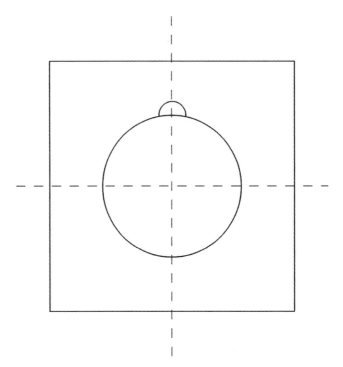

FIGURE P11.6 Geometry for Problem 11.6.

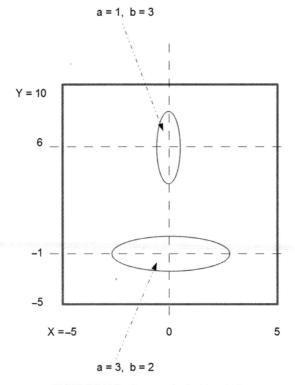

FIGURE P11.7 Structure for Problem 11.7.

11.6 Create a dataset for `ppt4.m` of a circle of radius 10 centered in an outer square of side(s) 40. Add a semicircle of radius 1 "sitting" on the original circle, centered at (0,10) as shown in Figure P11.6.

Set the outer square to 0 V and the circle and semicircle to 1.0 V. Calculate the voltage at several arbitrary points.

11.7 Starting with the program `ppt4.m`, add the capability for describing an ellipse in terms of its center and its major and minor axes. Since a general ellipse description requires four geometry parameters and the existing data structure allows for only three geometry parameters, we'll simplify the situation and require that the center of the ellipse be on the Y axis ($X = 0$). This guarantees backward capability with earlier datasets.

For the structure shown in Figure P11.7, set the outer box at 0 V and the ellipses at 1 V, and calculate the $V(y)$ at $X = 0$.

REFERENCES

1. *Brownian Motion*, www.wikipedia.com.

2. *Diffusion, Diffusion Equation*, www.wikipedia.com.

3. R. Geulusee, Probabilistic potential theory applied to electrical engineering problems, *Proc. IEEE* **61**(4): (April 1973).

4. L. Golnick and W. Smith, *The Cartoon Guide to Statistics*, HarperPerennial, New York, 1993.

5. L. Dworsky, *Probably Not*, Wiley, Hoboken, NJ, 2008.

12

The Finite Element Method (FEM)

12.1 INTRODUCTION

The finite element method (FEM) is probably the most widely used numerical approximation technique for solving electrostatic problems. This is true, as well as magnetostatic, general electromagnetic, structural (elastic body), and other physical system analyses.

The finite element method is similar to the FD method for electrostatics in that the space where the electric field exists is discretized (*meshed*, or "zoned up") and the electrodes, with applied voltages, are the boundary conditions. It is similar to the MoM in that it is readily amenable to complex geometries.

Because of its abilities to handle complex geometries and boundary conditions, the FEM is uniquely suited to structural analysis (airplane wings, bridges, vibrating structures, etc.). Historically, much of the formal development of the FEM occurred in the mechanical engineering community.[1] While electrostatics problems typically have fairly simple boundary conditions, they can have very complex structures, so the applicability and popularity of the FEM is well deserved.

The FEM formalizes and expands several topics that were foreshadowed in previous chapters:

1 The MoM technique employed both square and triangular cells over which the charge was simply approximated. The FD technique employed square cells over which voltage was simply approximated, with the approximation based on voltages at nodes common to several cells. The FEM may use many different shapes, in one, two and three dimensions, with different approximation functions tied to the voltages at nodes that are common to several (finite) elements.

Introduction to Numerical Electrostatics Using MATLAB, First Edition. Lawrence N. Dworsky.
© 2014 John Wiley & Sons, Inc. Published 2014 by John Wiley & Sons, Inc.

2 The FD method was derived by simple numerical approximation of Laplace's equation, but the concept of minimization of stored energy was introduced. This latter concept is central to the use of the FEM in electrostatics and it will be used as the beginning of the study of the topic.

The body of FEM literature is enormous. The breadth and depth of the subject also is enormous. The application of the FEM to electrostatics requires dealing only with a small subset of overall technique. The material below is not intended as a general introduction to FEM analysis; it only treats what is necessary in order to proceed to computer programs for problem setup and solving of Laplace's equation.

Introducing the FEM method in one dimension allows us to minimize the calculation detail that might obscure the concepts involved. On the other hand, one-dimensional (1d) electrostatic (Laplace's equation) problems can usually be solved analytically in the first place, so there really is no need for a numerical technique. Also, the matrix equations that result from 1dimensional FEM analysis using linear approximation functions are the same matrix equations that resulted from the FD analysis. This can be frustrating to the student because it appears to entail a lot of work which leads nowhere new. On balance, however, the simplicity that 1d derivations bring outweigh the delay their treatment brings in getting to useful results because of the stage setting they make possible.

12.2 SOLVING LAPLACE'S EQUATION BY MINIMIZING STORED ENERGY

We are looking for an approximate solution to Laplace's equation in one dimension

$$\frac{d^2V}{dx^2} = 0 \tag{12.1}$$

in the region $a \leq x \leq b$ where $V(a) = V_a$, $V(b) = V_b$ are the specified boundary conditions.

Begin by approximating $V(x)$ with the function

$$\phi(x) = V(x) + \varepsilon\mu(x) \tag{12.2}$$

where ε is a number and $\mu(x)$ is a twice-differentiable function satisfying $\mu(a) = \mu(b) = 0$.

For ε small, $\varphi(x)$ approximates $V(x)$ and at all times $\varphi(x)$ satisfies the same boundary conditions as $V(x)$.

Define $F(V)$ as

$$F(V) = \int_a^b \left(\frac{dV}{dx}\right)^2 dx \tag{12.3}$$

Except for the proper dimensional scaling factor, F is the energy stored in the electric field in the region $a \leq x \leq b$.

If we replace V in equation (12.3) with φ, then, of course, F will change and we may write its new value as

$$F + \delta F = \int_a^b \left(\frac{d\phi}{dx}\right)^2 = \int_a^b \left(\frac{d(V + \varepsilon\mu)}{dx}\right)^2 dx \tag{12.4}$$

Expanding equation (12.4), we have

$$F + \delta\mathrm{F} = \int \left(\frac{d(V + \varepsilon\mu)}{dx} \right)^2 dx = \int \left(\frac{dV}{dx} \right)^2 dx + 2\varepsilon \int \frac{d(\mu V)}{dx} dx + \varepsilon^2 \int \left(\frac{d\mu}{dx} \right)^2 dx \qquad (12.5)$$

The second term on the right is

$$2\varepsilon \int \frac{d(\mu V)}{dx} dx = 2\varepsilon \left[\mu V \right]_a^b = 0 \qquad (12.6)$$

because $\mu(a) = \mu(b) = 0$.

This leaves us with

$$F + \delta F = F(V) + \varepsilon^2 F(\mu) \qquad (12.7)$$

Equation (12.7) shows several things:

1   By definition, F is always positive. Therefore, $F + \delta F$ is never less than F. This demonstrates the assumption that was reasoned through previously that an approximate expression for V (that satisfies the correct boundary conditions) will always yield a stored energy, and therefore a capacitance for the structure, that is greater than the exact energy (and capacitance).

2   If the approximate expression for F has an adjustable parameter (or adjustable parameters), setting these parameters to minimize $F + \delta F$ gets us as close to the exact solution as our approximation will allow.

3   Because of the dependence of error in the energy approximation on ε^2 rather than simply on ε, for ε small enough, the approximation of the energy will be much more accurate than the approximation of V. Results that depend on the energy calculation, including capacitance and inductance, can be approximated very accurately.

There is much more information on this topic than has been presented here. Kwon and Bang give a detailed discussion of the various techniques for minimizing residual errors in approximate solutions to partial differential equations.[2] Although not explicitly discussed above, equations (12.1)–(12.7) provide a simple example of a variational calculation. Chari and Salon devote several chapters to variational methods and finite element formulation.[3] Both of these references derive the full multidimensional extension of this simple calculation.

12.3   A SIMPLE ONE-DIMENSIONAL EXAMPLE

As explained above, one-dimensional (1d) calculations and examples are a good way to introduce the FEM. The solution to Laplace's equation in 1d rectangular coordinates, with fixed voltages at the endpoints (boundaries), is a straight line, so developing approximate solutions is a comic exercise.

Instead, we will work with Laplace's equation in 1d cylindrical (i.e., circular) coordinates. We can still (and, indeed, will) find the exact solution, but now this exact solution is useful for evaluating approximate solution characteristics.

Figure 12.1 details the structure being considered.

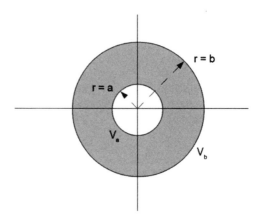

FIGURE 12.1 Concentric circle example structure.

The two concentric circles shown, in Figure 12.1, including the cross sections of two long concentric cylinders, have inner and outer radii a and b. These circles are conductors that are set at V_a and V_b, respectively. The space between these conductions, $a < r < b$, is filled with a uniform dielectric (possibly air).

Using Gauss's law, the charge on the inner circle is

$$Q = 2\pi \varepsilon r E_r \tag{12.8}$$

where $a < r < b$, ε is the dielectric constant of the region, and E_r is the electric field in the region.

The electric field is then

$$E_r = \frac{-dV}{dr} = \frac{Q}{2\pi \varepsilon r} \tag{12.9}$$

The voltage is

$$V(r) = \int E_r \, dr = \frac{-Q}{2\pi \varepsilon} \ln(r) + V_0 \tag{12.10}$$

Evaluating the first boundary condition, we obtain

$$V(a) = v_a = \frac{-Q}{2\pi \varepsilon} \ln(a) + V_0 \tag{12.11}$$

and substituting this back, we have

$$V(r) = \frac{-Q}{2\pi \varepsilon} \ln\frac{r}{a} + V_a \tag{12.12}$$

Evaluating the second boundary condition, we obtain

$$V(b) = V_b = \frac{-Q}{2\pi \varepsilon} \ln\frac{b}{a} + V_a \tag{12.13}$$

and therefore

$$\frac{Q}{2\pi\varepsilon} = \frac{V_b - V_a}{\ln(b/a)} \tag{12.14}$$

which results in

$$V(r) = (V_b - V_a)\frac{\ln(r/a)}{\ln(b/a)} + V_a \tag{12.15}$$

From equation (12.14), we have

$$\frac{Q}{|V_b - V_a|} = C = \frac{2\pi\varepsilon}{\ln(b/a)} \tag{12.16}$$

Redoing the above calculation with respect to the stored energy, we calculate the electric field using equation (12.15), specifically

$$-E_r = \frac{dV}{dr} = \frac{V_b - V_a}{\ln(b/a)}\frac{1}{r} \tag{12.17}$$

from which

$$U = \frac{\varepsilon}{2}\int_0^{2\pi}\int_a^b E_r^2\, r\, dr\, d\theta = \frac{\pi\varepsilon(V_b - V_a)^2}{\ln(b/a)} \tag{12.18}$$

and finally

$$C = \frac{2U}{(V_b - V_a)^2} = \frac{2\pi\varepsilon}{\ln(b/a)} \tag{12.19}$$

Equations (12.16) and (12.19), of course, agree; both derivations are shown for completeness.

The first approximation to examine is the simplest approximation that can satisfy the boundary conditions, a straight line connecting the two boundaries:

$$\phi(x) = V_a\left(\frac{b-r}{b-a}\right) + V_b\left(\frac{r-a}{b-a}\right) \tag{12.20}$$

Figure 12.2 shows $V(r)$ and $\phi(r)$ results for $a = 1$, $b = 100$, $V_a = 1$, and $V_b = 2$. The simple approximation for $V(r)$ doesn't appear to be a very realistic approximation at all.

Equation 12.20 could have been factored and/or expanded and then presented in any of several different but equivalent forms. The form chosen will prove to be very useful going forward. In this example $\phi(r)$, the approximation to $V(r)$, extends over the entire region of the structure, $a \le r \le b$. It is expressed in terms of the voltages at the ends of the region (nodes), which in this example happen to be the boundary conditions. Each node voltage

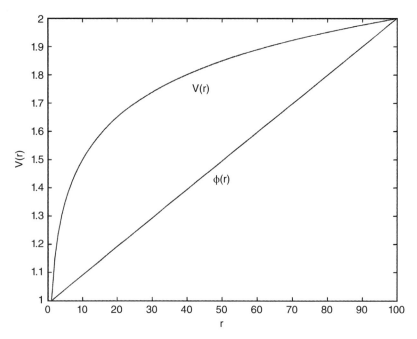

FIGURE 12.2 $V(r)$ for concentric circular electrodes and a straight-line approximation to $V(r)$.

(V_a and V_b) is multiplied by *basis functions*, $H_a(r)$ and $H_b(r)$, respectively. In this example these basis functions are simple linear equations; it will soon be shown that this is not a necessary condition. The basis functions have several properties:

1   $H_a = 1$ at $r = a$, $H_a = 0$ at $r = b$
2   $H_b = 1$ at $r = b$, $H_b = 0$ at $r = a$
3   $H_a + H_b = 1$ for all r

Continuing, we have

$$E_r = \frac{d\phi}{dr} = \frac{V_b - V_a}{b - a} \tag{12.21}$$

and then

$$U = \frac{\varepsilon}{2} \int_0^{2\pi} \int_a^b E_r^2 \, dr \, d\theta = \frac{\pi\varepsilon}{2}(V_b - V_a)^2 \frac{b/a + 1}{b/a - 1} \tag{12.22}$$

and

$$C = \pi\varepsilon \frac{b/a + 1}{b/a - 1} \tag{12.23}$$

Both the exact capacitance [equation (12.19)] and this approximate capacitance are functions only of the ratio b/a. The ratio of the approximate to the exact capacitance varies from 1.001 at $b/a = 1.1$ to 2.3 at $b/a = 100$, as shown in Figure 12.3.

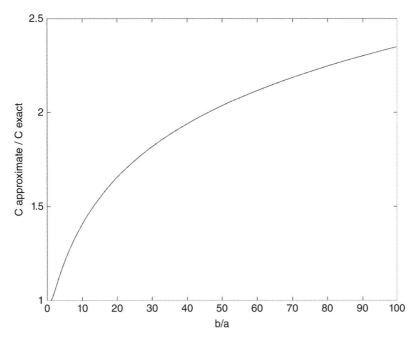

FIGURE 12.3 Ratio of *C* using a linear approximation to exact *C* versus *b/a*.

The approximation is fairly accurate for the former case, but poor for the latter case. In this latter case, inspection of Figure 12.2 shows the root of the problem, the actual *V(r)* function just doesn't look like a straight line!

12.4 A VERY SIMPLE FINITE ELEMENT APPROXIMATION

As a first attempt to try to improve the approximation, pick a point *c* midway between *a* and *b*, at a voltage V_c. Now, *V(x)* will be approximated by two lines, the first from *a* to *b* and the second from *b* to *c*. Specifically, we have

Region 1: $a \leq r \leq c = (a + b)/2$:

$$\phi_1 = V_a \frac{c-r}{c-a} + V_c \frac{r-a}{c-a} \tag{12.24}$$

Region 2: $c \leq r \leq b$:

$$\phi_2 = V_c \frac{b-r}{b-c} + V_b \frac{r-c}{b-a} \tag{12.25}$$

This approximation function is shown in Figure 12.4. Note that for the purpose of this figure, V_c has been chosen arbitrarily.

Each of these approximations are written using basis functions that have the same characteristics as described in the previous example. Each region has its own basis functions.

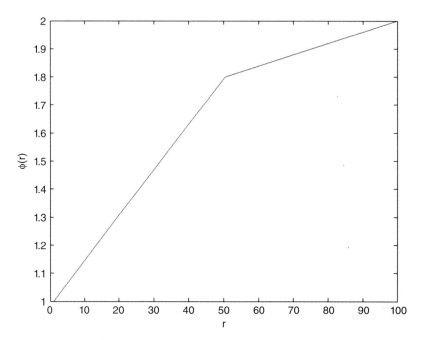

FIGURE 12.4   Two-line voltage approximation function.

The electric field and the stored energy in each region are found as in equations (12.21) and (12.22). The total stored energy is simply the sum of the energies in each region:

$$U = \frac{\pi\varepsilon}{2}\left[(V_c - V_a)^2\frac{c+a}{c-a} + (V_b - V_c)^2\frac{b+c}{b-c}\right] \tag{12.26}$$

At this point, V_c is still an arbitrary variable. We give it a value by invoking the principle that we want to choose V_c so as to minimize the approximate energy U:

$$\frac{dU}{dV_c} = (Vc - V_a)\frac{c+a}{c-a} + (Vc - V_b)\frac{b+c}{b-c} = 0 \tag{12.27}$$

and therefore

$$V_c = \frac{V_a[(c+a)/(c-a)] + V_b[(b+c)/(b-c)]}{[(c+a)/(c-a)] + [(b+c)/(b-c)]} \tag{12.28}$$

Figure 12.5 is a repeat of Figure 12.2 with the two-line approximate voltage added. While it's more like the exact $V(r)$ than is the single-line approximation, it's still not a very good approximation.

Figure 12.6 is a repeat of Figure 12.3 with the capacitance ratios for the two-line approximation added. While the results still are not very impressive for the larger values of b, the two-line approximation is significantly better than is the single-line approximation.

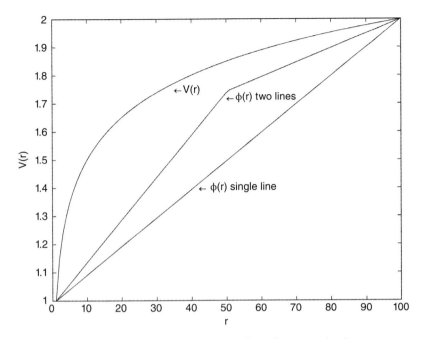

FIGURE 12.5 Exact single-line and two-line voltage approximations.

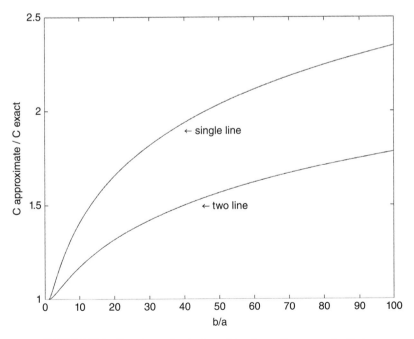

FIGURE 12.6 Capacitance for single- and two-line voltage approximations.

12.5   ARBITRARY NUMBER OF LINES APPROXIMATION

The next step in the process of improving the approximate solution to the concentric circle problem is to break the region down into a large number of line segments. Mathematically this is a direct extension of the procedure used in Section 12.4. To circumvent the need to write many sets of equations explicitly and also to allow for an arbitrary number of line segments, it is necessary to introduce some structured notation into the process.

Figure 12.7 shows the notation for the setup of this problem. In a 1d problem correlating the node number and the node location is relatively easy — this won't be the case in unstructured 2d and 3d problems.

Node 1 is at location $r_1 = a$. Node n is at location $r_n = b$. In general, node j is at location r_j. Region (element) j extends from r_j to r_{j+1}, $r_j \leq r \leq r_{j+1}$ and the size (length) of element j is $h_j = r_{j+1} - r_j$.

For element j it is convenient to define a *local* coordinate system u, namely, $0 \leq u \leq h_j$. In this region, then

$$V(u) = V_j \left(1 - \frac{u}{u_j} \right) + V_{j+1} \left(\frac{u}{u_j} \right) \tag{12.29}$$

$$-E = \frac{dV}{du} = \frac{1}{h_j} \left(V_{j+1} - V_j \right) \tag{12.30}$$

$$E^2 = \frac{1}{h_j^2} \left(V_{j+1} - V_j \right)^2 \tag{12.31}$$

Since we are in cylindrical coordinates, we must return to the global system before we integrate (12.31) to get the stored energy in this element. Since (12.31) is a constant expression, we have

$$U_j = \frac{\varepsilon}{2} \left(\frac{V_{j+1} - V_j^2}{h_j^2} \right) \int_{r_j}^{r_{j+1}} r \, dr = \left(V_{j+1} - V_j \right)^2 \frac{r_{j+1} + r_j}{h_j} \tag{12.32}$$

The total energy in the system is the sum of all of the element energies:

$$U = \sum_{j=1}^{n-1} U_j = \frac{\varepsilon}{2} (V_2 - V_1)^2 \frac{r_1 + r_2}{h_1} + \frac{\varepsilon}{2} (V_3 - V_2)^2 \frac{r_2 + r_3}{h_2} + \cdots \tag{12.33}$$

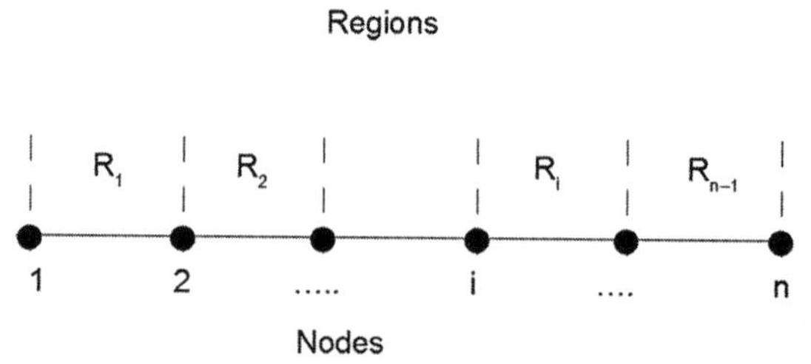

Regions

FIGURE 12.7   Node and region (element) layout for one-dimensional (1d) example.

To minimize this energy, we set the derivatives to each node voltage equal to zero:

$$0 = \frac{dU}{V_1} = (V_1 - V_2)\frac{r_1 + r_2}{h_1} \equiv a_{12}(V_1 - V_2) \tag{12.34}$$

$$0 = \frac{dU}{V_2} = (V_2 - V_1)\frac{r_1 + r_2}{h_1} + (V_2 - V_3)\frac{r_2 + r_3}{h_2} \equiv a_{12}(V_2 - V_1) + a_{23}(V_2 - V_3) = \cdots \tag{12.35}$$

This leads directly to the full set of linear equations:

$$\begin{bmatrix} a_{12} & -a_{12} & 0 & 0 & \cdots & 0 \\ -a_{12} & a_{12}+a_{23} & -a_{23} & 0 & \cdots & 0 \\ 0 & -a_{23} & a_{23}+a_{34} & -a_{34} & \cdots & 0 \\ \cdots & \cdots & \cdots & \cdots & \cdots & \cdots \\ 0 & 0 & 0 & 0 & -a_{n-1,n} & a_{n-1,n} \end{bmatrix} \begin{bmatrix} V_1 \\ V_2 \\ V_3 \\ \cdots \\ V_n \end{bmatrix} = \begin{bmatrix} 0 \\ 0 \\ 0 \\ \cdots \\ 0 \end{bmatrix} \tag{12.36}$$

Before this can be solved, the boundary conditions $V_1 = V_a$, $V_n = V_b$ must be inserted:

$$\begin{bmatrix} 1 & 0 & 0 & 0 & \cdots & 0 \\ -a_{12} & a_{12}+a_{23} & -a_{23} & 0 & \cdots & 0 \\ 0 & -a_{23} & a_{23}+a_{34} & -a_{34} & \cdots & 0 \\ \cdots & \cdots & \cdots & \cdots & \cdots & \cdots \\ 0 & 0 & 0 & 0 & 0 & 1 \end{bmatrix} \begin{bmatrix} V_1 \\ V_2 \\ V_3 \\ \cdots \\ V_n \end{bmatrix} = \begin{bmatrix} V_a \\ 0 \\ 0 \\ \cdots \\ V_b \end{bmatrix} \tag{12.37}$$

Figure 12.8 shows the results of solving equation (12.37), the *uniformly spaced nodes* curve ($a = 1$, $b = 100$). A higher number of nodes, meaning improved resolution, definitely produces better results with the ratio of the calculated to exact capacitance = 1.18 at $n = 12$.

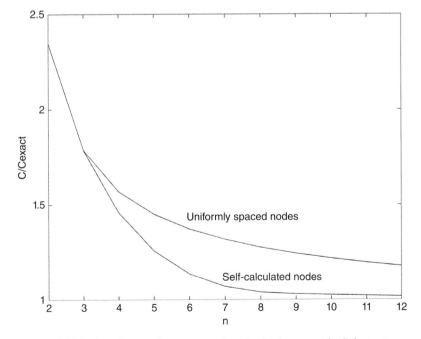

FIGURE 12.8 Capacitance ration versus number of nodes for concentric circle structure.

This calculation was performed by MATLAB program `fem1d1.m`.

```
% fem1d1.m   Repeat of the previous problem but using an organized system
% to create linear 1d elements
close

eps0 = 8.854;
n = 2;                  % Starting number of nodes
a = 1; b = 100;
ri = linspace(a, b, n);
h = ri(2:n) - ri(1:n-1);
Va = 0; Vb = 1;

% top of loop for adding regions

max_iters = 11;    % Number of subdivision cycles
for iter = 1 : max_iters
  array = spalloc (n, n, 3);
  f = zeros(n,1);

% coefficient array
  array(1, 1) = (ri(1) + ri(2))/h(1); array(n,n) = (ri(n-1) + ri(n))/h
(n-1);
  for i = 2 : n-1
    array(i,i) = (ri(i-1) + ri(i))/h(i-1) + (ri(i) + ri(i+1))/h(i);
  end

  for i = 1 : n-1
    array(i,i+1) = -(ri(i) + ri(i+1))/h(i);
  end

  for i = 2 : n
    array(i,i-1) = -(ri(i-1) + ri(i))/h(i-1);
  end

% bcs
  array(1,:) = 0; array(1, 1) = 1; f(n) = Va;
  array(n,:) = 0; array(n,n) = 1; f(n) = Vb;

% solve the equation set
  V = array\f;

% calculate the capacitance
  Utot = 0;
  for i = 1 : n-1
    Utot = Utot + (V(i+1) - V(i))^2*(ri(i+1) + ri(i))/h(i);
  end

  C = pi*eps0*Utot;
  C_exact = 2*pi*eps0/log(b/a);
  C/C_exact;
```

```
fprintf (' %d %f %f %f \n', n, C, C_exact, C/C_exact)

% Show the voltages
  if iter == max_iters-1
    V_exact = log(ri/a)/log(b/a);
    plot(ri, V, 'kx', ri, V_exact, 'k')
    xlabel ('r')
    ylabel ('V(r)')
  end

% Find the element with the largest voltage change
  dV = zeros(1,n-1);
  dV = (V(2:n) - V(1:n-1)); % ./h';
  [dVmax,index] = max(dV);

%subdivide the region - modify h and ri
  h = [h(1:index-1), h(index)/2, h(index)/2, h(index + 1:end)];
  n = n + 1;
  ri = [a];
  for i = 1 : n-1
    ri = [ri, ri(end) + h(i)];
  end

end
```

MATLAB program fem1d1.m has one additional capability, the results of which are also shown in Figure 12.8: the *self-calculated nodes* curve. Starting with some number of nodes (n in the program), the program solves the equation set and then examines the resulting voltage vector (V), looking for the two nodes with the greatest voltage difference between them. It then subdivides the element divided by these two nodes into two equally spaced elements. This is repeated max_iters times. The program could have been more efficiently written by having it subdivide, say, half of the existing elements — selecting those elements with the largest voltage step, at each iteration. The program as it is written, however, clearly emphasizes the point that higher resolution where the voltage is changing rapidly produces better results (for the same total number of nodes) than does equal-size elements.

Figure 12.9 shows the voltage profile produced by fed1d1.m for $n = 12$, using the self-calculated nodes. The calculated voltages at the nodes are superimposed on the exact solution evaluated at the node locations. The agreement is excellent.

In order to compare this 1d FEM calculation to the earlier FD calculation, look at any one of the (non-boundary-condition) lines in equation (12.37). For example, we have

$$-a_{23}V_2 + (a_{23} + a_{34})V_3 - a_{34}V_4 = 0 \tag{12.38}$$

Substituting the values of the coefficients and solving for V_3, we obtain

$$V_3 = \frac{V_2}{1 + [(r_3 + r_4)/(r_2 + r_3)]} + \frac{V_4}{1 + [(r_2 + r_3)/(r_3 + r_4)]} \tag{12.39}$$

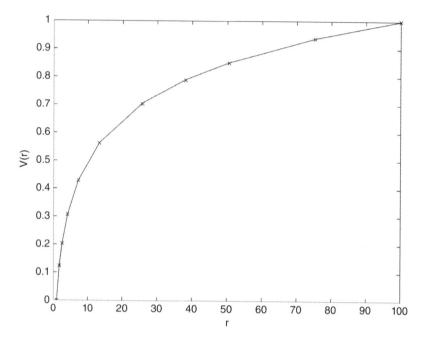

FIGURE 12.9   Voltage profile produced by `fed1d1.m` for $n = 12$.

If all h_i are the same $(= h)$, then $r_4 = r_3 + h$ and $r_2 = r_3 - h$, and therefore

$$V_3 = \frac{V_2}{1 + [(2r_3 + h)/(2r_3 - h)]} + \frac{V_4}{1 + [(2r_3 - h)/(2r_3 + h)]} = \frac{V_2}{2}\left[1 - \frac{1}{2r_3}\right] + \frac{V_4}{2}\left[1 + \frac{1}{2r_3}\right]$$

(12.40)

which is exactly the result presented in the FD derivation.

As predicted at the beginning of this chapter, 1d FEM derivations using linear elements yield interesting insights but no new results.

12.6   MIXED DIELECTRICS

One attribute of the FEM equation organization is the ease with which mixed dielectric regions are handled; the only caveat is that the dielectric is uniform in each element (region).

In going from equation (12.33) to equations (12.34) and (12.35), the factor $\varepsilon/2$ that appears in each term was simply dropped because the sum was set $= 0$. If each region is characterized by a dielectric constant

$$\varepsilon_j = k_j \varepsilon_0$$

(12.41)

then equations (12.34) and (12.35) become

$$0 = \frac{dU}{V_1} = k_1(V_1 - V_2)\frac{r_1 + r_2}{h_1} \equiv k_1 a_{12}(V_1 - V_2)$$

(12.42)

$$0 = \frac{dU}{V_2} = k_1(V_2 - V_1)\frac{r_1 + r_2}{h_1} + k_2(V_2 - V_3)\frac{r_2 + r_3}{h_2} \equiv k_1 a_{12}(V_2 - V_1) + k_2 a_{23}(V_2 - V_3)$$

$$(12.43)$$

The full matrix set [equation (12.36)], does not change if we simply update the definition of the a_{ij} terms to

$$a_{ij} \equiv k_i\frac{r_i + r_j}{h_i} \qquad (12.44)$$

12.7   A QUADRATIC APPROXIMATION

In Section 12.3 we used a straight line connecting the boundaries $r = a$ and $r = b$ as the approximate voltage function. In Section 12.4 this approximation was improved by splitting the line into two lines at an interior point c, specifically, $a < c < b$, and then adjusting V_c to minimize the total energy stored in the electric field. Another approach to finding an improved voltage approximation is to define a function that itself has one or more adjustable parameters, over the full region $a \le r \le b$.

Consider, for example, the function

$$V(r) = c_1 + c_2 r + c_3 r^2 \qquad (12.45)$$

This function has three parameters, so we will need to satisfy three conditions in order to fully specify it: $V(a) = V_a$, $V(b) = V_b$, and $V(c) = V_c$, where c again is an interior point, say, the middle of the region.

Once again we would like to express the approximate voltage function in terms of the node voltages and basis functions

$$V(r) = V_a f_a(r) + V_b f_b(r) + V_c f_c(r) \qquad (12.46)$$

where

$$\begin{aligned} f_a &= p_{a1} + p_{a2} r + p_{a3} r^2 \\ f_b &= p_{b1} + p_{b2} r + p_{b3} r^2 \\ f_c &= p_{c1} + p_{c2} r + p_{c3} r^2 \end{aligned} \qquad (12.47)$$

subject to

$$\begin{aligned} f_a(a) &= 1 & f_a(b) &= 0 & f_a(c) &= 0 \\ f_b(a) &= 0 & f_b(b) &= 1 & f_b(c) &= 0 \\ f_c(a) &= 0 & f_c(c) &= 0 & f_c(c) &= 1 \end{aligned} \qquad (12.48)$$

Inserting the appropriate three boundary conditions into each of the three equations, we obtain

$$\begin{bmatrix} 1 & a & a^2 \\ 1 & b & b^2 \\ 1 & c & c^2 \end{bmatrix} \begin{bmatrix} p_{a1} \\ p_{a2} \\ p_{a3} \end{bmatrix} = \begin{bmatrix} 1 \\ 0 \\ 0 \end{bmatrix} \qquad (12.49)$$

$$\begin{bmatrix} 1 & a & a^2 \\ 1 & b & b^2 \\ 1 & c & c^2 \end{bmatrix} \begin{bmatrix} p_{b1} \\ p_{b2} \\ p_{b3} \end{bmatrix} = \begin{bmatrix} 0 \\ 1 \\ 0 \end{bmatrix} \qquad (12.50)$$

$$\begin{bmatrix} 1 & a & a^2 \\ 1 & b & b^2 \\ 1 & c & c^2 \end{bmatrix} \begin{bmatrix} p_{c1} \\ p_{c2} \\ p_{c3} \end{bmatrix} = \begin{bmatrix} 0 \\ 0 \\ 1 \end{bmatrix} \qquad (12.51)$$

Solving these three sets of equations for $a = 1$, $b = 100$ and putting the results back into equation (12.47), we get the (three) basis functions shown in Figure 12.10.

These functions have the same properties as the previous basis functions; namely, they each equal one at their "home" nodes and zero at all other nodes, and the sum of all (three) functions is equal to one everywhere.

The steps to complete this solution follow the steps of the previous solution, but the details of the arithmetic are much more tedious.

First we find the electric field

$$-E_r = V_a \frac{df_a}{dr} + V_b \frac{df_b}{dr} + V_c \frac{df_c}{dr} V_a(p_{a2} + 2p_{a3}r) + V_b(p_{b2} + 2p_{b3}r) + V_c(p_{c2} + 2p_{c3}r) \quad (12.52)$$

then

$$U = \frac{\varepsilon}{2} \int_0^{2\pi} \int_a^b e_r^2 r\, dr\, d\theta \qquad (12.53)$$

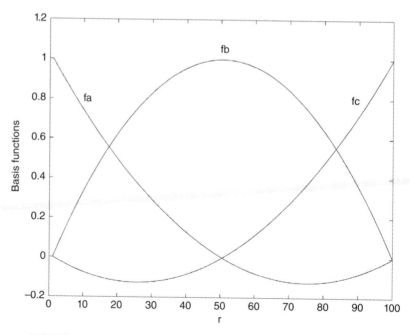

FIGURE 12.10   Basis functions for one element using a quadratic approximation.

and

$$\frac{dU}{V_c} = 0 \qquad\qquad (12.54)$$

We find V_c in terms of V_a, V_b, and the geometry and substitute this back into equation (12.46) for the voltage, equation (12.53) for the energy, and then the capacitance. The details of these calculations are shown in program `fem1d2.m` — quadratic element solution of the concentric circle problem:

```
% playing #3.  Repeat of the previous problem but using an organized
system
% to create 1 quadratic element
close

eps0 = 8.854;
a = 1; b = 100; c = (a+b)/2;
Va = 0;   Vb = 1;

ar = ones(3,3);
ar(1,2) = a; ar(1,3) = a*a;
ar(2,2) = b; ar(2,3) = b*b;
ar(3,2) = c; ar(3,3) = c*c;

f = [1;0;0];  pa = ar\f
f = [0;1;0];  pb = ar\f
f = [0;0;1];  pc = ar\f

r_plt = a : (b-a)/50 : b;
fa_plt = pa(1) + pa(2)*r_plt + pa(3)*r_plt.^2;
fb_plt = pb(1) + pb(2)*r_plt + pb(3)*r_plt.^2;
fc_plt = pc(1) + pc(2)*r_plt + pc(3)*r_plt.^2;

figure(1)
plot(r_plt, fa_plt, 'k', r_plt, fb_plt, 'k', r_plt, fc_plt, 'k')
xlabel('r')
ylabel('Basis Functions')
text(10, .8, 'fa')
text(50, 1.05, 'fb')
text(88, .8, 'fc')

% the first 3 terms will be needed for finding Vc
k1 = pc(2)^2*(b^2-a^2)/2 + 4/3*pc(2)*pc(3)*(b^3-a^3) + pc(3)^2*(b^4-a^4);

k2 = pa(2)*pc(2)*(b^2-a^2)/2 + (pa(2)*pc(3) + pc(2)*pa(3))*(b^3-a^3)*2/3 ...
       + pa(3)*pc(3)*(b^4-a^4);

k3 = pb(2)*pc(2)*(b^2-a^2)/2 + (pb(2)*pc(3) + pc(2)*pb(3))*(b^3-a^3)*2/3 ...
       + pb(3)*pc(3)*(b^4-a^4);

Vc = - (k2*Va + k3*Vb)/k1
```

```
k4 = pa(2)^2*(b^2-a^2)/2 + 4/3*pa(2)*pa(3)*(b^3-a^3) + pa(3)^2*(b^4-a*4);
k5 = pb(2)^2*(b^2-a^2)/2 + 4/3*pb(2)*pb(3)*(b^3-a^3) + pb(3)^2*(b^4-a*4);

k6 = pa(2)*pb(2)*(b^2-a^2)/2 + (pa(2)*pb(3) + pb(2)*pa(3))*(b^3-a^3)*2/3 ...
        + pa(3)*pb(3)*(b^4-a^4);

Utot = k1*Vc^2 + 2*k2*Va*Vc + 2*k3*Vb*Vc + k4*Va^2 + k5*Vb^2 + 2*k6*Va*Vb;

V_plt = Va*fa_plt + Vb*fb_plt + Vc*fc_plt;
figure(2)
plot(r_plt, V_plt)

C = 2*pi*eps0*Utot;
C_exact = 2*pi*eps0/log(b/a);
C/C_exact;
fprintf(' %d   %f   %f   %f   \n', n, C, C_exact, C/C_exact)
```

This program calculates a capacitance ratio of 1.60. This is better (lower) than the linear calculation for three nodes but not quite as good as the linear node calculation for four nodes.

One attribute of the FEM method is that it is possible to mix and match approximations as desired. In other words, some of the elements in a problem can be linear, some can be quadratic, and so forth. In 2d and 3d problems different-shape elements can also be mixed together, along with different approximation functions attached to different elements. The bookkeeping required to track this is significant, but computers do this kind of thing very well.

12.8 A SIMPLE TWO-DIMENSIONAL FEM PROGRAM

A 2d FEM program based on rectangles in a uniform grid doesn't add any new capabilities to what has already been presented. It is, however, a good learning tool because all of the steps necessary to create the FEM coefficient matrix are present. On a uniform grid there is a logical, simple way to number the nodes to make it easier to keep track of the node locations (the corners of the rectangles). The get_ijk.m function presented in Chapter 8 handles this job readily. This convenience disappears when different element shapes and orientations are allowed and/or the system becomes unstructured.

In one dimension, a linear approximation function FEM analysis generated the same coefficient matrix as did the 1d FD analysis. This will not repeat in two dimensions, as will be shown shortly.

Figure 12.11 shows a section of a square grid.

Consider the element shown shaded in Figure 12.11. Each such element in the grid will be referred to by the (i,j) value of its lower left corner, and the node numbers of the four vertices will be labeled according to the convention used in Chapter 8 (and also in the get_ijk.m function). Rectangle (i,j) has an X-dimension of 1 and a Y-dimension of 1. The four nodes (vertices) shown are labeled $V_1 = V(x,y)$, and so on, as shown, for convenience.

Consider a local coordinate system aligned with the grid and with its origin at point (i,j). In this coordinate system, the voltage $V(x,y)$ in the gray (shaded) rectangle is approximated by

$$V(x,y) = V_1(1-x)(1-y) + V_2(x)(1-y) + V_3(x)(y) + V_4(1-x)(y) \tag{12.55}$$

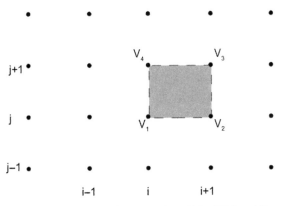

FIGURE 12.11 Section of a rectangular grid for FEM modeling.

The procedure at this point is straightforward:

1 Using equation (12.55), calculate the electric field components, E_x and E_y.

2 Calculate the stored energy in the electric field:

$$\frac{2U_e}{\varepsilon} = \int_0^1\int_0^1 \left(E_x^2 + E_y^2\right) dx\, dy \tag{12.56}$$

3 Calculate the (four) partial derivatives of the energy with respect to the node voltages:

$$0 = \frac{dU_e}{dV_1} = 4V_1 - V_2 - 2V_3 - V_4 \tag{12.57}$$

$$0 = \frac{dU_e}{dV_2} = -V_1 + 4V_2 - V_3 - 2V_4 \tag{12.58}$$

$$0 = \frac{dU_e}{dV_3} = -2V_1 - V_2 + 4V_3 - V_4 \tag{12.59}$$

$$0 = \frac{dU_e}{dV_4} = -V_1 - 2V_2 - V_3 + 4V_4 \tag{12.60}$$

4 For each element in the grid, translate the four nodes in each of equations (12.57)–(12.60) to the global (grid) node numbering. Then, for the row in the coefficient matrix corresponding to the voltage index on the LHS of the equation, add the corresponding terms on the RHS of the equation to the corresponding column in the coefficient matrix.

5 Bring in the boundary conditions and complete the solution process.

MATLAB program femrect.m—a simple 2d structured FEM program—is a rewrite of fd2.m, replacing the FD coefficient terms lines with the preceding lines:

```
%fmrect.m   A simple FEM example to show assembly procedure
clear
```

```
ns = 3;        % scaling factor for resolution studies
Imin = -20*ns; Imax = -Imin ; Jmin = -4*ns; Jmax = -Jmin; % definition
of array
Ic1 = -2*ns; Ic2 = -Ic1; Jc = 0; % inner conductor
[i,j,Kmax] = get_ijk(Imin,Imax,Jmin, Imax,Jmax,0); % number of variables

a = spalloc(Kmax,Kmax,9*Kmax); % allocate the coefficient array
b = zeros(Kmax,1);    % allocate the bcs array

% initialize a as if every variable is free
for j = Jmin: Jmax
  for i = Imin : Imax
    [i,j,k0] = get_ijk(Imin,Imax,Jmin, i,j,0);

% first equation
    [i,j,n1] = get_ijk(Imin,Imax,Jmin, i,j,0);
    n2 = n1 + 1;
    n3 = n2 + (Imax - Imin) + 1;
    n4 = n3 - 1;
    a(n1,n1) = a(n1,n1) + 4;
    if i < Imax, a(n1,n2) = a(n1,n2) - 1; end;
    if i < Imax & j < Jmax,   a(n1,n3) = a(n1,n3) -2; end;
    if j < Jmax, a(n1,n4) = a(n1,n4) -1; end;

% second equation

    [i,j,n2] = get_ijk(Imin,Imax,Jmin, i,j,0);
    n1 = n2 - 1;
    n3 = n2 + (Imax - Imin) + 1;
    n4 = n3 - 1;
    a(n2,n2) = a(n2,n2) + 4;
    if i > Imin, a(n2,n1) = a(n2,n1) - 1; end;
    if i < Imax & j < Jmax, a(n2,n3) = a(n2,n3) -1; end;
    if j < Jmax, a(n2,n4) = a(n2,n4) -2; end;

    % third equation

    [i,j,n3] = get_ijk(Imin,Imax,Jmin, i,j,0);
    n4 = n3 - 1;
    n1 = n4 - (Imax - Imin) - 1;
    n2 = n1 + 1;
    a(n3,n3) = a(n3,n3) + 4;
    if i > Imin & j > Jmin, a(n3,n1) = a(n3,n1) - 2; end;
    if j > Jmin, a(n3,n2) = a(n3,n2) -1; end;
    if i > Imin, a(n3,n4) = a(n3,n4) -1; end;

    % fourth equation

    [i,j,n4] = get_ijk(Imin,Imax,Jmin, i,j,0);
    n3 = n4 + 1;
    n1 = n4 - (Imax - Imin) - 1;
    n2 = n1 + 1;
```

```
    a(n4,n4) = a(n4,n4) + 4;
    if j > Jmin , a(n4,n1) = a(n4,n1) - 1; end;
    if i < Imax & j > Jmin, a(n4,n2) = a(n4,n2) -2; end;
    if i > Imin & j < Jmax, a(n4,n3) = a(n4,n3) -1; end;
  end

end

% Perimeter (v = 0) bcs
% clean out the appropriate row, replace it w/ 1 on diagonal
for i = Imin : Imax
    [i,j,k] = get_ijk(Imin,Imax,Jmin, i,Jmin,0);
    a(k,:) = 0; a(k,k) = 1;
    [i,j,k] = get_ijk(Imin,Imax,Jmin, i,Jmax,0);
    a(k,:) = 0; a(k,k) = 1;
end
for j = Jmin + 1 : Jmax - 1
    [i,j,k] = get_ijk(Imin,Imax,Jmin, Imin,j,0);
    a(k,:) = 0; a(k,k) = 1;
    [i,j,k] = get_ijk(Imin,Imax,Jmin, Imax,j,0);
    a(k,:) = 0; a(k,k) = 1;
end

% inner strip (v = 1) bcs
% clean out the appropriate row, replace it w/ 1 on diagonal
% put 1 in b array
for i = Ic1 : Ic2
    [i,j,k] = get_ijk(Imin,Imax,Jmin, i,Jc,0);
    a(k,:) = 0;
    a(k,k) = 1;
    b(k) = 1;
end

% solve for voltages
v = a\b;

% list non-zero voltages by row, column, & variable nrs
fprintf (' i  j  k Volts \n \n')
for k = 1 : Kmax
    volts = v(k);
    if volts > 0
      [i,j,k] = get_ijk(Imin,Imax,Jmin, 0,0,k);
      fprintf (' %3d  %3d  %3d  %8.3f \n', i, j, k, volts)
  end
end

% We'll need an xy voltage array for ongoing calcs

Sx = Imax - Imin + 1; Sy = Jmax - Jmin + 1;
Volts = zeros(Sx,Sy);
for ii = Imin:Imax
    i = ii - Imin + 1;
    for jj = Jmin:Jmax
```

```
        j = jj - Jmin + 1;
        [ii,jj,k] = get_ijk(Imin,Imax,Jmin, ii,jj,0);
        Volts(i,j) = v(k);
    end
end

C = C_Energy(Volts)
```

The difference in the results between programs fd2.m and femrect.m is not striking. The FEM calculation is better, that is, the capacitance is lower for all levels of resolution (ns). However, the improvement is only tenths of a percent. While fd2.m is a useful exercise in that in demonstrates taking local element results and assembling the coefficient matrix from it, it offers no real value in the quality of the results when a FD calculation can be easily written.

On the other hand, the 1d FEM formation yielded exactly the same results (and the same coefficient matrix) as did the FD formulation. In this 2d case, the fact that there is a small difference in results means that the coefficient matrices are not the same. This is obvious in that the FEM coefficient matrix has nine nonzero entries per (non-boundary-condition) rows, while the FD coefficient matrix has five.

Taking any of the non-boundary-condition rows of the coefficient matrix and rewriting it so that the references are to (i,j) coordinates rather than row numbers, we get (see Figure 12.12 for notation)

$$V_{i,j} = \frac{V_{i+1,j} + V_{i+1,j+1} + V_{i,j+1} + V_{i-1,j+1} + V_{i-1,j} + V_{i-1,j-1} + V_{i,j-1} + V_{i+1,j-1}}{8} \quad (12.61)$$

This equation tells us that, for a square grid, the FEM result is that the voltage at a node is calculated by averaging the voltages at the eight nearest nodes. This, according to the simple example above, yields slightly better results than does the FD calculation, which were that

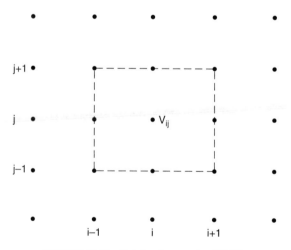

FIGURE 12.12   Notation for equation (12.61).

the voltage at a node is the average of the voltages at the four nearest nodes. If this is all it had to offer, the FEM calculation would not be worth the trouble.

In Chapter 13 we begin a discussion of unstructured systems and triangular element shapes. The value of the FEM calculations will soon become apparent.

PROBLEMS

12.1 The structure shown in Figure P12.1 is identical to the structure shown in Figure 9.16, rotated 45 degrees. Needless to say, we would expect the capacitance to be the same, regardless of which figure the calculation is based on. Create a simple voltage function that satisfies all boundary conditions, calculate the capacitance, and compare it to the exact result (90.6 pF/m) quoted in Section 9.7.

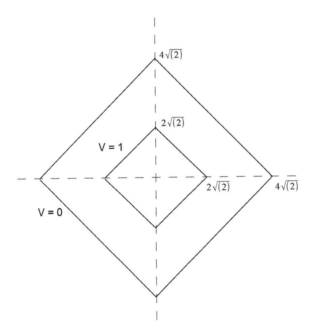

FIGURE P12.1   Square coaxial cable cross section discussed in Section 9.7.

12.2 Consider a stripline with an upper–lower box separation of b and a thin center conductor of width w. The sidewalls are infinitely far away. The simplest formula for the capacitance (per unit length), which should be reasonable for $w >> b$, is the ideal parallel plate capacitor relationship.

(a) For $w = 1$, let $b = 1$, and then $b = 10$, and compare this simple approximation to an online stripline capacitance calculator (assume an air dielectric).

(b) The approximate capacitances are lower than the published capacitances. How is this possible?

12.3 Improve the model of Problem 12.2 by creating a simple approximate voltage function that satisfies all boundary conditions, and compare the results to the published values. Are the approximate capacitances now higher than the published capacitances?

12.4 Figure P12.3 depicts a compound ideal parallel-plate capacitor structure; three electrodes are infinite in extent in Y and Z and separated by 1 m in x. Region 1 with $0 < x < 1$, is filled with an air dielectric and region 2, with $1 < x < 2$, is filled with a uniform dielectric of (relative) dielectric constant k.

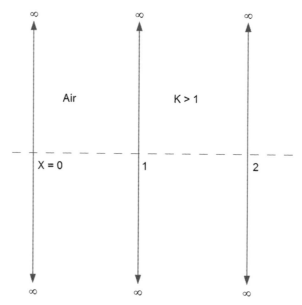

FIGURE P12.3 Structure for Problem 12.4.

Using the approximation tools of this chapter, find the exact voltage distribution $V(x)$ for $0 < x < 2$. (*Hint*: Since this is a one-dimensional rectangular coordinate problem, linear approximation functions are also the exact solution functions.)

REFERENCES

1. O. C. Zienkiewicz and R. L. Taylor, *The Finite Element Method*, 6th ed., Elsevier Butterworth-Heinemann, Burlington, MA, 2005.

2. Y. W. Kwon and H. Bang, *The Finite Element Method Using MATLAB*, 2nd ed., CRC Press, Boca Raton, FL, 2000.

3. M. V. K. Chari and S. J. Salon, *Numerical Methods in Electromagnetism*, Academic Press, San Diego, 2000.

13

Triangles and Two-Dimensional Unstructured Grids

13.1 INTRODUCTION

Because the finite element method (FEM) is capable of handling complex geometries and different boundary conditions, much of its development occurred in the mechanical engineering community.[1] Electrostatic applications of FEM are rarely as demanding in either the structural complexity or the boundary condition arenas as are many mechanical (stress–strain) problems, so the full menu of capabilities of FEM analysis will not be considered in this book.

The flexibility of using unstructured triangular cells was demonstrated using the method of moments (MoM) technique discussed earlier. In this chapter the use of unstructured triangular cells for planar two-dimensional (2d) FEM models will be developed. For MoM models, there was a good argument for working with right triangles — the barycenter of every triangle was of particular interest, and using right triangles facilitated the analyses. In FEM analysis there is nothing special about a right triangle, so triangles of any shape will be allowed.

Unstructured FEM grids are bookkeeping intensive. The computer program needs three sets of information:

1 *A List of Nodes.* This list assigns a node number to each node and specifies its location.

2 *A List of Elements.* When the elements are triangles, the list specifies each triangle and the (three) nodes that are their vertices.

Introduction to Numerical Electrostatics Using MATLAB, First Edition. Lawrence N. Dworsky.
© 2014 John Wiley & Sons, Inc. Published 2014 by John Wiley & Sons, Inc.

3   *A List of Boundary Conditions.* This list specifies which nodes are fixed or along planes of symmetry and, in the former case, the voltages at which these nodes are fixed.

There is no absolute "best way" to organize these lists. Lists 1 and 3 can be combined or kept separate.

Other than for very simple demonstration structures, FEM models are generated by computer. The computer software chosen to accomplish this task guides the design of the structure of the data file(s), as we will see shortly.

When calculating capacitance of an unstructured model, the total charge calculation is rarely used. This is due to the difficulty of drawing a satisfactory Gaussian surface. On the other hand, the stored energy calculation is very easy to implement.

Before looking at data structures, we need to derive the coefficient matrix entries for an arbitrary triangle.

13.2   ASIDE: THE AREA OF A TRIANGLE

This formula will be useful going forward. Figure 13.1 shows an arbitrary triangle (*T*) in the *X*–*Y* plane, inscribed in a rectangle.

The rectangle's sides are parallel to the *X* and *Y* axes, and the vertices of triangle *T* is as shown. The area of *T* is simply the area of the rectangle minus the areas of right triangles *A*, *B*, and *C*:

$$A_T = (x_2 - x_1)(y_3 - y_1) - \tfrac{1}{2}[(x_2 - x_1)(y_2 - y_1) + (x_2 - x_3)(y_3 - y_2) + (x_3 - x_1)(y_3 - y_1)] \quad (13.1)$$

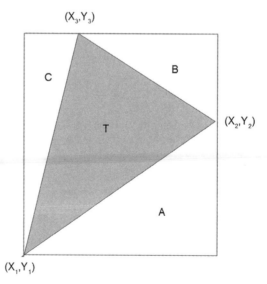

FIGURE 13.1   Arbitrary triangle inscribed in a rectangle for area calculation.

After some manipulation, We obtain

$$A_T = \frac{1}{2} \begin{vmatrix} 1 & x_1 & y_1 \\ 1 & x_2 & y_2 \\ 1 & x_3 & y_3 \end{vmatrix} \tag{13.2}$$

The choice of which vertex is vertex 1, and so on is arbitrary, provided that the vertices are numbered counterclockwise. If they are numbered clockwise, the equation (13.2) calculates the negative of the area.

We will adopt the convention that all element vertices are numbered counterclockwise.

13.3 THE COEFFICIENT MATRIX

Figure 13.2 shows the same triangle as in Figure 13.1, but without the surrounding rectangle.

The three vertices are set to voltages V_1, V_2, and V_3, as shown. The voltage anywhere in the triangle is given by the function

$$V(x,y) = V_1 f_1(x,y) + V_1 f_2(x,y) + V_3 f_3(x,y) \tag{13.3}$$

where the f_i are the basis functions

$$\begin{aligned} f_1 &= a_1 + b_1 x + c_1 y \\ f_2 &= a_2 + b_2 x + c_2 y \\ f_3 &= a_3 + b_3 x + c_3 y \end{aligned} \tag{13.4}$$

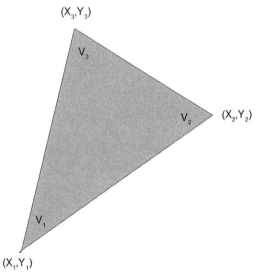

FIGURE 13.2 General triangle for analysis.

and a_i, b_i, and c_i are determined by the three sets of conditions

$$\begin{bmatrix} 1 & x_1 & y_1 \\ 1 & x_2 & y_2 \\ 1 & x_3 & y_3 \end{bmatrix} \begin{bmatrix} a_1 \\ b_1 \\ c_1 \end{bmatrix} = \begin{bmatrix} 1 \\ 0 \\ 0 \end{bmatrix} \tag{13.5}$$

$$\begin{bmatrix} 1 & x_1 & y_1 \\ 1 & x_2 & y_2 \\ 1 & x_3 & y_3 \end{bmatrix} \begin{bmatrix} a_2 \\ b_2 \\ c_2 \end{bmatrix} = \begin{bmatrix} 0 \\ 1 \\ 0 \end{bmatrix} \tag{13.6}$$

$$\begin{bmatrix} 1 & x_1 & y_1 \\ 1 & x_2 & y_2 \\ 1 & x_3 & y_3 \end{bmatrix} \begin{bmatrix} a_3 \\ b_3 \\ c_3 \end{bmatrix} = \begin{bmatrix} 0 \\ 0 \\ 1 \end{bmatrix} \tag{13.7}$$

The electric field components are

$$-E_x = \frac{\partial V}{\partial x} = V_1 \frac{\partial f_1}{\partial x} + V_2 \frac{\partial f_2}{\partial x} + V_3 \frac{\partial f_3}{\partial x} = V_1 b_1 + V_2 b_2 + V_3 b_3$$

$$-E_y = \frac{\partial V}{\partial y} = V_1 \frac{\partial f_1}{\partial y} + V_2 \frac{\partial f_2}{\partial y} + V_3 \frac{\partial f_3}{\partial y} = V_1 c_1 + V_2 c_2 + V_3 c_3 \tag{13.8}$$

and then

$$E^2 = E_x^2 + E_y^2 = (V_1 b_1 + V_2 b_2 + V_3 b_3)^2 + (V_1 c_1 + V_2 c_2 + V_3 c_3)^2 \tag{13.9}$$

Since the energy is not a function of x or y (in this approximation), we obtain

$$U = \frac{\varepsilon}{2} \iint_T E^2 \, dx \, dy = E^2 A_T \tag{13.10}$$

where A_T is as given by equation (13.2).

The terms that must be added to the coefficient matrix (also called *assembling* the matrix) are obtained by setting the derivatives of equation (13.10), using (13.9), to zero:

$$0 = \frac{\partial E^2}{\partial V_1} = A_T \left[V_1 (b_1^2 + c_1^2) + V_2 (b_1 b_2 + c_1 c_2) + V_3 (b_1 b_3 + c_1 c_3) \right]$$

$$0 = \frac{\partial E^2}{\partial V_2} = A_T \left[V_1 (b_1 b_2 + c_1 c_2) + V_2 (b_2^2 + c_2^2) + V_3 (b_2 b_3 + c_2 c_3) \right] \tag{13.11}$$

$$0 = \frac{\partial E^2}{\partial V_3} = A_T \left[V_1 (b_1 b_3 + c_1 c_3) + V_2 (b_2 b_3 + c_2 c_3) + V_3 (b_3^2 + c_3^2) \right]$$

Comparing equations (13.10) and (13.11), it appears that the triangle area A_T can drop out because of the 0 on the left side of equations (13.11). Remember, however, that here we are looking at only one triangle of an entire structure. We will sum the energy contributions from all triangles and then take derivatives and set them equal to zero. Except in the (possible but unlikely) case that all the triangles' areas are the same, the area terms will not drop out.

The coefficient matrix for an actual n node structure will have n rows, each row corresponding to a node. If the three nodes of the structure corresponding to the three nodes of the triangle T (Figure 13.2) are n_i for the node at (x_i, y_i), then equations (3.11) become

$$0 = \frac{\partial E^2}{\partial V_{n1}} = A_T \left[V_{n1} \left(b_1^2 + c_1^2 \right) + V_{n2} (b_1 b_2 + c_1 c_2) + V_{n3} (b_1 b_3 + c_1 c_3) \right]$$

$$= \frac{\partial E^2}{\partial V_{n2}} = A_T \left[V_{n1} (b_1 b_2 + c_1 c_2) + V_{n2} \left(b_2^2 + c_2^2 \right) + V_{n3} (b_2 b_3 + c_2 c_3) \right] \qquad (13.12)$$

$$= \frac{\partial E^2}{\partial V_{n3}} = A_T \left[V_{n1} (b_1 b_3 + c_1 c_3) + V_{n2} (b_2 b_3 + c_2 c_3) + V_{n3} \left(b_3^2 + c_3^2 \right) \right]$$

As an example, suppose that we have an eight-node system. The coefficient matrix a, is an 8×8 square matrix. Assume that we examine our node list and find that, for triangle 7, n_1 is node 1, n_2 is node 3, and n_3 is node 4. We add the terms from equation (13.12) to the coefficient matrix as equal to

$$a_{new} = a_{old} + \begin{bmatrix} A_7 \left(b_1^2 + c_1^2 \right) & 0 & A_7 (b_1 b_2 + c_1 c_2) & A_7 (b_1 b_3 + c_1 c_3) & 0 & 0 & 0 & 0 \\ 0 & 0 & 0 & 0 & 0 & 0 & 0 & 0 \\ A_7 (b_1 b_2 + c_1 c_2) & 0 & A_7 \left(b_2^2 + c_2^2 \right) & A_7 (b_2 b_3 + c_2 c_3) & 0 & 0 & 0 & 0 \\ A_7 (b_1 b_3 + c_1 c_3) & 0 & A_7 (b_2 b_3 + c_2 c_3) & A_7 \left(b_3^2 + c_3^2 \right) & 0 & 0 & 0 & 0 \\ 0 & 0 & 0 & 0 & 0 & 0 & 0 & 0 \\ 0 & 0 & 0 & 0 & 0 & 0 & 0 & 0 \\ 0 & 0 & 0 & 0 & 0 & 0 & 0 & 0 \\ 0 & 0 & 0 & 0 & 0 & 0 & 0 & 0 \end{bmatrix} \qquad (13.13)$$

After this has been done for all of the triangles, it is necessary only to bring in the boundary conditions (in exactly the same way as was done for FD equations) and then solve the system for the node voltages.

Once the node voltages have been found, the voltage everywhere is known through equations (13.3)–(13.7), the electric field is known through equation (13.8) (using the coefficients found above). The energy stored in the electric field of each triangle is known through equations (13.10) and (13.9) (again, using the coefficients found above), and the total energy in the system is found by summing all of the individual triangles' energies.

13.4 A SIMPLE EXAMPLE

Figure 13.3 shows a very simple triangular mesh (of a triangular structure). This structure is of no particular interest for its electrostatic properties; its use is for illustrating the FEM file information and matrix assembly technique while being as simple as possible. It does have all of the characteristics of more complicated (and more interesting) structures. Examining Figure 13.3, we note that

1 The structure is made up of nine triangles with seven common vertices, or nodes. The nodes are numbered using large numbers; the triangles are numbered using smaller, encircled, numbers.

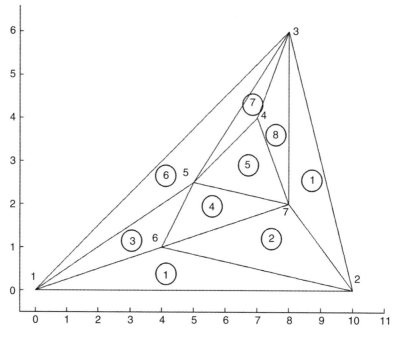

FIGURE 13.3   A very simple triangular mesh structure.

2   Each node has a number. The choice of node number (i.e., which node has which number) is arbitrary as long as each node is represented by a unique number, starting at 1, and no numbers are skipped.

3   There is a closed outer boundary. These are the lines connecting nodes 1–2, 2–3, and 3–1.

4   The nodes on the outer boundary will be set to a boundary condition (typically 0 V).

5   There are four internal nodes. At least one of these will be set to a boundary condition other than 0 V. The remaining node voltages will be the unknown variables that we wish to find. We can, of course, set all of the internal nodes to a boundary condition, but then the voltage distribution is fixed and there are no voltages to find. We can still find the stored energy if it is of interest.

6   All of the triangle locations are arbitrary so long as no triangle crosses over another triangle and no vertex of any triangle sits along a leg of another triangle.

7   Each triangle has a unique number. The numbering choice is arbitrary, with the same constraints as the node numbering. There is, in general, no formula relating the number of triangles to the number of nodes or linking a triangle's number to the node numbers of that triangle.

Our first task is to gather all the information about this structure, along with the boundary condition information, and put it in a form that's easily readable by a computer program. Since we want to write programs that are reusable for multiple structures, this means that we want to create data files with all of the necessary information. There is no absolute best algorithm for doing this. The choice presented here is to create three files to carry the node location information, the triangle descriptions, and the boundary conditions.

TABLE 13.1   Node Information

Node	X	Y
1	0	0
2	10	0
3	8	6
4	7	4
5	5	2.5
6	4	1
7	8	2

TABLE 13.2   Triangle Information

Triangle	Nodes		
1	1	2	6
2	6	2	7
3	1	6	5
4	6	7	5
5	5	7	4
6	5	3	1
7	5	4	3
8	4	7	3
9	7	2	3

TABLE 13.3   Boundary Condition Information

Ignore	Node	bc
1	1	0
1	2	0
1	3	0
1	5	1
1	6	1

Table 13.1 is a listing of the node information of this structure, `nodes_debug.txt`. The first column (italicized) is not actually part of the file. It is simply a counting list (row number) of the nodes. The computer program gets this information automatically as it reads the file; there is no need to explicitly show the counting in the data file. Columns 2 and 3, the actual data file, show the X and Y locations of each node.

Table 13.2 is a listing of the triangle information for this structure, `triangles_debug.txt`.

Again, the first (italicized) column is not actually part of the file; it is simply a counting list (indicating row number of the triangles). Columns 2–4 list the three nodes that are the vertices of the triangles. The order of the node numbering for each triangle is arbitrary as long as they are ordered counterclockwise for each triangle.

Table 13.3 is a listing of the boundary condition (bc) information file `bcs_debug.txt`. The format of this file is, at this point, not totally logical. Column 1 (italicized) conveys no necessary information; it is included for forward compatibility with the automatic meshing capability to be introduced in Chapter 14. For the moment, simply ignore column 1.

Columns 2 and 3 comprise a list of the nodes that are to be boundary conditions and the voltage to which these nodes are set, respectively. Referring to Figure 13.3, the three outer boundary nodes (1, 2, and 3) are set to 0 V and two of the internal nodes (5 and 6) are set to 1 V.

13.5 A TWO-DIMENSIONAL TRIANGULAR MESH PROGRAM

The MATLAB program `fem_2d_1.m` reads the data files described above, assembles the matrices, finds the node voltages and the structure's capacitance, and produces some useful graphics. It was written "looking forward" to provide compatibility with the more complicated problems presented in Chapter 14, so some pieces of the code are at this point unnecessary; this will be pointed out as needed:

- Program `fem_2d_1.m`—2d triangular mesh FEM program:

```
% 2d FEM with triangles

close all

filename = input ('Generic name for input files: ', 's');
nodes_file_name = strcat ('nodes_', filename, '.txt');
trs_file_name = strcat ('trs_', filename, '.txt');
bcs_file_name = strcat ('bcs_', filename, '.txt');

% get node data
dataArray = load (nodes_file_name);
nds.x = dataArray(:, 1); nds.y = dataArray(:, 2); nds.z = 0;
nr_nodes = length(nds.x)

% get triangles data
trs = (load(trs_file_name))';
nr_trs = length(trs)

% get bcs data
bcs = get_bcs(bcs_file_name);
dataArray = load(bcs_file_name);
bcs.node = dataArray(:,2); bcs.volts = dataArray(:,3);
nr_bcs = length(bcs.node)

% allocate the matrices
a = sparse(nr_nodes, nr_nodes);
f = zeros(nr_nodes,1);
b = zeros(3,nr_nodes); c = zeros(3,nr_nodes);

% generate the matrix terms
d = ones(3,3);
for tri = 1: nr_trs % Walk through all the triangles
  n1 = trs(1,tri); n2 = trs(2,tri); n3 = trs(3,tri);
  d(1,2) = nds.x(n1); d(1,3) = nds.y(n1);
  d(2,2) = nds.x(n2); d(2,3) = nds.y(n2);
  d(3,2) = nds.x(n3); d(3,3) = nds.y(n3);
```

```
  Area = det(d)/2;

  temp = d\[1; 0; 0]; b(1,tri) = temp(2); c(1,tri) = temp(3);
  temp = d\[0; 1; 0]; b(2,tri) = temp(2); c(2,tri) = temp(3);
  temp = d\[0; 0; 1]; b(3,tri) = temp(2); c(3,tri) = temp(3);

  a(n1,n1) = a(n1,n1) + Area*(a(n1,n1) + b(1,tri)^2 + c(1,tri)^2);
  a(n1,n2) = a(n1,n2) + Area*(a(n1,n2) + b(1,tri)*b(2,tri) +
c(1,tri)*c(2,tri));
  a(n1,n3) = a(n1,n3) + Area*(a(n1,n3) + b(1,tri)*b(3,tri) +
c(1,tri)*c(3,tri));

  a(n2,n2) = a(n2,n2) + Area*(a(n2,n2) + b(2,tri)^2 + c(2,tri)^2);
  a(n2,n1) = a(n2,n1) + Area*(a(n2,n1) + b(2,tri)*b(1,tri) +
c(2,tri)*c(1,tri));
  a(n2,n3) = a(n2,n3) + Area*(a(n2,n3) + b(2,tri)*b(3,tri) +
c(2,tri)*c(3,tri));

  a(n3,n3) = a(n3,n3) + Area*(a(n3,n3) + b(3,tri)^2 + c(3,tri)^2);
  a(n3,n1) = a(n3,n1) + Area*(a(n3,n1) + b(3,tri)*b(1,tri) +
c(3,tri)*c(1,tri));
  a(n3,n2) = a(n3,n2) + Area*(a(n3,n2) + b(3,tri)*b(2,tri) +
c(3,tri)*c(2,tri));

end

% bcs
for i = 1: nr_bcs
  n = bcs.node(i);
  a(n,:) = 0; a(n,n) = 1; f(n) = bcs.volts(i);
end

% Non-mesh nodes: Set up phony voltage just for matrix form
for i = 1 : nr_nodes
  if a(i,i) == 0
    a(i,:) = 0; a(i,i) = 1; f(i) = 0;
  end
end

v = a\f;

% calculate capacitance
C = get_tri2d_cap(v, nr_nodes, nr_trs, nds, trs)

% plot the mesh
show_mesh(nr_nodes, nr_trs, nr_bcs, nds, trs, bcs, v)
```

- Program get_tri2d_cap.m:

```
function C = get_tri2d_cap(v, nr_nodes, nr_trs, nds, trs)
% Calculate the capacitance of a 2d triangular structure
% Simple linear approx. function is assumed
```

```
eps0 = 8.854;

U = 0;
d = ones(3,3);
for tri = 1 : nr_trs
  n1 = trs(1,tri); n2 = trs(2,tri); n3 = trs(3,tri);
  d(1,2) = nds.x(n1); d(1,3) = nds.y(n1);
  d(2,2) = nds.x(n2); d(2,3) = nds.y(n2);
  d(3,2) = nds.x(n3); d(3,3) = nds.y(n3);
  temp = (d\[1; 0; 0]); b1 = temp(2); c1 = temp(3);
  temp = (d\[0; 1; 0]); b2 = temp(2); c2 = temp(3);
  temp = (d\[0; 0; 1]); b3 = temp(2); c3 = temp(3);

  Area = det(d)/2;
  if Area < = 0, disp 'Area < = 0 error'; end;

  U = U + Area*((v(n1)*b1 + v(n2)*b2 + v(n3)*b3)^2 ...
       +       (v(n1)*c1 + v(n2)*c2 + v(n3)*c3)^2);
end

U = U*eps0/2;
C = 2*U; % Potential difference of 1 volt assumed

end
```

- Program `show_mesh.m`:

```
function show_mesh(nr_nodes, nr_trs, nr_bcs, nds, trs, bcs, v)
% Take a look at the input data and the solution

  x_min = min(nds.x)*1.1;
  if x_min == 0, x_min = -.5; end;
  x_max = max(nds.x)*1.1;
  y_min = min(nds.y)*1.1;
  if y_min == 0, y_min = -.5; end;
  y_max = max(nds.y)*1.1;
  figure(1)
  axis([x_min, x_max, y_min, y_max]);
  hold on

  for i = 1 : nr_trs
    n1 = trs(1,i); n2 = trs(2,i); n3 = trs(3,i);
    x = [nds.x(n1), nds.x(n2), nds.x(n3), nds.x(n1)];
    y = [nds.y(n1), nds.y(n2), nds.y(n3), nds.y(n1)];
    plot(x,y,'k')
  end

  hold on

  for i = 1 : nr_bcs
    n = bcs.node(i);
    x = nds.x(n); y = nds.y(n);
    text(x, y, num2str(bcs.volts(i)))
  end
```

```
figure(2)
axis([x_min, x_max, y_min, y_max]);

for i = 1 : nr_trs
  n1 = trs(1,i); n2 = trs(2,i); n3 = trs(3,i);
  x = [nds.x(n1); nds.x(n2); nds.x(n3)];
  y = [nds.y(n1); nds.y(n2); nds.y(n3)];
  z = [v(n1); v(n2); v(n3)];
  colormap(gray)   % Delete this line for color graphics
  fill(x,y,z)
  hold on
end

end
```

The program begins by requesting the name for the data files, in this case debug. It then reads the three data files described in Section 13.4.

Matrix assembly and equation solution follow the analysis directly. The boundary conditions are inserted in exactly the same manner as was done in the FD analysis.

The lines beginning with the comment (fake bcs ...) do nothing in this example. The purpose of these lines of code will be explained in Chapter 14.

The capacitance of the structure is calculated by the function get_tri2d_cap.m. The stored energy is calculated directly from the equations above.

Function show_mesh.m produces two graphs. The first of these, Figure 13.4, is a sketch of the structure as understood by the program. This is useful for verifying that

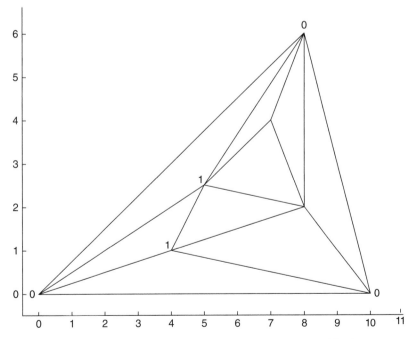

FIGURE 13.4 Sketch of structure and boundary conditions as understood by the computer.

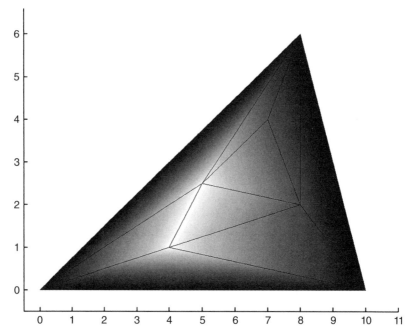

FIGURE 13.5 Sketch of structure and grayscale map of voltage profile.

the data files are correct. Also, the function puts the voltage boundary conditions at the appropriate nodes. Again, this is useful for verifying that the problem is being correctly set up.

The second graph produced is again a sketch of the structure, but this time the voltage profile is interpreted as shades of gray or color. Deleting line noted in the function show_mesh.m changes the MATLAB color map from gray shades that were needed for black-and-white Figure 13.5 (black = 0 V, white = 1 V) to more visually interesting colors, which vary from blue = 0 V to red = 1 V.

Inspection of these figures should verify the boundary condition voltages and provide a visual image of the voltage profile that both passes a commonsense test and gives some insight (when dealing with real structures) of the voltages involved.

PROBLEMS

13.1 Write a simple MATLAB script to create the data files for the structure shown in Figure P13.1 (or use a spreadsheet, do it manually, etc.). Run fem1_2d.m and verify that the data files are correct by inspecting the graphics and the resulting capacitance. Does the voltage profile generated look reasonable?

13.2 Modify the data files produced for Problem 13.1 to create the structure shown in Figure P13.4. Comment on the results of using this data file in Prob13_1.m (as compared to the previous results).

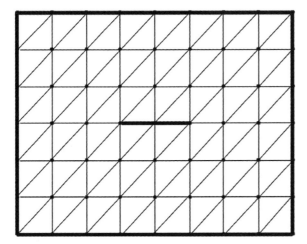

FIGURE P13.1   Structure for Problem 13.1.

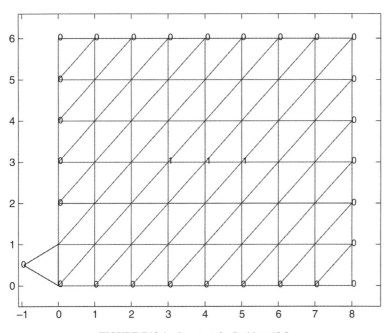

FIGURE P13.4   Structure for Problem 13.2.

13.3 Modify the data files produced for Problem 13.2 to create the structure shown in Figure P13.5.

13.4 Program `fem2d_1.m` will provide the basis of the working programs for the next few chapters. Readers intending to use the programs to solve real problems, or as the bases of programs to solve real problems, and who are familiar with vectorizing MATLAB program techniques, should consider vectorizing `fem2d_1.m` as a good problem, or project.

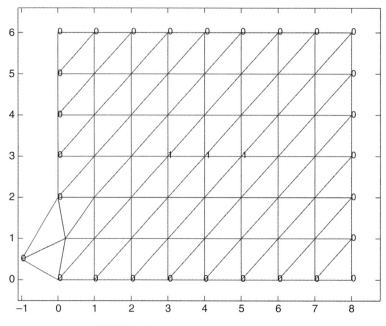

FIGURE P13.5 Structure for Problem 13.3.

REFERENCE

1. O. C. Zienkiewicz and R. L. Taylor, *The Finite Element Method*, 6th ed., Elsevier Butterworth-Heinemann, Burlington, MA, 2005.

14

A Zoning System and Some Examples

14.1 GENERAL INTRODUCTION

The program `fem_2d_1.m` presented in Chapter 13 is a general two-dimensional (2d) FEM program. It can handle sophisticated geometries. On the other hand, starting with a structure description and creating the *zoning* (also called the *meshing* or the *tessellation*), which generates the input files for this program, is a complex problem in its own right. Although generating an unstructured grid for a complex geometry by hand is not literally impossible, it is so difficult and the results would probably be so erroneous, that it would be a pointless quest. Computer-aided mesh generation is necessary.

Fortunately, for both academic and commercial purposes, many software packages have been developed. Some relatively simple packages were presented in the MoM unstructured grid discussed earlier. As was also discussed earlier, an in depth analysis or development of one of these programs is outside the scope of this book. On the other hand, we can't proceed with FEM examples or study the electrostatic properties of structures that FEM modeling would enable without a meshing system.

In this chapter we'll introduce the `gmsh` package; `gmsh` is a free three-dimensional (3d) FEM grid generator that may be downloaded along with very extensive documentation and tutorials (`hhtp://geuz.org/gmsh/`).[1] It is distributed under the GNU General Public License.[2] In the introduction to `gmsh` that follows, only a very small subset of `gmsh` capabilities is presented. The program(s) shown that read `gmsh` output files and parse the data for use by our MATLAB FEM programs were written only to handle the uses of `gmsh` discussed here. In other words, if you choose to extend your understanding and use of `gmsh` beyond what is presented, be prepared to extend the `gmsh` file parsing programs and possibly the FEM analysis programs to include these previously unconsidered capabilities.

Introduction to Numerical Electrostatics Using MATLAB, First Edition. Lawrence N. Dworsky.
© 2014 John Wiley & Sons, Inc. Published 2014 by John Wiley & Sons, Inc.

14.2 INTRODUCTION TO gmsh

The gmsh program is available in Windows, Macintosh, or Linux formats at http://geuz.org/gmsh/. The gmsh installation includes a full documentation manual. This manual, in turn, includes a full tutorial. There are also many other tutorials and examples available online.

A gmsh structure description may be constructed and modified interactively, by directly editing a data file, or by a combination of both of these techniques. The procedures for these approaches are described in the gmsh tutorials. In this chapter only data file input will be considered. This is not limiting in any way — all lines entered into a file using a text editor become part of the structure, and all information added interactively to the structure become lines in the file.

Starting gmsh, moving and sizing the windows as necessary and bringing up a drawing description file (xx.geo) in a text editor should be learned from the documentation/and/or tutorials. Any American Standard Code for Information Interchange (ASCII) text editor will work for editing gmsh files. An editor that displays line numbers is preferable for these discussions because referring to line numbers is an excellent and easy way to discuss the file's contents. The struct14_1.geo file — a simple gmsh rectangular plane description — is as follows:

```
// Gmsh structure file struct14_1.geo

lc = 1.;

// create a rectangular surface loop

Point(1) = {-2, -1, 0, lc};
Point(2) = {2, -1, 0, lc};
Point(3) = {2, 1, 0, lc};
Point(4) = {-2, 1, 0, lc};

Line(1) = {1, 2};
Line(2) = {2, 3};
Line(3) = {3, 4};
Line(4) = {4, 1};

Line Loop(3) = {1, 2, 3, 4};
Plane Surface(4) = {3} ;
```

The file struct14_1.geo displays the basic syntax of gmsh operation. Line 1 is a comment line. The // in a line denotes that everything following on that line is a comment; as in most programming languages, comments are ignored by the program.

Lines 2 (and lines 4, 6, etc.) are blank lines that are ignored by the program. Their purpose is to make groupings of content lines easier for a reader's eye to follow.

Line 3 defines the variable lc and sets it equal to one. (lc means *characteristic length* and will be discussed further below). The actual name of the variable is arbitrary.

Lines 7–10 each define a point. Each point has a unique number (1–4 in this example). The choice of numbers is arbitrary insofar as gmsh is concerned. Certain numbering conventions will be defined below to pass information to the MATLAB functions. (This is not part of gmsh or MATLAB, but rather "invented" conventions for our programs.) Each point has three coordinates and a characteristic length. In this example all four lines have the same characteristic length. This is not a necessary restriction — other variables could have been defined, and/or values have been entered directly into each point's description.

Lines 12–15 each define a line connecting two points. Again, line numbers are unique arbitrary choices. Note that the line numbers are independent of the point numbers. The order of the points in each line definition is arbitrary, but will be relevant when lines are connected.

Line 17 defines a *line loop*, namely, a polygon, in this case a closed rectangle. If the numbering of the line segments is taken as {tail, head}, then the head of one segment should connect to the tail of the next in the loop. If line 2 is defined as {3, 2} instead of {2, 3}, then line 17 would have to be modified to {1, −2, 3, 4} to preserve the definition of the simple rectangle.

Line 18 defines a plane surface as the area enclosed by line loop 3.

Figure 14.1 shows the structure described by struct14_1.geo. This is the structure that should be visible in the gmsh display window. When the gmsh display window is the active window, hovering the cursor over one of the corners of the rectangle (a defined point) will cause the point number and its vertices to be listed at the bottom of the window. Similarly, hovering the cursor over one of the lines will produce a listing of the line number and its defining points. The dashed lines inside the rectangle reflect the fact that this is a defined plane, and hovering the cursor over a dashed line will produce the plane number and its defining lines.

If the upper choice box in the gmsh control window is switched to mesh (it should have been at its opening choice of geometry) and the 2D choice is selected, the rectangle is meshed into triangles (Figure 14.2).

Since the boundaries of a rectangle are straight lines, no compromises had to be made – the surface is entirely filled with triangles. If we want higher-resolution meshing, all we have to do is reduce the value of lc and repeat the meshing operation.

Figure 14.3 shows the meshing for lc reduced to 0.25.

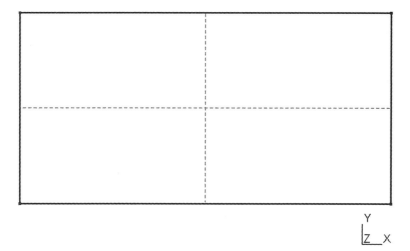

FIGURE 14.1   The gmsh structure described by struct14_1.geo.

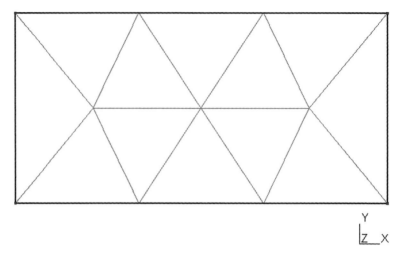

FIGURE 14.2   struct14_1.geo showing 2d meshing (lc = 1).

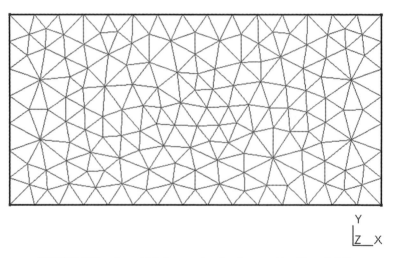

FIGURE 14.3   struct14_1.geo showing finer meshing (lc = 0.25).

The use of one or more characteristic lengths is now apparent. This number determines how finely gmsh will subdivide a surface into a mesh. Also, note that now there is no regularity to the location of the nodes. This is an unstructured meshing; there is no simple formula that can calculate a node number from the node's location, or vice versa.

Here is gmsh file struct14_2.geo — a gmsh transmission line cross sectional structure:

```
// Gmsh structure struct14_2.geo

lc = .5;
// exterior loop
Point(1) = {-2, -1, 0, lc};
```

```
Point(2) = {2, -1, 0, lc};
Point(3) = {2, 1, 0, lc};
Point(4) = {-2, 1, 0, lc};

Line(1) = {1, 2};
Line(2) = {2, 3};
Line(3) = {3, 4};
Line(4) = {4, 1};

// interior surface which will become a hole
Point(5) = {-1, -.05, 0, lc};
Point(6) = {1, -.05, 0, lc};
Point(7) = {1, .05, 0, lc};
Point(8) = {-1, .05, 0, lc};

Line(105) = {5, 6};
Line(106) = {6, 7};
Line(107) = {7, 8};
Line(108) = {8, 5};

Line Loop(1) = {105, 106, 107, 108};
Plane Surface(200) = {1};

// exterior surface
Line Loop(3) = {1, 2, 3, 4};
Plane Surface(4) = {3, 1} ;
```

Program `struct14_2.geo` begins as a repeat of `struct14_1.geo`, up through line 13. The line loop and plane surface commands of `struct14_1.geo`, however, have been removed.

Then `struct14_2.geo` defines four new points (lines 16–19), and then four new lines (lines 21–24), which form a small rectangle, inside the first rectangle. Lines 26 and 27 then create a line loop and define a plane surface.

Line 30 defines line loop 3 for the first rectangle. It could have been left where it was originally, but was moved to keep it with it corresponding plane surface command, line 31. This plane surface command has two arguments: {3, 1}. The first number, 3, refers to line loop 3. This is the outer rectangle's surface area. The second number, 1 (and any other numbers, if they are present), refers to line loops that are internal to line loop 3, and are "holes" in that surface.

Figure 14.4 shows the structure that has been defined. It is the structure of the cross section of a rectangular transmission line with a rectangular center conductor.

The lines of the inner rectangle are numbered by integers between 100 and 199. The plane surface defined by this line loop is numbered by an integer between 200 and 299. These choices are not gmsh-based. They are chosen, as will be seen in Section 14.3, to facilitate using this structural definition as an electrostatics description — that is, as an easy way to tell upcoming software how to identify an inner conductor (inside the outer boundary).

Figure 14.5 shows the result of the 2d meshing of `struct14_2.geo`.

This figure shows that gmsh is smart enough to understand that the characteristic length chosen for all the line segments is too great for the short sides of the inner rectangle.

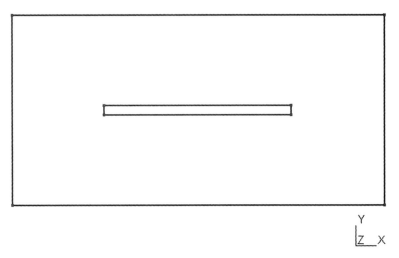

FIGURE 14.4   Structure defined by struct14_2.geo.

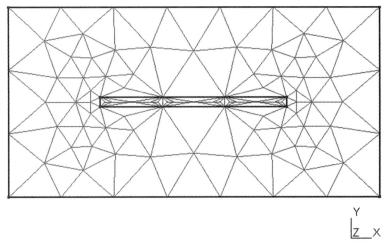

FIGURE 14.5   Two-dimensional meshing of struct14_2.geo.

It adjusts triangle sizes automatically and produces a mesh that is very usable for electrostatic modeling. The overall characteristic length could, of course, always be reduced and/or individual regions tweaked as desired.

The last thing we need to do with this structure in gmsh is to choose Save in the mesh dropdown list gmsh will create the file struct14_2.msh.

14.3   TRANSLATING THE gmsh.msh FILE

The file struct14_2.msh is an ASCII text file. While computers will have no trouble reading it, it is a difficult file to read by eye; floating-point numbers are 15 digits long and a single space separates columns. The file struct14_2_edited.msh.txt is, as its name implies, an edited version of `struct14_2.msh`; the floating-point numbers are rounded to three digits and the columns have tab separators. Here is `struct14_2_edited.msh.txt`— the `struct14_2.msh` file reformatted for easier reading:

```
$  MeshFormat
2.2    0    8
$EndMeshFormat
$Nodes
99
1      -2        -1         0
2       2        -1         0
3       2         1         0
4      -2         1         0
5      -1       -0.05       0
6       1       -0.05       0
7       1        0.05       0
8      -1        0.05       0
9      -1.429              -1          0
10     -0.857   -1          0
11     -0.286   -1          0
12      0.286   -1          0
13      0.857   -1          0
14      1.429   -1          0
15      2       -0.333      0
16      2        0.333                  0
17      1.429    1          0
18      0.857    1          0
19      0.286    1          0
20     -0.286    1          0
21     -0.857    1          0
22     -1.429               1          0
23     -2        0.333      0
24     -2       -0.333      0
25     -0.333              -0.05       0
26      0.333   -0.05       0
27      0.333    0.05       0
28     -0.333    0.05       0
29     -1.599               0          0
30      1.599    0          0
31      1.316    0.474      0
32     -1.316   -0.474      0
33     -1.316    0.474      0
34      1.316   -0.474      0
35     -0.84    -0.547      0
36      0.84     0.547      0
37      0.84    -0.547      0
38     -0.84     0.547      0
39      1.269    0          0
40     -1.269    0          0
41     -1.185              -0.256      0
42      1.185               0.256      0
```

43	-1.185	0.256	0	
44	1.185	-0.256	0	
45	0.94	-0.27	0	
46	-0.94	0.27	0	
47	-0.94	-0.27	0	
48	0.94	0.27	0	
49	0	-0.536	0	
50	0	0.536	0	
51	1.109	0	0	
52	-1.109	0	0	
53	-1.609	-0.643	0	
54	1.609	0.643	0	
55	1.609	-0.643	0	
56	-1.609	0.643	0	
57	-0.462	0.576	0	
58	0.462	-0.576	0	
59	-0.462	-0.576	0	
60	0.462	0.576	0	
61	-1.469		-0.255	0
62	1.469	0.255	0	
63	-1.469	0.255	0	
64	1.469	-0.255		0
65	1.103	0.121	0	
66	-1.103	-0.121	0	
67	-1.103	0.121	0	
68	1.103	-0.121	0	
69	-1.09	-0.464	0	
70	1.09	0.464	0	
71	-1.09	0.464	0	
72	1.09	-0.464	0	
73	0.646	-0.295	0	
74	0.746	-0.15	0	
75	-0.646	0.295	0	
76	-0.746	0.15	0	
77	0.646	0.295	0	
78	0.746	0.15	0	
79	-0.646		-0.295	0
80	-0.746	-0.15	0	
81	-1.187		-0.718	0
82	1.187	0.718	0	
83	-1.187	0.718	0	
84	1.187	-0.718	0	
85	0.667	0	0	
86	-0.667	0	0	
87	0	0	0	
88	-0.211	0	0	
89	0.455	0	0	
90	-0.455	0	0	
91	0.878	0	0	
92	-0.878	0	0	
93	0.211	0	0	
94	-0.282	0	0	

```
95        -0.385        0          0
96         0.385        0          0
97         0.948        0          0
98        -0.948        0          0
99         0.282        0          0
$EndNodes
$Elements
212
1     15    2    0    1      1
2     15    2    0    2      2
3     15    2    0    3      3
4     15    2    0    4      4
5     15    2    0    5      5
6     15    2    0    6      6
7     15    2    0    7      7
8     15    2    0    8      8
9      1    2    0    1      1      9
10     1    2    0    1      9     10
11     1    2    0    1     10     11
12     1    2    0    1     11     12
13     1    2    0    1     12     13
14     1    2    0    1     13     14
15     1    2    0    1     14      2
16     1    2    0    2      2     15
17     1    2    0    2     15     16
18     1    2    0    2     16      3
19     1    2    0    3      3     17
20     1    2    0    3     17     18
21     1    2    0    3     18     19
22     1    2    0    3     19     20
23     1    2    0    3     20     21
24     1    2    0    3     21     22
25     1    2    0    3     22      4
26     1    2    0    4      4     23
27     1    2    0    4     23     24
28     1    2    0    4     24      1
29     1    2    0  105      5     25
30     1    2    0  105     25     26
31     1    2    0  105     26      6
32     1    2    0  106      6      7
33     1    2    0  107      7     27
34     1    2    0  107     27     28
35     1    2    0  107     28      8
36     1    2    0  108      8      5
37     2    2    0    4      7     65     48
38     2    2    0    4      5     66     47
39     2    2    0    4      8     46     67
40     2    2    0    4      6     45     68
41     2    2    0    4     31     62     54
42     2    2    0    4     32     61     53
43     2    2    0    4     33     56     63
44     2    2    0    4     34     55     64
```

45	2	2	0	4	15	30	64
46	2	2	0	4	23	29	63
47	2	2	0	4	16	62	30
48	2	2	0	4	24	61	29
49	2	2	0	4 6	51	7	
50	2	2	0	4 5	8	52	
51	2	2	0	4	15	64	55
52	2	2	0	4	23	63	56
53	2	2	0	4	16	54	62
54	2	2	0	4	24	53	61
55	2	2	0	4 8	76	46	
56	2	2	0	4 6	74	45	
57	2	2	0	4 7	48	78	
58	2	2	0	4 5	47	80	
59	2	2	0	4	13	37	58
60	2	2	0	4	21	38	57
61	2	2	0	4	18	60	36
62	2	2	0	4	10	59	35
63	2	2	0	4	23	24	29
64	2	2	0	4	15	16	30
65	2	2	0	4	32	53	81
66	2	2	0	4	31	54	82
67	2	2	0	4	33	83	56
68	2	2	0	4	34	84	55
69	2	2	0	4	39	42	65
70	2	2	0	4	40	41	66
71	2	2	0	4	40	67	43
72	2	2	0	4	39	68	44
73	2	2	0	4 1	53	24	
74	2	2	0	4 3	54	16	
75	2	2	0	4 4	23	56	
76	2	2	0	4 2	15	55	
77	2	2	0	4	12	13	58
78	2	2	0	4	20	21	57
79	2	2	0	4	18	19	60
80	2	2	0	4	10	11	59
81	2	2	0	4	11	12	49
82	2	2	0	4	19	20	50
83	2	2	0	4	17	18	82
84	2	2	0	4 9	10	81	
85	2	2	0	4	21	22	83
86	2	2	0	4	13	14	84
87	2	2	0	4	25	49	26
88	2	2	0	4	27	50	28
89	2	2	0	4	41	47	66
90	2	2	0	4	42	48	65
91	2	2	0	4	43	67	46
92	2	2	0	4	44	68	45
93	2	2	0	4	39	65	51
94	2	2	0	4	40	66	52
95	2	2	0	4	40	52	67
96	2	2	0	4	39	51	68

97	2	2	0	4	13	84	37
98	2	2	0	4	21	83	38
99	2	2	0	4	10	35	81
100	2	2	0	4	18	36	82
101	2	2	0	4	1	9	53
102	2	2	0	4	3	17	54
103	2	2	0	4	2	55	14
104	2	2	0	4	4	56	22
105	2	2	0	4	36	70	82
106	2	2	0	4	35	69	81
107	2	2	0	4	37	84	72
108	2	2	0	4	38	83	71
109	2	2	0	4	5	52	66
110	2	2	0	4	7	51	65
111	2	2	0	4	8	67	52
112	2	2	0	4	6	68	51
113	2	2	0	4	9	81	53
114	2	2	0	4	17	82	54
115	2	2	0	4	22	56	83
116	2	2	0	4	14	55	84
117	2	2	0	4	26	49	58
118	2	2	0	4	28	50	57
119	2	2	0	4	25	59	49
120	2	2	0	4	27	60	50
121	2	2	0	4	12	58	49
122	2	2	0	4	20	57	50
123	2	2	0	4	19	50	60
124	2	2	0	4	11	49	59
125	2	2	0	4	40	61	41
126	2	2	0	4	39	62	42
127	2	2	0	4	40	43	63
128	2	2	0	4	39	44	64
129	2	2	0	4	27	77	60
130	2	2	0	4	25	79	59
131	2	2	0	4	26	58	73
132	2	2	0	4	28	57	75
133	2	2	0	4	32	41	61
134	2	2	0	4	31	42	62
135	2	2	0	4	33	63	43
136	2	2	0	4	34	64	44
137	2	2	0	4	45	74	73
138	2	2	0	4	46	76	75
139	2	2	0	4	48	77	78
140	2	2	0	4	47	79	80
141	2	2	0	4	26	73	74
142	2	2	0	4	28	75	76
143	2	2	0	4	27	78	77
144	2	2	0	4	25	80	79
145	2	2	0	4	41	69	47
146	2	2	0	4	42	70	48
147	2	2	0	4	43	46	71
148	2	2	0	4	44	45	72

149	2	2	0	4	37	73	58
150	2	2	0	4	38	75	57
151	2	2	0	4	36	60	77
152	2	2	0	4	35	59	79
153	2	2	0	4	37	45	73
154	2	2	0	4	38	46	75
155	2	2	0	4	36	77	48
156	2	2	0	4	35	79	47
157	2	2	0	4	29	61	40
158	2	2	0	4	30	62	39
159	2	2	0	4	30	39	64
160	2	2	0	4	29	40	63
161	2	2	0	4	32	81	69
162	2	2	0	4	31	82	70
163	2	2	0	4	33	71	83
164	2	2	0	4	34	72	84
165	2	2	0	4	32	69	41
166	2	2	0	4	31	70	42
167	2	2	0	4	34	44	72
168	2	2	0	4	33	43	71
169	2	2	0	4	38	71	46
170	2	2	0	4	37	72	45
171	2	2	0	4	36	48	70
172	2	2	0	4	35	47	69
173	2	2	0	4	26	74	6
174	2	2	0	4	27	7	78
175	2	2	0	4	28	76	8
176	2	2	0	4	5	80	25
177	2	2	0	200	25	94	28
178	2	2	0	200	26	96	27
179	2	2	0	200	25	28	95
180	2	2	0	200	5	98	8
181	2	2	0	200	6	7	97
182	2	2	0	200	26	27	99
183	2	2	0	200	91	97	7
184	2	2	0	200	27	93	99
185	2	2	0	200	92	8	98
186	2	2	0	200	5	92	98
187	2	2	0	200	85	91	7
188	2	2	0	200	28	94	88
189	2	2	0	200	28	90	95
190	2	2	0	200	26	99	93
191	2	2	0	200	86	8	92
192	2	2	0	200	27	96	89
193	2	2	0	200	27	87	93
194	2	2	0	200	26	6	85
195	2	2	0	200	26	89	96
196	2	2	0	200	25	90	86
197	2	2	0	200	27	89	85
198	2	2	0	200	27	85	7
199	2	2	0	200	5	86	92
200	2	2	0	200	28	86	90

```
201   2   2   0   200    85    6    91
202   2   2   0   200    26   85    89
203   2   2   0   200    25   87    88
204   2   2   0   200    25   95    90
205   2   2   0   200    25   88    94
206   2   2   0   200    25   26    87
207   2   2   0   200    28   88    87
208   2   2   0   200    91    6    97
209   2   2   0   200    27   28    87
210   2   2   0   200    26   93    87
211   2   2   0   200    28    8    86
212   2   2   0   200     5   25    86
$EndElements
```

The program `parse_msh_file.m` translates the gmsh mesh file `struct14_2.
msh` to the three data files that `fem_2d_1.m` requires as discussed in the last chapter. File
`parse_msh_file.m` is not a general-purpose gmsh–to–MATLAB conversion program.
It handles only those gmsh capabilities discussed in these chapters. It performs limited
validity checking, and, while it prompts for an output file name, it does not check to see
whether that name is already in use.

If `parse_msh_file.m` succeeds, it returns the three files `nodes_filename.txt`,
`trs_filename.txt`, and `bcs_filename.txt`, where `filename` is a user-supplied
name. If the program fails for any reason, the three output files are generated, but they are
empty files.

File `parse_msh_file.m` parses `struct14_2.msh` using the following
algorithms:

1 Lines 10 and 11. Get the name of the `.msh` file to read; open the file (for reading).
2 Lines 13–31. Verify the header information.
3 Lines 33–59. Read node location information and store in nodes array.
4 Lines 60–91. Read triangle and boundary condition information and store in `trs` and
   `bcs` arrays. The Column 2 in the `.msh` file in this section indicates an element type:
   $15 = $ point, $1 = $ line, $2 = 3$ node triangle. Points entries are ignored. Triangle entries,
   except for excluded regions, are added to the `trs` array. Line entries are added to the
   `bcs` array.
5 Line 92. Close the input file
6 Lines 94–103. List the `bcs` file, prompt for lines to become 1-V boundary conditions
   (the default is 0 V).
7 Lines 105–117. Prompt for the output file name and write the files to disk.

Program `parse_msh_file.m` — MATLAB script for converting gmsh output for
FEM analysis usage — is as follows:

```
% parse_msh_file.m
% Read the mesh file generated by gmsh and parse it
% This is a very limited routine, based only on the
%     capabilities of gmsh that are being used

% Element types handled are 1 = line, 2 = 3 node triangle, 15 = point
```

```
clear

filename = uigetfile('*.msh');   % read the file
fid = fopen(filename, 'r');

tline = fgetl(fid);   % read the first file line
if (strcmp(tline, '$MeshFormat')) == 0   % Verify file type
  'Problem with this file'
  nodes = []; trs = []; bcs = [];
  fclose (fid);
  return
end

tline = fgetl(fid);   % look for version number
[header] = sscanf(tline, '%f %d %d');
if (header(1) ~ = 2.2), 'Possible gmsh version incompatibility', end;

tline = fgetl(fid);   % Look for $EndMeshFormat
if (strcmp(tline, '$EndMeshFormat')) == 0
  'No $EndMeshFormat Line found'
  nodes = []; trs = []; bcs = [];
  fclose (fid);
  return
end

tline = fgetl(fid);   % Look for $Nodes
if (strcmp(tline, '$Nodes')) == 0
  'No $Nodes Line found'
  nodes = []; trs = []; bcs = [];
  fclose (fid);
  return
end

% NOTE: Nodes is being read as a 3-d file even though only 2-d
%         usage is anticipated in this version

nr_nodes = fscanf(fid, '%d', 1);   % get number of nodes
nodes = [0,0,0]; % zeros(nr_nodes, 3);   % allocate nodes array
for i = 1 : nr_nodes   % get nodes information
  temp = fscanf(fid, '%g', 4);
  n = temp(1);
  nodes(n, :) = temp(2:4);
end
fgetl(fid);      % Clear the line

tline = fgetl(fid);      % Look for $EndNodes
if (strcmp(tline, '$EndNodes')) == 0
  'No $EndNodes Line found'
  nodes = []; trs = []; bcs = [];
  fclose (fid);
  return
end

tline = fgetl(fid);   % Look for $Elements
```

```
if (strcmp(tline, '$Elements')) == 0
  'No $Nodes Line found'
  nodes = []; trs = []; bcs = [];
  fclose (fid);
  return
end

trs = []; bcs = [];
nr_elements = fscanf(fid, '%d', 1);   % get number of elements
for i = 1: nr_elements                  % process elements information
  temp = fscanf(fid, '%d %d', 2);       % get element type
  if temp(2) == 15                    % defined point
    fgetl(fid);                         % clear the line
  elseif temp(2) == 1               % defined line
    temp2 = fscanf(fid, '%d %d %d %d %d', 5);
    bcs = [bcs; [temp2(3:4)', 0]];
    fgetl(fid);
  elseif temp(2) == 2               % created mesh triangle
    temp3 = fscanf(fid, '%d %d %d %d %d %d', 6);
    if temp3(3) < 200                % ignore excluded triangles
      trs = [trs; temp3(4:6)'];
    end
    fgetl(fid);
  else
    'Unknown type in mesh file'
    nodes = []; trs = []; bcs = [];
    fclose (fid);
    return
  end
end
fclose (fid);

nr_bc_lines = length(bcs);
while 1 == 1
  'bcs file: '
  bcs
  line_fix = input('Enter line # to set 1 volt bc, 0 to conclude: ');
  if line_fix == 0, break, end;
  for i = 1 : nr_bc_lines
    if bcs(i,1) == line_fix, bcs(i,3) = 1; end;
  end
end

filename = input('Name for output files, NO OVERWRITE CHECKING: ', 's');
nodes_file_name = strcat('nodes_', filename, '.txt');
trs_file_name = strcat('trs_', filename, '.txt');
bcs_file_name = strcat('bcs_', filename, '.txt');
fid = fopen (nodes_file_name, 'w');
fprintf (fid, '%g\t%g\t%g\r\n', nodes');
fclose (fid);
fid = fopen (trs_file_name, 'w');
fprintf (fid, '%d\t%d\t%d\r\n', trs');
fclose (fid);
fid = fopen (bcs_file_name, 'w');
fprintf (fid, '%d\t%d\t%d\r\n', bcs');
fclose (fid);
```

Item 6 in the list preceding this program deserves some discussion. The program first presents the boundary condition file as listed in the following table; only the last 12 lines of the file are shown in this table:

Line Number	Node Number	Boundary Condition
3	22	0
4	4	0
4	23	0
4	24	0
105	5	0
105	25	0
105	26	0
106	6	0
107	7	0
107	27	0
107	28	0
108	8	0

The first column in the table (i.e., in the file) is the line number from the original. geo file. Since the data being processed are of the fully meshed structure, there is typically more than one line section (triangle side) associated with each of these line numbers. The second column contains one of the nodes associated with the line segment. If we restrict ourselves to conductors having a closed surface and nonzero enclosed area, then only one node number is necessary. The last column shows the boundary condition associated with this node. The original, default, value as shown for all nodes is 0 V.

The program prompt at this point reads Enter line # to set 1 volt bc, 0 to conclude: The response to this prompt could be any of the line numbers shown in the table. The reason for numbering internal electrodes with line numbers > 100 is now apparent; these line segments appear at the end of the list and are easy to identify.

A response of 105 (with a carriage return) causes the file to be modified and reprinted as follows:

Line Number	Node Number	Boundary Condition
3	22	0
4	4	0
4	23	0
4	24	0
105	5	1
105	25	1
105	26	1
106	6	0
107	7	0
107	27	0
107	28	0
108	8	0

As this table shows, all nodes associated with the original line 105 have been set to 1 V. Repeating this procedure, entering 106, 107, and 108 (in any order) produces the desired file, as listed in the following table:

Line Number	Node Number	Boundary Condition
3	22	0
4	4	0
4	23	0
4	24	0
105	5	1
105	25	1
105	26	1
106	6	1
107	7	1
107	27	1
107	28	1
108	8	1

The entire inner conductor surface has now been set to 1 V and the file is ready for use. A response of 0 terminates this part of the program; the program then moves on to writing the newly generated data files to disk.

14.4   RUNNING THE FEM ANALYSIS

The data files are now ready for `fem_2d_1.m` (described and listed in Chapter 13). The program produces Figures 14.6 and 14.7.

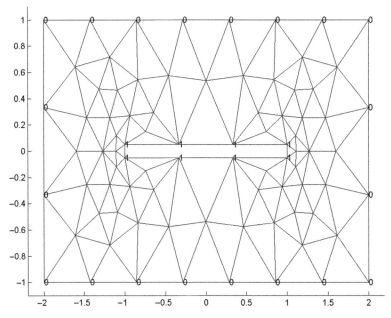

FIGURE 14.6   gmsh zoning of `struct14_2.geo` showing boundary conditions.

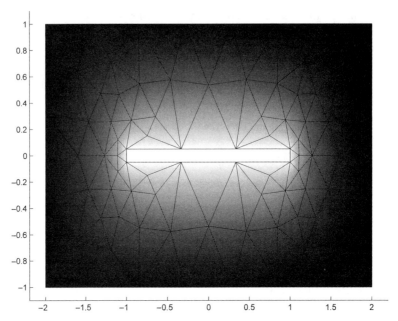

FIGURE 14.7   Graphical output of `fem_2d_1` showing voltage profile.

Repeating a note from Chapter 13, removing the line near the bottom of the `show_mesh.m` function `colormap (gray)` will produce an on-screen version of Figure 14.7 with color interpretation of the voltage distribution.

Program `fem_2d_1.m` also calculates the capacitance of the structure, $C = 58.97$ pF/m. An online calculator of stripline impedance[3] predicts $C = 55.6$ pF/m. Even with the sides of the box as close as they are to the center conductor, the FEM prediction is high.

Increasing the outer box sidewall positions to 3 m and decreasing the characteristic length (lc) to 0.1 reduces the capacitance error to $< 0.5\%$.

14.5   MORE `gmsh` FEATURES AND EXAMINING THE ELECTRIC FIELD

File `struct14_3.geo` illustrates several additional gmsh features, all of which require no modification of `parse_msh_file.m` or `fem_2d_1.m` for use:

```
// gmsh structure file struct14_3.geo

lc = .1;
// exterior loop
Point(1) = {-1.5, -1.5, 0, lc};
Point(2) = {1.5, -1.5, 0, lc};
Point(3) = {1.5, 1.5, 0, lc};
Point(4) = {-1.5, 1.5, 0, lc};

Line(1) = {1, 2};
Line(2) = {2, 3};
Line(3) = {3, 4};
Line(4) = {4, 1};
```

```
// interior surface which will become a hole
Point(5) = {-.5, -.5, 0, lc};
Point(6) = {.5, -.5, 0, lc};
Point(7) = {.5, 0, 0, lc};
Point(8) = {-.5, 0, 0, lc};
Point(9) = {0, 0, 0, lc};

Line(105) = {8, 5};
Line(106) = {5, 6};
Line(107) = {6, 7};

Circle(108) = {7, 9, 8};

// rotate the inner conductor about the z axis
a = 3.1416/4;
Rotate {{0,0,1},{0,0,0},a}{Point{5};Point{6};Point{7};Point{8};
Point{9};}

Line Loop(1) = {105, 106, 107, 108};
Plane Surface(200) = {1};

// exterior surface
Line Loop(3) = {1, 2, 3, 4};
Plane Surface(4) = {3, 1} ;
```

The structure produced by struct14_3.geo (with the boundary conditions tabulated and the meshing shown) is shown in Figure 14.8.

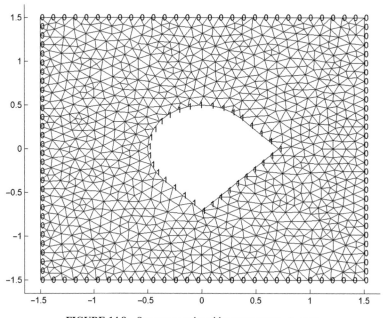

FIGURE 14.8   Structure produced by struct14_3.geo.

Referring to the `struct14_3.geo` listing, points 1–4 and lines 1–4 define an outer box. Points 5–8 define a rectangle, but lines 105–108 only connect three sides of this rectangle. Point 9 defines a fifth point that lies midway between points 7 and 8. Circle 108 defines the arc of a circle connecting points 7 and 9, with its center at point 8. The rotate instruction rotates points 5–9 about the Z axis by $45°$ ($\pi/4$ radians). This is an arbitrary rotation used only to show the capability.

There is also a command for translating a group of points. At first glance this seems unnecessary — why not just define the points at their final locations? Translation, however, is very valuable for facilitating parameter studies, such as moving the inner conductor about and studying the results on the structure's fields, capacitance, and other properties. Translation, together with an available duplication command, allows for placement of multiple complex electrode shapes in a structure at arbitrary locations. Combining all of these capabilities (translation, duplication, and rotation) allows for easy generation and manipulation of complex structures.

Returning to Figure 14.8, the boundary conditions are 0 V on the outer (square) boundary and 1 V on the inner conductor. The former boundary condition is set by default; the latter is exactly the same as in the previous examples. Following the prompts in `parse_msh_file.m`, line 105–107 and circle 108 must be manually set to 1 V.

Figure 14.9 shows the voltage distribution for `struct14_3.geo` as generated by `show_mesh.m`.

Calling the Matlab function `get_tri2d_E.m` at the end of `fem_2d_1.m` results in Figure 14.10, the electric field magnitude distribution of `struct14_3.geo`. The calculation of the electric field is identical to this calculation in `get_tri2d_cap.m` and there would have been some programming efficiency in including the electric field calculation and graphics generation into this latter function. MATLAB program `get_tri2d_E.m` — showing the electric field profile — is written as a separate function only for the convenience of the reader:

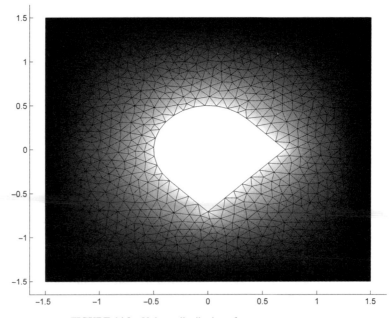

FIGURE 14.9   Voltage distribution of `struct14_3.geo`.

```
function get_tri2d_E(v, nr_nodes, nr_trs, nds, trs)
%  Calculate and show the field magnitude profile
%  E_mag calculation is identical to that in capacitance calculation
%  function

  eps0 = 8.854;

  d = ones(3,3); E_mag = zeros(nr_trs,1); Ex = E_mag; Ey = E_mag;
  for tri = 1 : nr_trs
    n1 = trs(1,tri); n2 = trs(2,tri); n3 = trs(3,tri);
    d(1,2) = nds.x(n1); d(1,3) = nds.y(n1);
    d(2,2) = nds.x(n2); d(2,3) = nds.y(n2);
    d(3,2) = nds.x(n3); d(3,3) = nds.y(n3);
    temp = (d\[1; 0; 0]); b1 = temp(2); c1 = temp(3);
    temp = (d\[0; 1; 0]); b2 = temp(2); c2 = temp(3);
    temp = (d\[0; 0; 1]); b3 = temp(2); c3 = temp(3);

    E_x(tri) = v(n1)*b1 + v(n2)*b2 + v(n3)*b3;
    E_y(tri) = v(n1)*c1 + v(n2)*c2 + v(n3)*c3;
    E_mag(tri) = E_x(tri).^2 + E_y(tri).^2;

  end

  E_peak = sqrt(max(E_mag))
  E_mag = sqrt(E_mag/max(E_mag));   % normalize
  E_x = abs(E_x)/max(E_x);
  E_y = abs(E_y)/max(E_y);

  E_show = E_mag;   % pick E_x, E_y or E_mag to display

  figure(3)

  x_min = min(nds.x)*1.1;
  if x_min == 0, x_min = -.5; end;
  x_max = max(nds.x)*1.1;
  y_min = min(nds.y)*1.1;
  if y_min == 0, y_min = -.5; end;
  y_max = max(nds.y)*1.1; axis([x_min, x_max, y_min, y_max]);
  axis([x_min, x_max, y_min, y_max]);
  axis square;
  hold on

  for tri = 1 : nr_trs
    n1 = trs(1,tri); n2 = trs(2,tri); n3 = trs(3,tri);
    x = [nds.x(n1); nds.x(n2); nds.x(n3)];
    y = [nds.y(n1); nds.y(n2); nds.y(n3)];
    E = [E_show(tri); E_show(tri); E_show(tri)];
    colormap(gray) % Delete this line for color graphics
    fill(x,y,E)
  end
end
```

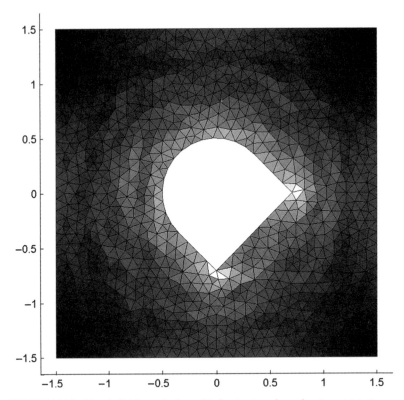

FIGURE 14.10 Electric field magnitude profile for structure shown in `struct14_3.geo`.

Figure 14.10 shows the electric field magnitude profile produced by `get_tri2d_E.m`. Once again, the color figure produced on a computer display allows more insight than the black-and-white text rendition. In either case, it is clear that the electric field peaks at the sharp corners of the structure. Modifying `get_tri2d_E.m` to show field components, to tabulate field values, and so on, is very straightforward.

14.6 MULTIPLE ELECTRODES

(*Note*: This section relies heavily on the results of Chapter 10. The electrostatic concepts and circuital analysis of a multiple electrode structure are the same regardless of whether FD or FEM analysis is utilized to find the parameters of the structure. Even if the reader has no interest in FD analysis, reading Chapter 10 before reading this section is recommended.)

The `gmsh` file `struct14_4.geo` describes a simple rectangular two-electrode structure.

```
// Gmsh structure file struct14_4.geo

lc = .1;
// exterior loop
Point(1) = {-2, -1.5, 0, lc};
Point(2) = {2, -1.5, 0, lc};
Point(3) = {2, 1.5, 0, lc};
Point(4) = {-2, 1.5, 0, lc};
```

```
Line(1) = {1, 2};
Line(2) = {2, 3};
Line(3) = {3, 4};
Line(4) = {4, 1};

// interior surface which will become a hole
Point(5) = {-1, -.7, 0, lc};
Point(6) = {-.2, -.8, 0, lc};
Point(7) = {-.2, .45, 0, lc};
Point(8) = {-1, .5, 0, lc};
Line(105) = {5, 6};
Line(106) = {6, 7};
Line(107) = {7, 8};
Line(108) = {8, 5};

Line Loop(1) = {105, 106, 107, 108};
Plane Surface(200) = {1};

// second inner surface
Point(15) = {0., -.5, 0, lc};
Point(16) = {1, -.5, 0, lc};
Point(17) = {.8, .5, 0, lc};
Point(18) = {.2, .5, 0, lc};
Line(115) = {15, 16};
Line(116) = {16, 17};
Line(117) = {17, 18};
Line(118) = {18, 15};

Line Loop(2) = {115, 116, 117, 118};
Plane Surface(210) = {2};

// exterior surface
Line Loop(3) = {1, 2, 3, 4};
Plane Surface(4) = {3, 1, 2} ;
```

The structure described by struct14_4.geo is shown in Figure 14.11.

In this structure, two obviously different, nontouching, electrodes are contained in the same outer boundary. File struct14_4.geo defines both of these electrodes as excluded regions inside the outer boundary. The left-hand electrode is bounded by lines 105–108; the right-hand electrode, by lines 115–118.

The following questions might be posed:

1 What are the various capacitances of this structure?
2 If the left-hand electrode (electrode 1) is set as a boundary condition while the right-hand electrode (electrode 2) is left floating, what voltage will it float to?

As described in Chapter 10, these questions may be answered using a circuit model of the system. Figure 10.1 shows the three-capacitor model that describes this system.

If we set both electrodes to 1 V (by setting the boundary condition on lines 105–108 and lines 115–118 = 1), we find that fem_2d_1.m calculates C_a = the sum of C_1 and C_2 = 75.66 pF/m.

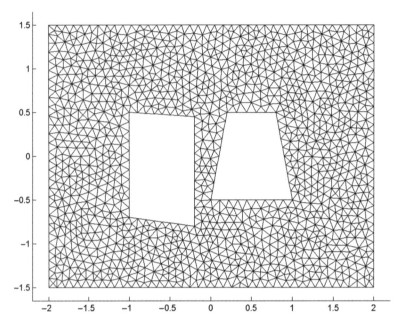

FIGURE 14.11   Two-electrode structure described by `struct 14_14.geo`.

If we set electrode 1 to 1 V and electrode 2 to 0 V by setting only lines 105–108 = 1 V, then lines 115–118 remain at 0 V by default, and we calculate $C_a =$ the sum of C_1 and $C_{12} =$ 85.98 pF/m.

If we set electrode 2 to 1 V and electrode 1 to 0 V, we calculate $C_b =$ the sum of C_2 and $C_{12} = 77.43$ pF/m.

Equations (10.4), (10.5), and (10.6) invert these relationships and give us the capacitance values $C_1 = 42.1$, $C_2 = 33.6$, and $C_{12} = 43.9$ pF/m, respectively.

If we set electrode 1 to 1 V and let electrode 2 float, then (again referring to the derivation in Chapter 10), we obtain

$$V = (1\,V)\frac{C_{12}}{C_2 + C_{12}} = \frac{43.9}{33.6 + 43.9} = 0.57\,V \tag{14.1}$$

As described in Chapter 10, the voltage distribution resulting from the first FEM calculation above is identically the sum of the voltage distributions of the the the last two FEM calculations above, and this relationship may be used to eliminate one FEM calculation.

Utilization of a symmetric structure as described in Chapter 10 to eliminate one FEM calculation is also a valid option. Since we are using unstructured grids for the FEM calculation, however, we have no a priori guarantee that the latter two FEM calculations above will yield identical results, even though we are using a symmetric structure. In this case the calculation that yields the lower capacitance is the more accurate one, but we have no way of knowing which calculation will yield the lower capacitance until we do both calculations, and at that point there is no longer any value in "eliminating" one calculation because both calculations have already been done — we simply choose the lower result to use in both cases.

PROBLEMS

14.1 Modify `get_tri2d_E.m` so that it can display E_x and E_y as well as E_{mag}. Show these results for `struct14_4` with the right electrode set to 1 V and the left electrode set to 0 V.

14.2 Write a MATLAB function that takes the voltage profile from a `fem2d_1.m` run and creates a graph of equipotential contours.

14.3 Create a `gmsh` structure file (`xxx.geo`) for two concentric squares of sides 2 and 4. Mesh the inner region finer than the outer region. Rotate the inner square in 5° steps; create the mesh and the boundary conditions, and then find the capacitance and peak electric field magnitude as a function of the rotation angle.

14.4 Using any text editor, edit the boundary condition file of the previous example (at 0 rotation) and remove all entries for the right-hand wall of the outer square. (Re)run the FEM analysis and comment on the capacitance and voltage distribution as compared to the previous results. (*Hint*: This strategy will be formally presented in Chapter 15.)

REFERENCES

1. C. Geuzaine and J. F. Remacle, Gmsh: A three-dimensional finite element mesh generator with built-in pre- and post-processing facilities, *Int. J. Num. Methods Eng.*, **79**(11): 1309–1331 (2009).

2. `http://geuz.org/gmsh/doc/LICENSE.txt`

3. `http://www.ideaconsulting.com/strip.htm`

15

Some FEM Topics

15.1 SYMMETRIES

One of the principal reasons for the popularity of the FEM is the ease with which boundary conditions are handled. All of the FEM examples presented thus far have utilized boundary conditions formed by setting boundary nodes to an assigned voltage. Since we have dealt exclusively with triangular shapes with simple linear shape functions, this means that the lines connecting these nodes (an edge of the triangle) are set to a linear interpolation between these two node voltages.

A second type of boundary condition occurs when the node voltages are not assigned but the electric flux along line edges is assigned. The *electric flux* at a region boundary is defined as the integral of the normal component of the electric field lines crossing that boundary (in this case the triangle's edge).

There are elegant derivations of how to translate these boundary conditions into FEM equations.[1] These derivations, in turn, require some background in variational calculus and vector calculus. We will consider only a subset of the general problem, so a simpler explanation of how to set up our equations will suffice.

A *symmetry boundary*, or a magnetic wall, is a boundary which has zero electric flux across it. That is, the electric field at the boundary is parallel to the boundary wall (and the normal field across the boundary wall is zero).

Figure 15.1 shows an example of a symmetric structure. It may also be regarded as a subset of a larger symmetric structure.

In this figure, the structure is symmetric about the labeled symmetry axis (the dashed line). Although it is not necessary for the zoning (node and triangle locations) to be symmetric about a symmetry axis, it is always possible to create them this way — if this were not possible, the structure cannot be symmetric about the symmetry axis.

Introduction to Numerical Electrostatics Using MATLAB, First Edition. Lawrence N. Dworsky.
© 2014 John Wiley & Sons, Inc. Published 2014 by John Wiley & Sons, Inc.

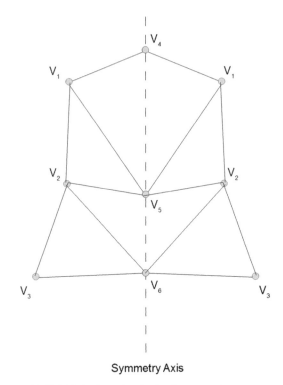

Symmetry Axis

FIGURE 15.1 Example of a simple symmetric structure.

In Figure 15.1, the node numbers are indicated in the subscript of each voltage at that node (V_1, V_2, etc.). Since everything is symmetric about the symmetry axis, the same node numbering (and voltages) are shown on both sides of this axis.

Now, suppose that we first create an FEM coefficient matrix and forcing function vector using only the nodes on and to the left of the symmetry axis. Then we create a second FEM coefficient matrix and forcing function vector using only the nodes on and to the right of the symmetry axis. The two coefficient matrices and the two forcing function vectors will be identical.

Finally, we create a third FEM coefficient matrix and forcing function vector using the entire structure. These will just be the sums of the first two coefficient matrices and forcing function vectors, respectively. Since the first two coefficient matrices are identical and the first two forcing function vectors are identical, all three sets will have the same solution vector (node voltages).

In other words, the FEM formulation algorithm for a structure with a symmetry axis is to create the matrices for only one side of the symmetry axis (including the nodes on the symmetry axis itself), leaving the nodes on the symmetry axis as variables. This conclusion, as a programming algorithm, may be explained simply as "Don't worry about it, everything will take care of itself."

15.2 A SYMMETRY EXAMPLE, INCLUDING A TWO-SIDED CAPACITANCE ESTIMATE

The MATLAB programs in Chapters 13 and 14 may be updated to accommodate symmetry axes with several simple modifications.

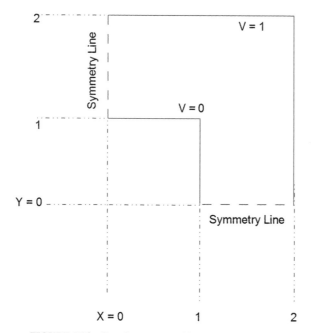

FIGURE 15.2   Sample structure with two symmetry lines.

Figure 15.2 shows a quarter section of a square within a square. This is the same example that was used in Chapter 10. The gmsh coding for this consists simply of the six-sided polygon shown in Figure 15.2; there are no excluded interior regions.

The first code modifications needed are to parse_msh_file.m. As written, parse_msh_file.m considered only conductor boundaries that were closed polygons. In this situation a full list of the nodes containing the endpoints of all the line segments would contain each node twice (once on each end of each line segment). To avoid this redundancy, only one end of the line was examined for its node number.

As Figure 15.2 shows, we now have to consider some nodes that form the junction of a conductor and a symmetry line — with the boundary condition (bc) on the conductor taking precedence. Adding one extra line after line 77 accomplishes this task. When a structure contains connected conductor line segments, the program does generate some redundant information. This information doesn't cause any trouble, so to keep things as simple as possible, no extra code was written to "search and destroy" these extra lines.

Next, we need a way to explain our new bc information to the programs. In Problem 14.4, this was done by manually editing the bcs file. This time we'll update the parse_msh_file program to add this capability.

The bc data entry system is now as follows:

- The program will loop through all of the defined lines of the structure.
- For each line, set any desired bc voltage except -9999, which will be used as a flag to indicate a line of symmetry.

MATLAB program parse_mesh_file2.m is as follows:

```
% parse_msh_file2.m
```

```
% Read the mesh file generated by gmsh and parse it
% This is a very limited routine, based only on the
%   capabilities of gmsh that are being used
% This is an extension parse_msh_file.m, to allow for symmetries

%   Element types handled are 1 = line, 2 = 3 node triangle, 15 = point

  filename = uigetfile('*.msh');              % read the file
  fid = fopen(filename, 'r');

  tline = fgetl(fid);                              % read the first file line
  if (strcmp(tline, '$MeshFormat')) == 0    % Verify file type
    'Problem with this file'
    nodes = []; trs = []; bcs = [];
    fclose (fid);
    return
  end

  tline = fgetl(fid);                         % look for version number
  [header] = sscanf(tline, '%f %d %d');
  if (header(1) ~= 2.2), 'Possible gmsh version incompatibility', end;

  tline = fgetl(fid);                         % Look for $EndMeshFormat
  if (strcmp(tline, '$EndMeshFormat')) == 0
    'No $EndMeshFormat Line found'
    nodes = []; trs = []; bcs = [];
    fclose (fid);
    return
  end

  tline = fgetl(fid);                         % Look for $Nodes
  if (strcmp(tline, '$Nodes')) == 0
    'No $Nodes Line found'
    nodes = []; trs = []; bcs = [];
    fclose (fid);
    return
  end

  % NOTE: Nodes is being read as a 3-d file even though only 2-d
  %        usage is anticipated in this version

nr_nodes = fscanf(fid, '%d', 1);             % get number of nodes
nodes = [0,0,0]; % zeros(nr_nodes, 3);    % allocate nodes array
for i = 1 : nr_nodes % get nodes information
  temp = fscanf(fid, '%g', 4);
  n = temp(1);
  nodes(n,:) = temp(2:4);
end
fgetl(fid);                                  % Clear the line

  tline = fgetl(fid);                         % Look for $EndNodes
  if (strcmp(tline, '$EndNodes')) == 0
    'No $EndNodes Line found'
```

```
  nodes = [] ; trs = [] ; bcs = [] ;
  fclose (fid) ;
  return
end

tline = fgetl (fid) ;                       % Look for $Elements
if (strcmp(tline, '$Elements')) == 0
  'No $Nodes Line found'
  nodes = [] ; trs = [] ; bcs = [] ;
  fclose (fid) ;
  return
end

trs = [] ; bcs = [] ;
nr_elements = fscanf(fid, '%d', 1) ;        % get number of elements
for i = 1 : nr_elements                     % process elements information
  temp = fscanf(fid, '%d %d', 2) ;          % get element type
  if temp (2) == 15                         % defined point
    fgetl (fid) ;                           % clear the line
  elseif temp (2) == 1                      % defined line
    temp2 = fscanf(fid, '%d %d %d %d %d', 5) ;
    bcs = [bcs; [temp2 ( 3:4)', 0]] ;

% -------------------------------------------------
%   added line for symmetry calcs - get both nodes on bc line
%   this adds some redundancy to the bcs file, but no harm
    bcs = [bcs; [temp2 (3), temp2 (5), 0]] ;
% -------------------------------------------------

    fgetl (fid) ;
  elseif temp (2) == 2                      % created mesh triangle
    temp3 = fscanf(fid, ' %d  %d  %d  %d  %d  %d', 6) ;
    if temp3 (3) < 200                      % ignore excluded triangles
      trs = [trs; temp3 (4:6)'] ;
    end
    fgetl (fid) ;
  else
    'Unknown type in mesh file'
    nodes = [] ; trs = [] ; bcs = [] ;
    fclose (fid) ;
    return
  end
end
fclose (fid) ;

nr_bc_lines = length(bcs) ;
'bcs file: '
bcs
% ----- modified lines for symmetry calculations ---------------
disp ('Set bcs for all lines, -9999 for symmetry line') ;
while 1 == 1
  line_nr = input ('Enter line nr for bc: ')
```

```
      if line_nr == 0, break; end;
      j = input ('Enter bc: ')                    % decide on the bc
      for k = 1 : nr_bc_lines                     % set the bc
        if bcs (k,1) == line_nr, bcs (k,3) = j; end;
      end
    'bcs file: '
    bcs
  end
end
% -------------------------------------------------------------------

filename = input ('Name for output files, NO OVERWRITE CHECKING: ', 's');
nodes_file_name = strcat ('nodes_', filename, '.txt');
trs_file_name = strcat ('trs_', filename, '.txt');
bcs_file_name = strcat ('bcs_', filename, '.txt');
fid = fopen (nodes_file_name, 'w');
fprintf (fid, '%g\t%g\t%g\r\n', nodes');
fclose (fid);
fid = fopen (trs_file_name, 'w');
fprintf (fid, '%d\t%d\t%d\r\n', trs');
fclose (fid);
fid = fopen (bcs_file_name, 'w');
fprintf (fid, '%d\t%d\t%d\r\n', bcs');
fclose (fid);
```

Program parse_msh_file2.m is the revised version of this program.

Next, fem_2d.1.m must be modified to properly deal with the new information. The changes are so minor that the revised programs need not be fully listed. These changes are as follows:

1 In fem_2d_1.m, in the section beginning with the comment line %bcs, change the
   if statement to

   ```
   ifbcs.volts(i) ~ = -9999          % symmetry axis = -9999
   ```

2 In show_mesh.m, change the line before the text command to the same as above.

The first change above causes the FEM analysis to not set up any boundary conditions at edges specified as boundary conditions; in other words, just ignore them, as explained above. Edges are treated the same as internal nodes.

The second change above does the same for the show_mesh routine.

Referring to Figure 15.2, number the lines counterclockwise, beginning with the symmetry line at $Y = 0$. The responses to the bcs prompt in (the modified version of) show_mesh.m are then −9999, 1, 1, −9999, 0, and 0.

Figure 15.3 shows the same structure, including the meshing; Figure 15.4 shows the voltage distribution of the structure; and Figure 15.5 shows the electric field distribution.

The calculated capacitance of the structure is $4(22.77) = 91.1$ pF/M. As mentioned in Chapter 10, the exact capacitance of this structure is 90.6 pF/M. This FEM result is 0.5% high. This structure was zoned with 385 nodes, a fairly low-resolution zoning.

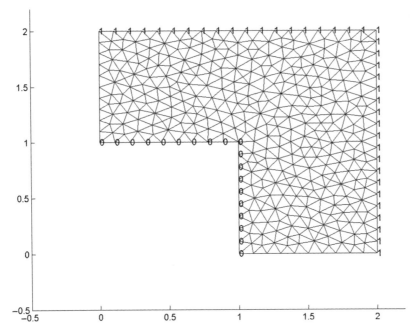

FIGURE 15.3   Structure shown in Figure 15.2 after meshing.

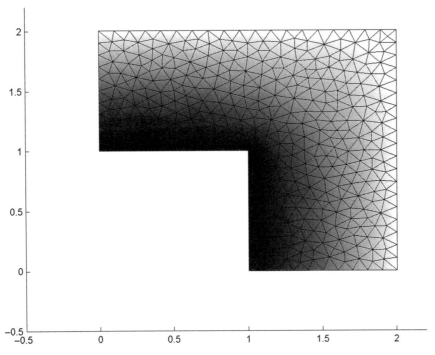

FIGURE 15.4   Structure from Figure 15.2 showing voltage distribution.

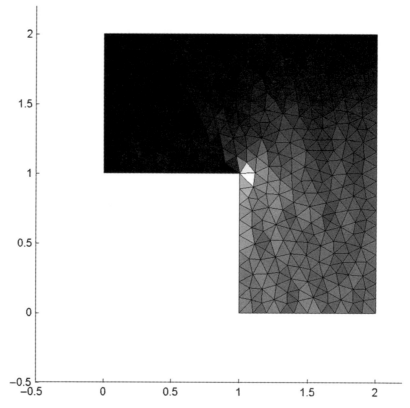

FIGURE 15.5   Structure from Figure 15.2 showing field distribution.

As was discussed in Section 9.7, the dual capacitance may be found by reversing the fixed and symmetry boundary lines. For this same structure (and same mesh file), this means that the bc responses to `parse_msh_file.m` are 0, −9999, −9999, 1, −9999, and −9999.

Examination of the figures produced by `fem_2d_1.m` show that the program has correctly interpreted the boundary conditions and is producing reasonable looking voltage and field distributions.

The capacitance produced by the program is $C_{\text{dual}} = 3.48$. Using equations (9.27) and (9.28) to calculate the electrical low-capacitance estimate, we obtain

$$C_{\text{low}} = \frac{4\varepsilon_0^2}{C_{\text{dual}}} = 90.1 \text{ pF/m} \tag{15.1}$$

This capacitance is 0.5% lower than the exact value. As discussed in Chapter 9, the best estimate is obtained by averaging the high- and low-capacitance estimates; the result is essentially the exact answer.

At this point the value of FEM modeling is becoming apparent. The ability to represent arbitrary structures and the ability to easily apply different boundary conditions clearly separates it from the FD and MoM techniques. Although FEM programs are more involved than either FD or MoM programs, from a user's perspective they are much more flexible.

Figure 15.6 shows another example of the flexibility of the FEM calculation.

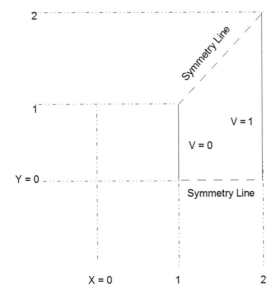

FIGURE 15.6 Structure with eightfold symmetry.

Figure 15.6 shows the same structure as in the previous example, but in this case only $\frac{1}{8}$th of the total square conductors is shown. The line from (1,1) to (2,2) is also a line of symmetry, or magnetic wall. It does not matter that the line is not parallel to either the *X* or the *Y* axis.

Both `fem_2d_1.m` and `parse_msh_file1.m`, including the modifications described above, treat this geometry correctly and produce good results for both the physical and the dual-structure capacitances, again with excellent accuracy for the average of the two capacitances, using only a modest number of nodes.

15.3 AXISYMMETRIC STRUCTURES

An *axisymmetric structure* is a structure that has cylindrical symmetry. The geometry is described, in cylindrical coordinates, in terms of *r* and *z* only — there are no angular variations. Since there are only two variables involved, these structures may be zoned in the same manner as 2d rectangular problems, but the interpretation of the drawing is very different.

For example, consider the example used in Section 15.2 (Figures 15.2 and 15.6). These figures described the cross section of an infinitely long (in *z*) pair of concentric squares. Figure 15.7 shows the various ways this structure may be zoned, either using the full cross section (a) or taking advantage of various axes of symmetry (b), (c), and (d).

Figure 15.8 shows a perspective view of the same concentric square transmission line. An important point to keep in mind here is that Figure 15.8 does not show the entire structure described. It just shows a section (in *z*) of the structure. The structure is infinitely long in Z, and we have been dealing only with a cross section of it. Since such a structure with no Z dependence (i.e., uniform) can be described entirely by voltages and fields that are only functions of *X* and *Y*, a 2d analysis finds everything there is to know about the structure.

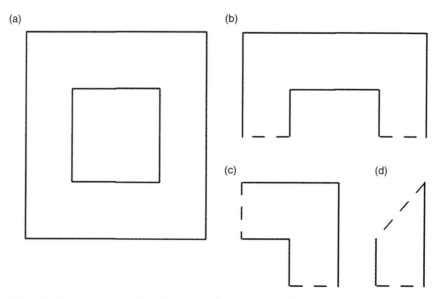

FIGURE 15.7 Cross section choices for concentric square transmission line: (a) Full cross section: (b) $\frac{1}{2}$ symmetry; (c) $\frac{1}{4}$ symmetry; (d) $\frac{1}{8}$ symmetry.

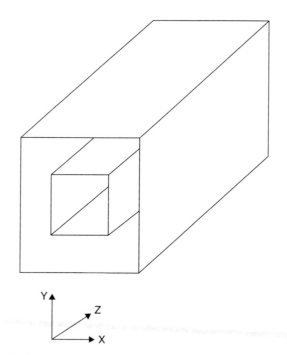

FIGURE 15.8 Perspective view of concentric square transmission line.

Voltages, electric field components and energy density are functions of X and/or Y only and are valid for any value of Z. Total energy and capacitance values are calculated as joules and farads per unit length.

Now consider Figure 15.9.

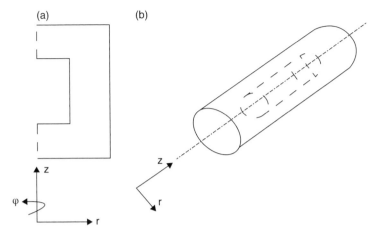

FIGURE 15.9   (a) Cross section between r and z and (b) perspective view of concentric cylinder structures.

Figure 15.9a is identical to Figure 15.7b except that the axes are now labeled (r,z) rather than (x,y). The structures represented, however, are not the same. Figure 15.9b shows the axisymmetric structure described by Figure 15.9a. There are two important points to make here:

1   Figure 15.9 shows the entire structure being modeled, not just a piece of it. It is a 3d structure shown in its entirety with both cylinders shorted at both ends. The capacitance between the inner and outer cylinders and the total stored energy (for some prescribed electrode voltages) are measured in farads and joules, respectively.

2   The location of $z = 0$ for the structure is arbitrary, as is the location of $x = 0$ and $y = 0$ in rectangular coordinates insofar as the properties of the structure are concerned. In other words, Figure 15.9b may be moved (caused to slide) up or down the z axis without changing any structure properties. On the other hand, the r axis is fixed. In order to retain the *axisymmetric* status, there can be no φ dependence, and the structure in Figure 15.9b must always be fully describable by its cross section (Figure 15.9a).

Deriving the formulas for the coefficient matrix for an axisymmetric structure using triangles with the shape functions defined in Chapter 13 duplicates the analysis of Chapter 13, except that X is replaced with r and Y is replaced with z, up to the energy expression [equation (13.10)]:

$$U = \frac{\varepsilon}{2} \iint_T E^2 \, dx \, dy = E^2 A_T \tag{15.2}$$

As in this equation, the energy is constant over the triangle. The integral, over the area of the triangle, however, is now

$$U = (2\pi) \frac{\varepsilon}{2} E^2 \iint_T r \, dr \, dz \tag{15.3}$$

Evaluating the integral in equation (15.3) is not a trivial task. Fortunately, it is an already accomplished and well-documented task. If we put aside the physics behind equation (15.3) and look only at the integral

$$\iint_T r \, dr \, dz = \iint_T x \, dx \, dy \tag{15.4}$$

we recognize this integral as the X component of the center of mass of an arbitrary triangle T in the X–Y plane.

The solution to this, as documented in many places,[2-4] is

$$\iint_T r \, dr \, dz = r_c A_T \tag{15.5}$$

where A_T is the triangle's area, as before, and r_c is the average of the three nodes' r values:

$$r_c = \frac{r_{n1} + r_{n2} + r_{n3}}{3} \tag{15.6}$$

Modifying the MATLAB code to study axisymmetric problems is very simple:

1 When calculating the coefficient matrix, simply multiply the area by r_c. In the code below, this multiplication is carried out every time A_T is used rather than before it is used. This is unnecessary, but hopefully the clearest way to present this change.

2 When calculating the capacitance, multiply the area by r_c, as above, and then multiply by an additional 2π, the integral over φ. This latter multiplication was not needed in the coefficient term calculation because, being the same for all terms, it drops out.

The MATLAB listings below show the code, including both the earlier changes in this chapter and the calculations corrected for axisymmetric structures:

- MATLAB program fem_2d.3.m — MATLAB script for FEM modeling of axisymmetric structures:

```
% 2d FEM with triangles
% fem_2d_3.m is for axisymmetric structures
% This includes all the modifications to fem_2d_1.m to handle
symmetries
% Just to avoid rewriting a lot of code, nodes are referred to as x & y
% These should be interpreted as r and z, respectively

close

filename = input ('Generic name for input files: ', 's');
nodes_file_name = strcat ('nodes_', filename, '.txt');
trs_file_name = strcat ('trs_', filename, '.txt');
bcs_file_name = strcat ('bcs_', filename, '.txt');

% get node data from file
dataArray = load (nodes_file_name);
nds.x = dataArray (:,1); nds.y = dataArray (:,2); nds.z = 0;
nr_nodes = length (nds.x)
```

```
% get triangles data from file
trs = (load(trs_file_name))';
nr_trs = length(trs)

% get bcs data from file
dataArray = load(bcs_file_name);
bcs.node = dataArray(:,2); bcs.volts = dataArray(:,3);
nr_bcs = length(bcs.node)

% allocate the matrices
a = sparse(nr_nodes, nr_nodes);
f = zeros(nr_nodes,1);
b = zeros(3,nr_nodes); c = zeros(3,nr_nodes);

% generate the matrix terms
d = ones(3,3);
for tri = 1: nr_trs % Walk through all the triangles
  n1 = trs(1,tri); n2 = trs(2,tri); n3 = trs(3,tri);
  d(1,2) = nds.x(n1); d(1,3) = nds.y(n1);
  d(2,2) = nds.x(n2); d(2,3) = nds.y(n2);
  d(3,2) = nds.x(n3); d(3,3) = nds.y(n3);
  Area = det(d)/2;
  rc = (nds.x(n1) + nds.x(n2) + nds.x(n3))/3.;

  temp = d\[1; 0; 0]; b(1,tri) = temp(2); c(1,tri) = temp(3);
  temp = d\[0; 1; 0]; b(2,tri) = temp(2); c(2,tri) = temp(3);
  temp = d\[0; 0; 1]; b(3,tri) = temp(2); c(3,tri) = temp(3);

  a(n1,n1) = a(n1,n1) + rc*Area*(b(1,tri)^2 + c(1,tri)^2);
  a(n1,n2) = a(n1,n2) + rc*Area*(b(1,tri)*b(2,tri) + c(1,tri)*c
  (2,tri));
  a(n1,n3) = a(n1,n3) + rc*Area*(b(1,tri)*b(3,tri) + c(1,tri)*c
  (3,tri));

  a(n2,n2) = a(n2,n2) + rc*Area*(b(2,tri)^2 + c(2,tri)^2);
  a(n2,n1) = a(n2,n1) + rc*Area*(b(2,tri)*b(1,tri) + c(2,tri)*c
  (1,tri));
  a(n2,n3) = a(n2,n3) + rc*Area*(b(2,tri)*b(3,tri) + c(2,tri)*c
  (3,tri));

  a(n3,n3) = a(n3,n3) + rc*Area*(b(3,tri)^2 + c(3,tri)^2);
  a(n3,n1) = a(n3,n1) + rc*Area*(b(3,tri)*b(1,tri) + c(3,tri)*c
  (1,tri));
  a(n3,n2) = a(n3,n2) + rc*Area*(b(3,tri)*b(2,tri) + c(3,tri)*c
  (2,tri));

end

%bcs
for i = 1: nr_bcs
  if bcs.volts(i) >= 0
  n = bcs.node(i);
```

```
      a(n,:) = 0; a(n,n) = 1; f(n) = bcs.volts(i);
    end
  end

  % fake bcs for nodes in excluded region - assuming boundary at 1 volt
  % There might not be any excluded regions in symmetry problems
  % This is being left for backwards-compatibility of the program
  for i = 1 : nr_nodes
    if a(i,i) == 0
      a(i,:) = 0; a(i,i) = 1; f(i) = 1;
    end
  end

  v = a\f;

  % plot the mesh
  show_mesh2(nr_nodes, nr_trs, nr_bcs, nds, trs, bcs, v)

  % calculate capacitance
  C = get_tri2d_cap_2(v, nr_nodes, nr_trs, nds, trs)

  % show field profile
  get_tri2d_E_2(v, nr_nodes, nr_trs, nds, trs)

  % show voltage along chosen edge

  x_temp = []; v_temp = [];
  for i = 1 : nr_nodes
  %    if abs(nds.y(i)) < .1 %nds.y(i) == 0
    if nds.x(i) < .01 & nds.y(i) >= 0 %nds.y(i) == 0
  %      fprintf ('%d %d %f \n', i, nds.y(i), v(i))
      x_temp = [x_temp, nds.y(i)];
      v_temp = [v_temp, v(i)];
    end
  end

  %x_temp = sort(x_temp);
  %v_temp = sort(v_temp);
  v_exact = -log(x_temp)/log(2) + 1;
  % v_exact = x_temp - 1;
  figure(4)
  %plot(x_temp, v_temp, '.r', x_temp, v_exact, 'x')
  plot(sort(x_temp), sort(v_temp), 'k')
```

- MATLAB program get_tri2d_cap_2.m—MATLAB function for capacitance of axisymmetric structures:

```
  function C = get_tri2d_cap_2(v, nr_nodes, nr_trs, nds, trs)
  %   Calculate the capacitance of a 2d triangular structure
  %   Simple linear approx. function is assumed
  %   This version is for axisymmetric structures

    eps0 = 8.854;
```

```
U = 0;
d = ones(3,3);
for tri = 1 : nr_trs
  n1 = trs(1,tri); n2 = trs(2,tri); n3 = trs(3,tri);
  d(1,2) = nds.x(n1); d(1,3) = nds.y(n1);
  d(2,2) = nds.x(n2); d(2,3) = nds.y(n2);
  d(3,2) = nds.x(n3); d(3,3) = nds.y(n3);
  temp = (d\[1; 0; 0]); b1 = temp(2); c1 = temp(3);
  temp = (d\[0; 1; 0]); b2 = temp(2); c2 = temp(3);
  temp = (d\[0; 0; 1]); b3 = temp(2); c3 = temp(3);

  Area = det(d)/2;
  if Area <= 0, disp 'Area <= 0 error'; end;
  rc = (d(1,2) + d(2,2) + d(3,2))/3;

  U = U + 2*pi*rc*Area*((v(n1)*b1 + v(n2)*b2 + v(n3)*b3)^2 ...
      + (v(n1)*c1 + v(n2)*c2 + v(n3)*c3)^2);
end

  U = U*eps0/2;
  C = 2*U;        % Potential difference of 1 volt assumed

end
```

- MATLAB program `get_tri2d_E_2.m`—MATLAB function for E field distribution in axisymmetric structures:

```
function get_tri2d_E(v, nr_nodes, nr_trs, nds, trs)
%  Calculate and show the field magnitude profile
%  E_mag calculation is identical to that in capacitance calculation
%  function

eps0 = 8.854;

d = ones(3,3); E_mag = zeros(nr_trs,1);
for tri = 1 : nr_trs
    n1 = trs(1,tri); n2 = trs(2,tri); n3 = trs(3,tri);
    d(1,2) = nds.x(n1); d(1,3) = nds.y(n1);
    d(2,2) = nds.x(n2); d(2,3) = nds.y(n2);
    d(3,2) = nds.x(n3); d(3,3) = nds.y(n3);
    temp = (d\[1; 0; 0]); b1 = temp(2); c1 = temp(3);
    temp = (d\[0; 1; 0]); b2 = temp(2); c2 = temp(3);
    temp = (d\[0; 0; 1]); b3 = temp(2); c3 = temp(3);

    %  Use these two lines to look at E_magnitude
    E_mag(tri) = ((v(n1)*b1 + v(n2)*b2 + v(n3)*b3)^2 ...
        + (v(n1)*c1 + v(n2)*c2 + v(n3)*c3)^2);

    %  Use these two lines to only look at Ex
    %  E_mag(tri) = eps0/2*(v(n1)*b1 + v(n2)*b2 + v(n3)*b3)^2;

    %  Use these two lines to only look at Ey
```

```
    %  E_mag(tri) = eps0/2*(v(n1)*c1 + v(n2)*c2 + v(n3)*c3)^2;

end
E_peak = max(E_mag)

E_mag = sqrt(E_mag/max(E_mag)); % normalize

figure(3)

x_min = min(nds.x)*1.1;
if x_min == 0, x_min = -.5; end;
x_max = max(nds.x)*1.1;
y_min = min(nds.y)*1.1;
if y_min == 0, y_min = -.5; end;
y_max = max(nds.y)*1.1; axis([x_min, x_max, y_min, y_max]);
axis([x_min, x_max, y_min, y_max]);
axis square;
hold on

for tri = 1 : nr_trs
    n1 = trs(1,tri); n2 = trs(2,tri); n3 = trs(3,tri);
    x = [nds.x(n1); nds.x(n2); nds.x(n3)];
    y = [nds.y(n1); nds.y(n2); nds.y(n3)];
    E = [E_mag(tri); E_mag(tri); E_mag(tri)];
    fill(x,y,E)
end
```

A very simple example of an asymmetric structure is shown in Figure 15.10.

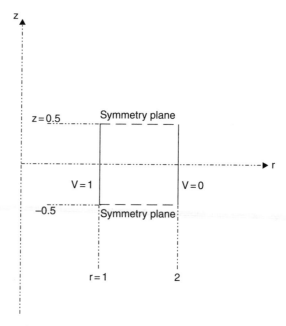

FIGURE 15.10   Infinite-length concentric cylinder structure.

Figure 15.10 shows two concentric cylinders, of radii 1 and 2. They are both 1 unit long, centered at $z = 0$. The use of symmetric planes at $z = -0.5$ and $z = 0.5$ means that the FEM calculation is effectively for a pair of infinitely long concentric cylinders. In other words, while we previously had been using a 2d cross section calculation to model uniformly infinitely long structures, now we are using an axisymmetric 3d calculation to do the same thing. This example adds nothing to our solution capability, but is instructional as to setting up structures for the axisymmetric calculation. Program `struct15_1.geo`—the gmsh file for the structure in Figure 15.10—is very simple:

```
//   Structure file struct15_1.geo
//   Symmetric coaxial line for 3d calc

lc = .05;

Point(1) = {1, -.5, 0, lc};
Point(2) = {2, -.5, 0, lc};
Point(3) = {2, .5, 0, lc};
Point(4) = {1, .5, 0, lc};

Line(1) = {1, 2};
Line(2) = {2, 3};
Line(3) = {3, 4};
Line(4) = {4, 1};

//   exterior surface
Line Loop(1) = {1, 2, 3, 4};
Plane Surface(2) = {1} ;
```

For this gmsh file, the bc prompts in `parse_msh_file2.m` are $-9999, 0, -9990$, and 1.

Program `fem_2d.3.m` calculates the capacitance of this structure to be 80.266 pF/m. The exact capacitance, as derived in Chapter 13, is

$$C = \frac{2\pi\varepsilon_0}{\ln(b/a)} = \frac{2\pi\varepsilon_0}{\ln(2)} = 80.259 \text{ pF/m} \tag{15.7}$$

This FEM calculation, with a 530-node, 980-triangle structure, predicts the capacitance with an error of 0.01%.

The gmsh file `struct15_2.geo`—using parametric data—creates the structure shown and discussed in Figures 15.11–15.13:

```
//   structure file struct15_2.geo
//   Finite length concentric cylinders

lc = .1;

h = 0.6;    //   overall structure 1/2 height
t = 0.5;    //   end cap thickness
h2 = h - t;
ra = 1;     //   inner radius
rb = 2;     //   outer radius
```

```
//  exterior loop
Point(1) = {0, -h, 0, lc};
Point(2) = {rb, -h, 0, lc};
Point(3) = {rb, h, 0, lc};
Point(4) = {0, h, 0, lc};
Point(5) = {0, h2, 0, lc};
Point(6) = {ra, h2, 0, lc};
Point(7) = {ra, -h2, 0, lc};
Point(8) = {0, -h2, 0, lc};

Line(1) = {1, 2};
Line(2) = {2, 3};
Line(3) = {3, 4};
Line(4) = {4, 5};
Line(5) = {5, 6};
Line(6) = {6, 7};
Line(7) = {7, 8};
Line(8) = {8, 1};

//  exterior surface
Line Loop(1) = {1, 2, 3, 4, 5, 6, 7, 8};
Plane Surface(2) = {1} ;
```

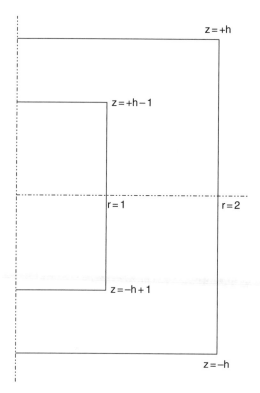

FIGURE 15.11   Structure defined using `struct15_2.geo`.

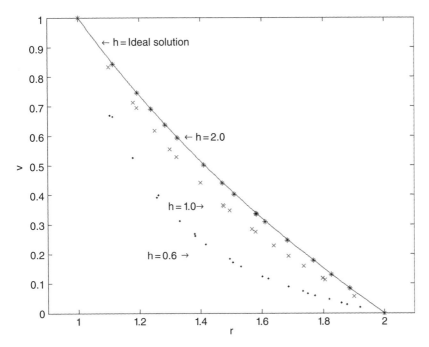

FIGURE 15.12   Results of FEM analysis of `struct8.geo`.

This file introduces a very convenient capability of `gmsh`, namely, the ability to perform calculations using defined variables, and refer to these variables for structural information. These variables are defined schematically in Figure 15.11.

Referring to Figure 15.11, when h is very large as compared to r_a, r_b, and t, we expect the voltage profile near $z = 0$ to approximate the voltage profile between two infinitely long cylinders, as calculated in Chapter 10:

$$V(r) = 1 - \frac{\ln(r)}{\ln(2)} \tag{15.8}$$

Figure 15.12 shows a plot of equation (15.8) and data for the FEM analysis of `struct15_2.geo`, using three different values of h. All other values are kept constant as in `struct15_2.geo`. When $h = 0.8$, the voltage profile looks nothing like the analytic solution except, of course, that the boundary conditions are satisfied. When $h = 1.0$, the voltage profile approximates the analytic solution. When $h = 2.0$, the voltage profile is indistinguishable from the analytic solution. The value of the ability to set parametric values in `gmsh` and then vary them to perform parameter studies on a given structure is considerable. To be complete, we should vary the resolution and make sure that it is high enough to ensure that the results are due to the fact that the parameters are deliberately varied and not noticeably due to subtle changes in zoning when a parameter is changed.

Figure 15.13 is a repeat of the structure of Figure 15.11, but only the $z \geq 0$ region is shown. The value $Z = 0$ is an axis of symmetry (along with $r = 0$), and the FEM analysis may be run this way. The results, of course, will be identical to these results.

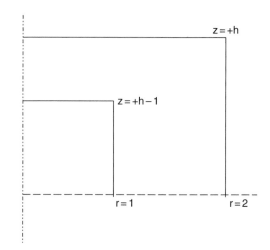

FIGURE 15.13   Structure from Figure 15.11 showing Z symmetry.

Again, note that in axisymmetric structures z-axis symmetries behave correctly. The $r = 0$ symmetry is necessary, but no other r-axis symmetries are meaningful.

15.4   THE GRADED-POTENTIAL BOUNDARY CONDITION

A situation often encountered in electrostatic analysis is an axisymmetric structure where the outer boundary bc is a potential that is graded (typically linearly) between two defined points. An example of such a structure is shown in Figure 15.14, showing a simplified cross section of a cathode ray tube (CRT). For about 50 years, before the advent of liquid crystal display (LCD) displays, the CRT was the mainstay of the television, oscilloscope, and (in the later years) the computer monitor industries.

A CRT is built using a sealed glass package, under high vacuum. The base of the package (near $z = 0$ in the figure) holds an electron emitter structure and control electrodes; these are not shown in this figure. The sidewalls of the package are coated with a highly resistive material so that voltages may be applied but very little current will flow. The sidewall tapers out (in r) from the base to the anode, or screen, of the CRT. The screen is coated with phosphor materials (materials that emit light when struck by energetic electrons). The screen is set to a high voltage, typically $\sim 15\,kV$.

The tapered sidewall of the package is set to a linearly graded voltage that is equal to 0 at the base and equal to the anode voltage at the anode. Electrons that have been emitted into the vacuum and aimed at particular locations on the anode (the screen) accelerate through the (approximately) uniform field of the graded region and strike the anode.

Incorporating the capability of setting one or more lines to a graduated voltage may be accomplished by modifying `parse_msh_file2.m`, modifying `fem_2d_3.m`, or adding a new and separate MATLAB script that the bc data file must be run through. This latter approach was chosen, not because of any performance incentives, but in the interest of keeping each piece of code as clear and unentangled as possible.

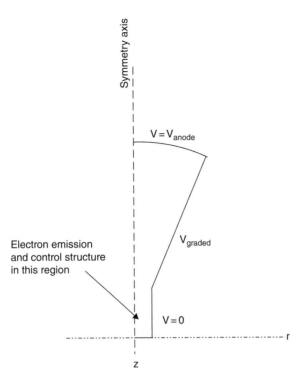

FIGURE 15.14 A simplified CRT outline.

MATLAB program `graded_bc.m` is as follows:

```
% graded_bc.m A short script to apply a graded boundary condition to an
% existing bc_XXX.txt file

filename = input ('Generic name for input files: ', 's');
bcs_file_name = strcat ('bcs_', filename, '.txt');
nodes_file_name = strcat ('nodes_', filename, '.txt');

bcs = load (bcs_file_name);
nds = load (nodes_file_name);

line_nr = 8888;   % get input information
while line_nr ~= 0
   bcs
   line_nr = input ('Enter line number for graded bc, 0 to terminate: ');
   if line_nr == 0
     break
   end
   v1 = input ('Enter starting voltage ')
   v2 = input ('Enter ending voltage ')

   % find the line number and get the start and stop points in the file
   for i = 1 : length (bcs)
```

```
      if bcs(i,1) == line_nr
        p1 = bcs(i,2);                    % start point
        break
      end
    end
    p2 = bcs(length(bcs),2);
    for j = i + 1 : length(bcs)
      if bcs(j,1) ~ = line_nr
        p2 = bcs(j-1,2);                  % stop point
        break
      end
    end

    %  get the length of the line
    line_length = sqrt((nds(p1,1) - nds(p2,1))^2 + (nds(p1,2) -
nds(p2,2))^2);

    for i = 1 : length(bcs)               % scale the voltage by length along the
      if bcs(i,1) == line_nr              % chosen line and set the bc
        p = bcs(i,2);
        bcs(i,3) = v1 + (v2-v1)*sqrt((nds(p,1) - nds(p1,1))^2 ...
          + (nds(p,2) - nds(p1,2))^2)/line_length;
      end
    end
end

%  replace the bcs file on disk
save (bcs_file_name, 'bcs', '-ascii')
```

Program `graded_bc.m` should be run after the `parse_msh_file2.m` has been run for a given dataset. In `parse_msh_file2.m`, the line whose bc is to be graded may be set to any convenient, temporary, value (0 is fine). Program `graded_bc.m` will prompt for a line (or lines) to grade and then ask for starting and stopping voltages for the gradation.

(*Important note*: Program `graded_bc.m` does not perform any validity checking on the entered data. A line with voltage graded between, say, 0 and 100-V should connect to a 0-V bc line at its beginning and to a 100-V bc line at its end. Any other conditions are not physical and will lead to useless results in the FEM analysis.)

The gmsh structure file `struct15_3.geo`—file for the CRT profile—is presented here:

```
//  structure file struct15_3.geo
//  Simple cylinder for demonstrating bias capability

lc = .5;

//  exterior loop
Point(1) = {0, 0, 0, lc};
Point(2) = {1, 0, 0, lc};
Point(3) = {1, 2, 0, lc};
Point(4) = {5, 10, 0, lc};
```

```
Point(5) = {0, 11., 0, lc};

Point(6) = {0, -.5, 0, lc}; // circle center point

Line(1) = {1, 2};
Line(2) = {2, 3};
Line(3) = {3, 4};
Circle(4) = {4, 6, 5};
Line(5) = {5, 1};

//  exterior surface
Line Loop(1) = {1, 2, 3, 4, 5};
Plane Surface(2) = {1} ;
```

The gmsh file struct15_3.geo generates the CRT outline structure shown in Figure 15.14. Using parse_msh_file2.m to set the fixed boundary conditions of 0 V everywhere except at the screen, 100 V at the screen; and then using graded_bc.m to set a linear taper from the base (neck) of the CRT to the screen, we get the FEM analysis model shown in Figure 15.15. Remember that the line at $r = 0$ with no bc numbers in this figure denotes the axisymmetric $r = 0$ boundary condition. Also, as may be seen, a relatively coarse meshing grid is shown; the intent is to demonstrate the file generation procedure and not to strive for a very accurate result.

Figure 15.16 shows the voltage profile along the $r = 0$ axis for the CRT structure.

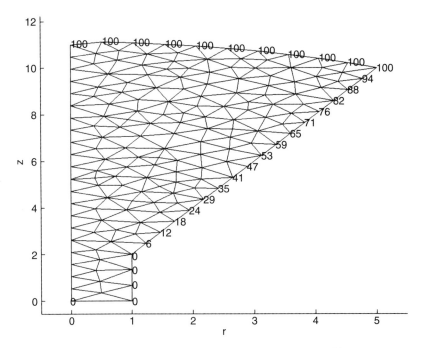

FIGURE 15.15   fem_2d.3.m figure showing CRT boundary condition setup.

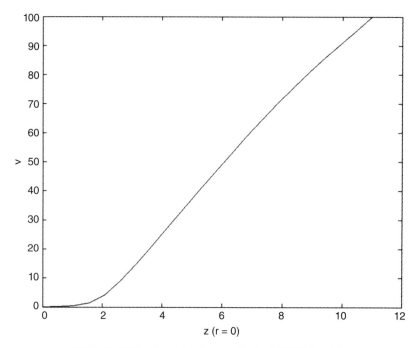

FIGURE 15.16   The $r = 0$ voltage profile for CRT FEM model.

The voltage profile shows that the neck of the CRT is essentially a field-free region; the voltage increases approximately linearly with z between $z = 2$ and the screen.

15.5   UNBOUNDED REGIONS

[*Note*: In this section we are no longer considering axisymmetric structures. We are (back) to 2d rectangular cross sections of uniform infinite (in z) structures. The MATLAB scripts and functions for the rectangular and axisymmetric cases are not very different — care should be taken that the correct programs are being used because the results are very different and using the wrong program will result in nonsensical (if any) results.]

Many structures are not well described as a layout with an enclosing boundary. In the physical world no structure is infinitely far away from everything else, but often the best physical approximation is that this structure is alone in the universe. Figure 15.17 is an example of such a structure.

Figure 15.17 shows the cross section of a parallel wire transmission line. In practice, these lines are built using periodic insulators to support the structure or the lines are embedded into a flat plastic continuous insulator. The latter structure is the 300-Ω twin lead brown TV antenna line that dominated home TV installations for many years. The former structure was popular for many years in open installations, such as antennas on large ships (before the advent of good low-loss dielectric materials) because it was the lowest-loss transmission line available.

The structure of Figure 15.17 is idealized in that there is no mechanical support. It is a good example for studying approximate analysis techniques because the exact capacitance (and hence transmission line parameters) may be calculated.[5] For $s = 3$ and $r = 0.5$, we obtain

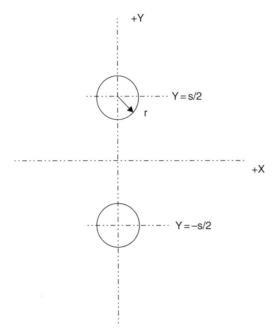

FIGURE 15.17 Cross section of a parallel wire transmission line.

$$C = \frac{\pi\epsilon}{\cosh^{-1}(s/2r)} = 15.78 \, \text{pF/m} \tag{15.9}$$

If we model this structure setting the electrodes at 0.5 and −0.5 V then we still maintain the convenience of dealing with a 1-V difference between the electrodes, and also since the average value of the voltages is zero, we can expect the voltage at infinity to be zero (see Chapter 3 for further discussion of this issue).

If we now want to build an FEM mesh to model this structure, we immediately see our problem — we need to create a mesh encompassing all space. This is clearly not a realistic task.

If the voltage goes to zero at infinity and we have centered our structure at or near the origin, then we should be able to create a $v = 0$ circular boundary far enough from the origin. This would give us a finite region to build a mesh. The question, of course, is what exactly — is meant by "far enough."

The ghmsh program struct15_4.geo—showing a parallel wire line region with a circular boundary creates just this situation:

```
//  structure file struct15_4.geo
//  First program for setting up semi-infinite boundaries

lc = 1.;

//  outer circle
rout = 6;
```

```
Point(10) = {rout, 0.0, 0, lc};
Point(11) = {0.0, 0.0, 0, lc};
Point(12) = {0.0, rout, 0, lc};
Point(13) = {-rout, 0.0, 0, lc};
Point(14) = {0, -rout, 0, lc};

Circle(20) = {10, 11, 12};
Circle(21) = {12, 11, 13};
Circle(22) = {13, 11, 14};
Circle(23) = {14, 11, 10};

sep = 3;
rin = 0.5;
//  upper inner circle
Point(30) = {rin, sep/2, 0, lc};
Point(31) = {0, rin + sep/2, 0, lc};
Point(32) = {-rin, sep/2, 0, lc};
Point(33) = {0, sep/2-rin, 0, lc};
Point(34) = {0, sep/2, 0, lc};

Circle(40) = {30, 34, 31};
Circle(41) = {31, 34, 32};
Circle(42) = {32, 34, 33};
Circle(43) = {33, 34, 30};
Line Loop(44) = {40, 41, 42, 43};
//Plane Surface(45) = {44};

//  lower inner circle
Point(50) = {rin, -sep/2, 0, lc};
Point(51) = {0, rin-sep/2, 0, lc};
Point(52) = {-rin, -sep/2, 0, lc};
Point(53) = {0, -sep/2-rin, 0, lc};
Point(54) = {0, -sep/2, 0, lc};

Circle(60) = {50, 54, 51};
Circle(61) = {51, 54, 52};
Circle(62) = {52, 54, 53};
Circle(63) = {53, 54, 50};

Line Loop(64) = {60, 61, 62, 63};
//Plane Surface(65) = {64};
Line Loop(24) = {20, 21, 22, 23};
Plane Surface(25) = {24, 44, 64};
```

The proper boundary conditions for this structure are 0.5 and −0.5 V for the inner circles and 0 V for the outer circle.

Figure 15.18 shows the results of this exercise.

At an outer boundary shell radius of approximately ≥20, the predicted capacitance is essentially the exact capacitance. The procedure works well, but its practicality is questionable. Simply increasing the outer shell increases the size of the coefficient matrix rapidly. Since this example was simply two circles, the necessary resolution was modest. If a more

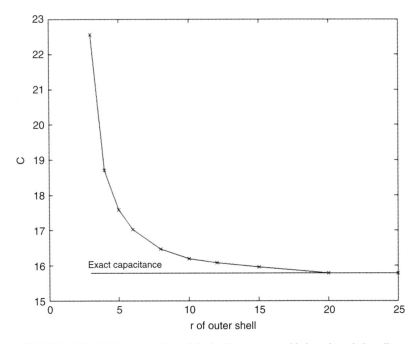

FIGURE 15.18 FEM analysis of parallel wire line versus outside boundary circle radius.

complex structure was involved and higher resolution was needed, this process could be impractical. Some other approaches are needed.

Chari and Salon present a survey of various techniques developed to deal with unbounded, or *open bounded regions*.[6] Simplified versions of two of these techniques will be presented here.

Figure 15.19 shows a section of the outer region of the structure described above (`struct15_4.geo`). The outer circle radius is R. One triangle with two nodes on the outer circle at voltages V_a and V_b is shown. This figure is exaggerated for clarity. The outer circle in an actual structure with a reasonable resolution mesh would be much closer to the outer triangle line than the figure implies.

Experience tells us that once we're a reasonable distance from an electrically neutral structure (such as our parallel wire line example), the voltage profile from a circle surrounding this structure out to infinity is not a very interesting function. There will be a an angular dependence (the angle from the X axis to a line from the origin to a field point), just as there is an angular dependence of the electric dipole voltage profile far from the dipole itself. Also, the voltage will have to fall off at least as rapidly as it does in the case of the electric dipole, that is, with the square of the distance from the origin.

In Figure 15.19, two lines from the origin (the center of the circle) have angles φ_1 and φ_2 with respect to the X axis, as shown.

Since V_b and V_a cannot be too different, and since the arc of the circle is close to the line connecting nodes a and b, we will approximate the voltage on the circle arc between these two nodes as follows:

$$V_0 \simeq \frac{V_a + V_b}{2} \tag{15.10}$$

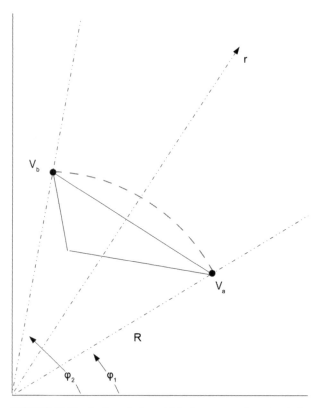

FIGURE 15.19   Section of `struct15_4.geo` near the outer circle.

Then, for $r \geq R$, we let

$$V(r) = \frac{a+b}{2}\frac{R^2}{r^2} \tag{15.11}$$

We have now created a new element for our structure. This element is bounded by the arc $r = R$ and the lines φ_1 and φ_2. It is unbounded as r goes to infinity. The voltage profile chosen [equation (15.11)] ensures that the energy in the region is finite. This voltage profile is also the voltage profile of an electric dipole (see Chapter 1).

Unfortunately, this approximation does not satisfy several other conditions that we have demanded of FEM structures up to this point. The voltage on the arc is not identically the voltage on the line connecting nodes a and b in Figure 15.20. There is a totally ignored region between this new element and the outer triangle that it connects to. The voltages along the lines from R to infinity are not the same from element to element around the circle. In other words, this is not a very good approximation. It does, however, give us a first pass at dealing with an unbounded region.

In this new element the electric field is

$$-E_r = \frac{\partial V}{\partial r} = \frac{V_a + V_b}{2}\left(\frac{-2R^2}{r^3}\right) = -(V_a + V_b)\frac{R^2}{r^3} \tag{15.12}$$

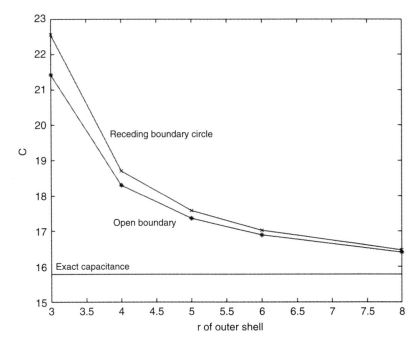

FIGURE 15.20   Results of open boundary condition approximation.

$$E^2 = (V_a + V_b)^2 \frac{R^4}{r^6} \tag{15.13}$$

The energy stored in this field is

$$U = \frac{\varepsilon}{2} \int\limits_{\phi_1}^{\phi_2} \int\limits_{R}^{\infty} E^2(r)\, r\, dr\, d\phi = 2\varepsilon(\phi_2 - \phi_2)(V_a + V_b)^2 \tag{15.14}$$

The terms that must be added to the coefficient matrix to include this energy are

$$
\begin{aligned}
0 = \frac{\partial U}{V_a} &= 4\varepsilon(\phi_2 - \phi_1)(V_a + V_b) \\
&= \frac{\partial U}{V_b} = 4\varepsilon(\phi_2 - \phi_1)(V_b + V_a)
\end{aligned}
\tag{15.15}
$$

Modifications to the MATLAB code to implement these equations are shown in the following listings:

- MATLAB script `fem_2d_4.m`—MATLAB script with simple open boundary element:

```
%  2d FEM with triangles
%  fem_2d_4.m is an extension of fem_2d_2.m,
%  for processing of semi-infinite approximation 1
```

```
close

clear

filename = input ('Generic name for input files: ', 's');
nodes_file_name = strcat ('nodes_', filename, '.txt');
trs_file_name = strcat ('trs_', filename, '.txt');
bcs_file_name = strcat ('bcs_', filename, '.txt');

%  get node data from file
dataArray = load (nodes_file_name);
nds.x = dataArray(:,1); nds.y = dataArray(:,2); nds.z = 0;
nr_nodes = length (nds.x)

%  get triangles data from file
trs = (load (trs_file_name))';
nr_trs = length (trs)

%  get bcs data from file
dataArray = load (bcs_file_name);
bcs.node = dataArray(:,2); bcs.volts = dataArray(:,3);
nr_bcs = length (bcs.node)

%  allocate the matrices
a = sparse (nr_nodes, nr_nodes);
f = zeros (nr_nodes,1);
b = zeros (3,nr_nodes); c = zeros (3,nr_nodes);

%  Find the outer radius
r_outer_sq = 0;
for i = 1 : nr_nodes
    r_sq = nds.x(i)^2 + nds.y(i)^2;
    if r_sq > r_outer_sq
      r_outer_sq = r_sq;
    end
end
fprintf ('Outer radius found to be %f \n', sqrt (r_outer_sq))
r_outer_sq = .9999*r_outer_sq; % getting around roundoff issues

%  generate the matrix terms
d = ones (3,3);
for tri = 1: nr_trs % Walk through all the triangles
  n1 = trs (1,tri); n2 = trs (2,tri); n3 = trs (3,tri);
  d(1,2) = nds.x(n1); d(1,3) = nds.y(n1);
  d(2,2) = nds.x(n2); d(2,3) = nds.y(n2);
  d(3,2) = nds.x(n3); d(3,3) = nds.y(n3);
  Area = det (d)/2;

  temp = d\ [1; 0; 0]; b(1,tri) = temp(2); c(1,tri) = temp(3);
  temp = d\ [0; 1; 0]; b(2,tri) = temp(2); c(2,tri) = temp(3);
  temp = d\ [0; 0; 1]; b(3,tri) = temp(2); c(3,tri) = temp(3);

  a(n1,n1) = a(n1,n1) + Area*( b(1,tri)^2 + c(1,tri)^2);
  a(n1,n2) = a(n1,n2) + Area*(b(1,tri)*b(2,tri) + c(1,tri)*c
(2,tri));
```

```
  a(n1,n3) = a(n1,n3) + Area*(b(1,tri)*b(3,tri) + c(1,tri)*c
(3,tri));

  a(n2,n2) = a(n2,n2) + Area*(b(2,tri)^2 + c(2,tri)^2);
  a(n2,n1) = a(n2,n1) + Area*(b(2,tri)*b(1,tri) + c(2,tri)*c
(1,tri));
  a(n2,n3) = a(n2,n3) + Area*(b(2,tri)*b(3,tri) + c(2,tri)*c
(3,tri));

  a(n3,n3) = a(n3,n3) + Area*(b(3,tri)^2 + c(3,tri)^2);
  a(n3,n1) = a(n3,n1) + Area*(b(3,tri)*b(1,tri) + c(3,tri)*c
(1,tri));
  a(n3,n2) = a(n3,n2) + Area*(b(3,tri)*b(2,tri) + c(3,tri)*c
(2,tri));

  % An outer bc triangle will have two nodes on the outer circle

  r1_sq = nds.x(n1)^2 + nds.y(n1)^2;
  r2_sq = nds.x(n2)^2 + nds.y(n2)^2;
  r3_sq = nds.x(n3)^2 + nds.y(n3)^2;
  if r1_sq > r_outer_sq & r2_sq > r_outer_sq
    % 'n3 inner bc node'
    theta1 = atan2(nds.y(n1),nds.x(n1));
    theta2 = atan2(nds.y(n2),nds.x(n2));
    t = 4*(theta2-theta1);
    a(n1,n1) = a(n1,n1) + t;
    a(n1,n2) = a(n1,n2) + t;
    a(n2,n2) = a(n2,n2) + t;
    a(n2,n1) = a(n2,n1) + t;
  elseif r2_sq > r_outer_sq & r3_sq > r_outer_sq
    % 'n1 inner bc node'
    theta1 = atan2(nds.y(n2),nds.x(n2));
    theta2 = atan2(nds.y(n3),nds.x(n3));
    % t = 4*(theta2-theta1); % here
    a(n2,n2) = a(n2,n2) + t;
    a(n2,n3) = a(n2,n3) + t;
    a(n3,n3) = a(n3,n3) + t;
    a(n3,n2) = a(n3,n2) + t;
  elseif r3_sq > r_outer_sq & r1_sq > r_outer_sq
    % 'n2 inner bc node'
    theta1 = atan2(nds.y(n3),nds.x(n3));
    theta2 = atan2(nds.y(n1),nds.x(n1));
    % t = 4*(theta2-theta1); % here
    a(n3,n3) = a(n3,n3) + t;
    a(n3,n1) = a(n3,n1) + t;
    a(n1,n1) = a(n1,n1) + t;
    a(n1,n3) = a(n1,n3) + t;
  end
end
%bcs
for i = 1: nr_bcs
    if bcs.volts(i) ~= -9999
      n = bcs.node(i);
```

```
      a(n, :) = 0; a(n,n) = 1; f(n) = bcs.volts(i);
   end
end

%  fake bcs for nodes in excluded region - assuming boundary at 1 volt
%  There might not be any excluded regions in symmetry problems
%  This is being left for backwards-compatibility of the program
for i = 1 : nr_nodes
  if a(i,i) == 0
    a(i, :) = 0; a(i,i) = 1; f(i) = 1;
  end
end

v = a\f;

%  plot the mesh
show_mesh2(nr_nodes, nr_trs, nr_bcs, nds, trs, bcs, v)

%  calculate capacitance
C = get_tri2d_cap4(v, nr_nodes, nr_trs, nds, trs)

%  show field profile
get_tri2d_E(v, nr_nodes, nr_trs, nds, trs)

%  show voltage along y = 0 edge

x_temp = []; v_temp = [];
for i = 1 : nr_nodes
  if nds.y(i) == 0
%    fprintf ('%d  %d  %f \n', i, nds.y(i), v(i))
    x_temp  =  [x_temp, nds.x(i)];
    v_temp  =  [v_temp, v(i)];
  end
end
figure(4)
plot(x_temp, v_temp, '.')
```

- MATLAB function `get_tri2d_cap4.m`—MATLAB function for capacitance of simple open boundary element:

```
function C = get_tri2d4_cap(v, nr_nodes, nr_trs, nds, trs)
%  Calculate the capacitance of a 2d triangular structure
%  Simple linear approx. function is assumed
%  This version is for first infinite boundary approximation

  eps0 = 8.854;

%  Find the outer radius
  r_outer_sq = 0;
  for i = 1 : nr_nodes
    r_sq = nds.x(i)^2 + nds.y(i)^2;
    if r_sq > r_outer_sq
```

```
      r_outer_sq = r_sq;
    end
  end
  fprintf ('Outer radius found to be %f \n', sqrt(r_outer_sq))%
  r_outer_sq = .9999*r_outer_sq; % getting around roundoff issues

  U = 0;
  d = ones(3,3);
  for tri = 1 : nr_trs
    n1 = trs(1,tri); n2 = trs(2,tri); n3 = trs(3,tri);
    d(1,2) = nds.x(n1); d(1,3) = nds.y(n1);
    d(2,2) = nds.x(n2); d(2,3) = nds.y(n2);
    d(3,2) = nds.x(n3); d(3,3) = nds.y(n3);
    temp = (d\[1; 0; 0]); b1 = temp(2); c1 = temp(3);
    temp = (d\[0; 1; 0]); b2 = temp(2); c2 = temp(3);
    temp = (d\[0; 0; 1]); b3 = temp(2); c3 = temp(3);

    Area = det(d)/2;
    if Area < = 0, disp 'Area < = 0 error'; end;

    U = U + Area*((v(n1)*b1 + v(n2)*b2 + v(n3)*b3)^2 ...
        + (v(n1)*c1 + v(n2)*c2 + v(n3)*c3)^2);

%   Add terms to the outer triangles
%   An outer bc triangle will have two nodes on the outer circle
    r1_sq = nds.x(n1)^2 + nds.y(n1)^2;
    r2_sq = nds.x(n2)^2 + nds.y(n2)^2;
    r3_sq = nds.x(n3)^2 + nds.y(n3)^2;
    if r1_sq > r_outer_sq & r2_sq > r_outer_sq
%    'n3 inner bc node'
        theta1 = atan2(nds.y(n1),nds.x(n1));
        theta2 = atan2(nds.y(n2),nds.x(n2));
        t = 4*(theta2-theta1);
        U = U + t*(v(n1) + v(n2))^2;
    elseif r2_sq > r_outer_sq & r3_sq > r_outer_sq
%    'n1 inner bc node'
        theta1 = atan2(nds.y(n2),nds.x(n2));
        theta2 = atan2(nds.y(n3),nds.x(n3));
        t = 4*(theta2-theta1);
        U = U + t*(v(n2) + v(n3))^2;
    elseif r3_sq > r_outer_sq & r1_sq > r_outer_sq
%    'n2 inner bc node'
      theta1 = atan2(nds.y(n3),nds.x(n3));
      theta2 = atan2(nds.y(n1),nds.x(n1));
      t = 4*(theta2-theta1);
      U = U + t*(v(n3) + v(n1))^2;
    end

  end

  U = U*eps0/2;
  C = 2*U;               % Potential difference of 1 volt assumed

end
```

In both of these listings, the program code was written to clearly show calculation of the new terms. Neither listing reflects coding for optimum performance.

The results of the FEM simulation presented above using the same structure as previously (`struct10.geo`) are shown in Figure 15.20. The open boundary approximation brings a definite although not a large amount of improvement to the calculation. The most improvement occurs at small values of outer-shell radius, as would be expected.

Another approach to dealing with the open boundary problem is exemplified by the gmsh file `struct15_5.geo`.

- File gmsh `struct15_5.geo`—layered resolution boundary region example:

```
//  structure file struct15_5.geo
//  2nd program for setting up semi-infinite boundaries

lc1 = 1.; lc2 = 1.5; lc3 = 2;
Point(1) = {0, 0, 0, lc1};      // Center of structure high res
Point(2) = {0, 0, 0, lc2};      // Center of structure low res
Point(5) = {0,0,0, lc3};        // center of structure very low res
sep = 3.; // Center - Center separation of inner circles
Point(3) = {0, sep/2, 0, lc1};
Point(4) = {0, -sep/2, 0, lc1};

//  3rd level outer
r3 = 6;
Point(50) = {r3, 0, 0, lc3};
Point(51) = {0, r3, 0, lc3};
Point(52) = {-r3, 0, 0, lc3};
Point(53) = {0, -r3, 0, lc3};
Circle(54) = {50, 2, 51};
Circle(55) = {51, 2, 52};
Circle(56) = {52, 2, 53};
Circle(57) = {53, 2, 50};
Line Loop(58) = {54, 55, 56, 57};

//  2nd level outer
r2 = 4;
Point(10) = {r2, 0, 0, lc2};
Point(11) = {0, r2, 0, lc2};
Point(12) = {-r2, 0, 0, lc2};
Point(13) = {0, -r2, 0, lc2};
Circle(14) = {10, 2, 11};
Circle(15) = {11, 2, 12};
Circle(16) = {12, 2, 13};
Circle(17) = {13, 2, 10};
Line Loop(18) = {14, 15, 16, 17};

//  Original Structure outer level
r1 = 3;
Point(20) = {r1, 0, 0, lc1};
Point(21) = {0, r1, 0, lc1};
```

```
Point(22) = {-r1, 0, 0, lc1};
Point(23) = {0, -r1, 0, lc1};
Circle(24) = {20, 1, 21};
Circle(25) = {21, 1, 22};
Circle(26) = {22, 1, 23};
Circle(27) = {23, 1, 20};
Line Loop(28) = {24, 25, 26, 27};


//  Upper inner circle
r0 = 0.5;
Point(30) = {r0, sep/2, 0, lc1};
Point(31) = {0, r0+sep/2, 0, lc1};
Point(32) = {-r0, sep/2, 0, lc1};
Point(33) = {0, sep/2-r0, 0, lc1};
Circle(34) = {30, 3, 31};
Circle(35) = {31, 3, 32};
Circle(36) = {32, 3, 33};
Circle(37) = {33, 3, 30};
Line Loop(38) = {34, 35, 36, 37};


//  Lower inner circle
r0 = 0.5;
Point(40) = {r0, -sep/2, 0, lc1};
Point(41) = {0, r0-sep/2, 0, lc1};
Point(42) = {-r0, -sep/2, 0, lc1};
Point(43) = {0, -sep/2-r0, 0, lc1};
Circle(44) = {40, 4, 41};
Circle(45) = {41, 4, 42};
Circle(46) = {42, 4, 43};
Circle(47) = {43, 4, 40};


Line Loop(48) = {44, 45, 46, 47};

Plane Surface(100) = {28, 38, 48};
Plane Surface (19) = {18, 28};
Plane Surface (59) = {58, 18};
```

Program `struct15_5.geo` is a repeat of `struct10.geo` using a $v = 0$ outer boundary, but the space inside the outer boundary consists of successive annular rings, with the resolution decreasing (`lc` value increasing) as the radii of the rings increases. In other words, use less resolution where less resolution is needed. This example, with an outer radius of 6, results in a capacitance calculation of 17.33 pF/m. This is not quite as accurate as the capacitance predicted for an outer radius of 6 with uniform high resolution, but it is accomplished using significantly fewer nodes (169 vs. 321—almost a factor of 2 improvement). These results may, of course, be extended using larger outer rings and further decreased resolution. The attractive features of this approach are that there are no further mathematical approximations or program modifications necessary and that `gmsh` quite easily handles all of the details.

FIGURE 15.21   mesh file `struct15_5.jpg`.

15.6 DIELECTRIC MATERIALS

It is very easy to modify the calculations and computer programs to correctly handle the effects of different dielectric constants in the structure, as long as the dielectric constant is uniform in each triangle. This isn't a limitation — simply define the triangles so that this condition is satisfied.

Going back to the derivations in Chapter 13, the total energy in the system is

$$U = \sum_T E_T^2 A_T \qquad (15.16)$$

The sum here is over all of the triangles in the structure.

In equation 15.16, the contribution from each triangle to U contains ε, the dielectric permittivity. Assume now that each triangle has a unique permittivity, $\varepsilon = k_T \varepsilon_0$, where k_T is the relative dielectric constant of triangle T.

We build the coefficient matrix row by row by taking the derivative of equation (15.16) with respect to each node voltage and set this derivative to zero. When we do this, $\varepsilon_0/2$ drops out but k_T remains for each triangle's (T) contribution.

From a programming perspective, we must add the relative dielectric constant information to the structure's description — either as a new column in the triangle's data file or possibly as a new file altogether. Repeating the earlier example of triangle T's contribution to the coefficient matrix, assume that we have an eight-triangle system with triangle T's vertices at nodes 1, 3, and 4. The contribution to the coefficient matrix of triangle T is now equal to (see equation 13.13 for comparison)

$$
\begin{bmatrix}
A_T k_T \left(b_1^2 + c_1^2 \right) & 0 & A_t k_T (b_1 b_2 + c_1 c_2) & A_T k_T (b_1 b_3 + c_1 c_3) & 0 & 0 & 0 & 0 \\
0 & 0 & 0 & 0 & 0 & 0 & 0 & 0 \\
A_T k_T (b_1 b_2 + c_1 c_2) & 0 & A_T k_T \left(b_2^2 + c_2^2 \right) & A_T k_T (b_2 b_3 + c_2 c_3) & 0 & 0 & 0 & 0 \\
A_T k_T (b_1 b_3 + c_1 c_3) & 0 & A_T k_T (b_2 b_3 + c_2 c_3) & +A_T k_T \left(b_3^2 + c_3^2 \right) & 0 & 0 & 0 & 0 \\
0 & 0 & 0 & 0 & 0 & 0 & 0 & 0 \\
0 & 0 & 0 & 0 & 0 & 0 & 0 & 0 \\
0 & 0 & 0 & 0 & 0 & 0 & 0 & 0 \\
0 & 0 & 0 & 0 & 0 & 0 & 0 & 0
\end{bmatrix}
$$

$$(15.17)$$

Updating the earlier calculation is easy; when creating each triangle's contribution to the coefficient matrix, simply multiply that triangle's area by its (relative) dielectric constant.

As an example, consider the gmsh file `struct15_6.geo`—structure file for microstrip line example:

```
//      structure file struct15_6.geo
//      structure for symmetry calculation
//      stripline / microstrip line example

lc = .2;            // background resolution

t = .1;             // thickness of center conductor
w = 1.0;            // 1/2 width of center conductor
h = 8;              // height of box
d = 2.;             // height of dielectric slab
b = 4;              // 1/2 width of outer box

Point(1) = {0, 0, 0, lc};
Point(2) = {b, 0, 0, lc};
Point(3) = {b, h, 0, lc};
Point(4) = {0, h, 0, lc};
Point(5) = {0, d + t, 0, lc};
Point(6) = {w, d + t, 0, lc/5}; // higher resolution near the conductor edges
Point(7) = {w, d, 0, lc/5};
Point(8) = {0, d, 0, lc};

Line(1) = {1, 2};
Line(2) = {2, 3};
Line(3) = {3, 4};
Line(4) = {4, 5};
Line(5) = {5, 6};
Line(6) = {6, 7};
Line(7) = {7, 8};
Line(8) = {8, 1};
```

(a)

(b)

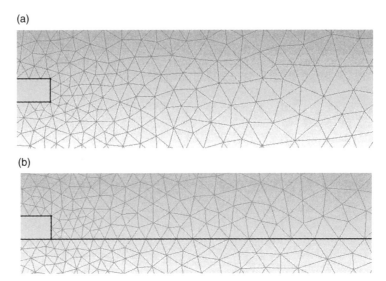

FIGURE 15.22 gmsh structure file with dielectric slab edge defined.

```
Line Loop(1) = {1, 2, 3, 4, 5, 6, 7, 8};
Plane Surface(1) = {1} ;
```

File `struct15_6.geo` utilizes *X* symmetry and creates a boxed microstrip cross section. It is set up with a relatively low-resolution meshing, except in the area of the edge of the center conductor. The resulting mesh file may be parsed through `parse_msh_file2.m` and then the FEM analysis performed using `fem_2d.m`.

The purpose of this example is to create a dielectric slab in the region below the center conductor ($y = d = 2.0$ in the `struct15_6.geo` file) and modify `fem_2d.m` to correctly set up and solve the FEM model.

Our first task arises, as may be seen by examining the mesh file Figure 15.22a is to create a straight edge along $y = d$ so that the dielectric slab region can be defined (Figure 15.22b). Since `struct15_6.geo` was created without this goal in mind, the resulting mesh triangle vertices are not along the desired interface line.

This issue is resolved in `struct15_7.geo`—`struct15_6` with dielectric edge delineated—by adding a line along what will be the dielectric surface:

```
// structure file struct15_7.geo
// structure for symmetry calculation
// stripline / microstrip line example
// dielectric edge line added

lc = .2;                    // background resolution

t = .1;                     // thickness of center conductor
w = 1.0;                    // 1/2 width of center conductor
h = 8;                      // height of box
d = 2.;                     // height of dielectric slab
b = 4;                      // 1/2 width of outer box
```

```
Point(1) = {0, 0, 0, lc};
Point(2) = {b, 0, 0, lc};
Point(3) = {b, h, 0, lc};
Point(4) = {0, h, 0, lc};
Point(5) = {0, d + t, 0, lc};
Point(6) = {w, d + t, 0, lc/5};    // higher resolution near the
conductor edges
Point(7) = {w, d, 0, lc/5};
Point(8) = {0, d, 0, lc};
Point(9) = {b, d, 0, lc};

Line(1) = {1, 2};
Line(2) = {2, 9};
Line(3) = {9, 7};
Line(4) = {7, 8};
Line(5) = {8, 1};

// Lower box
Line Loop(1) = {1, 2, 3, 4, 5};
Plane Surface(1) = {1} ;

Line(6) = {9, 3};
Line(7) = {3, 4};
Line(8) = {4, 5};
Line(9) = {5, 6};
Line(10) = {6, 7};

//upper box
Line Loop(2) = {-3, 6, 7, 8, 9, 10};
Plane Surface(2) = 2;
```

Two modifications to `fed_2d.m` are required in order to perform the dielectric edge calculation, and then a modification to the capacitance calculation function:

- MATLAB program `fem_2d_diel.m`—namely, `fem_2d_2.m` with dielectric region capability:

```
%  2d FEM with triangles
%  fem_2d_4.m is an extension of fem_2d_2.m, with dielectric regions

close
clear

filename = input('Generic name for input files: ', 's');
nodes_file_name = strcat('nodes_', filename, '.txt');
trs_file_name = strcat('trs_', filename, '.txt');
bcs_file_name = strcat('bcs_', filename, '.txt');

%  get node data from file
nds = get_nodes(nodes_file_name);
nr_nodes = length(nds.x)

% get triangles data from file
```

```
trs = get_triangles(trs_file_name);
nr_trs = length(trs)

% get bcs data from file
bcs = get_bcs(bcs_file_name);
nr_bcs = length(bcs.node)

edge_y = 2.0;

% allocate the matrices
a = sparse(nr_nodes, nr_nodes);
f = zeros(nr_nodes,1);
b = zeros(3,nr_nodes); c = zeros(3,nr_nodes);

% generate the matrix terms
d = ones(3,3);
all_er = [];
for tri = 1: nr_trs % Walk through all the triangles
  n1 = trs(1,tri); n2 = trs(2,tri); n3 = trs(3,tri);
  d(1,2) = nds.x(n1); d(1,3) = nds.y(n1);
  d(2,2) = nds.x(n2); d(2,3) = nds.y(n2);
  d(3,2) = nds.x(n3); d(3,3) = nds.y(n3);
  Area = det(d)/2;

% ---- Patch for dielectric region

th = 1.001*edge_y;        % Height of the dielectric layer
er = 20;                  % Dielectric constant
if nds.y(n1) < = th & nds.y(n2) < = th & nds.y(n3) < = th
  Area = Area*er;
  all_er = [all_er, er];
else
    all_er = [all_er, 1];
end

% ------------------------------------------------------------

temp = d\[1; 0; 0]; b(1,tri) = temp(2); c(1,tri) = temp(3);
temp = d\[0; 1; 0]; b(2,tri) = temp(2); c(2,tri) = temp(3);
temp = d\[0; 0; 1]; b(3,tri) = temp(2); c(3,tri) = temp(3);

a(n1,n1) = a(n1,n1) + Area*(b(1,tri)^2 + c(1,tri)^2);
a(n1,n2) = a(n1,n2) + Area*(b(1,tri)*b(2,tri) + c(1,tri)*c(2,tri));
a(n1,n3) = a(n1,n3) + Area*(b(1,tri)*b(3,tri) + c(1,tri)*c(3,tri));

a(n2,n2) = a(n2,n2) + Area*(b(2,tri)^2 + c(2,tri)^2);
a(n2,n1) = a(n2,n1) + Area*(b(2,tri)*b(1,tri) + c(2,tri)*c(1,tri));
a(n2,n3) = a(n2,n3) + Area*(b(2,tri)*b(3,tri) + c(2,tri)*c(3,tri));

a(n3,n3) = a(n3,n3) + Area*(b(3,tri)^2 + c(3,tri)^2);
a(n3,n1) = a(n3,n1) + Area*(b(3,tri)*b(1,tri) + c(3,tri)*c(1,tri));
a(n3,n2) = a(n3,n2) + Area*(b(3,tri)*b(2,tri) + c(3,tri)*c(2,tri));
```

```
end

%bcs
for i = 1: nr_bcs
  if bcs.volts(i) >= 0
    n = bcs.node(i);
    a(n,:) = 0; a(n,n) = 1; f(n) = bcs.volts(i);
  end
end

% fake bcs for nodes in excluded region - assuming boundary at 1 volt
% There might not be any excluded regions in symmetry problems
% This is being left for backwards-compatibility of the program
for i = 1 : nr_nodes
  if a(i,i) == 0
    a(i,:) = 0; a(i,i) = 1; f(i) = 1;
  end
end

v = a\f;

% plot the mesh
show_mesh2(nr_nodes, nr_trs, nr_bcs, nds, trs, bcs, v)

% calculate capacitance
C = get_tri2d_cap_diel(v, nr_nodes, nr_trs, nds, trs, all_er)

% show field profile
get_tri2d_E(v, nr_nodes, nr_trs, nds, trs)

% show voltage along y = 0 edge_y
x_temp = []; v_temp = [];
for i = 1 : nr_nodes
  if nds.y(i) == 0
%     fprintf ('%d %d %f \n', i, nds.y(i), v(i))
    x_temp = [x_temp, nds.x(i)];
    v_temp = [v_temp, v(i)];
  end
end
figure(4)
plot(x_temp, v_temp, '.')
```

- MATLAB program `get_tri2d_2d_cap_diel.m`—capacitance calculation with dielectric region capability:

```
function C = get_tri2d_cap_diel(v, nr_nodes, nr_trs, nds,
trs, all_er)
%      Calculate the capacitance of a 2d triangular structure
%      dielectric region added

  eps0 = 8.854;
```

```
U = 0;
d = ones(3,3);
for tri = 1 : nr_trs
    n1 = trs(1,tri); n2 = trs(2,tri); n3 = trs(3,tri);
    d(1,2) = nds.x(n1); d(1,3) = nds.y(n1);
    d(2,2) = nds.x(n2); d(2,3) = nds.y(n2);
    d(3,2) = nds.x(n3); d(3,3) = nds.y(n3);
    temp = (d\[1; 0; 0]); b1 = temp(2); c1 = temp(3);
    temp = (d\[0; 1; 0]); b2 = temp(2); c2 = temp(3);
    temp = (d\[0; 0; 1]); b3 = temp(2); c3 = temp(3);

    Area = det(d)/2;
    if Area < = 0, disp 'Area < = 0 error'; end;

    U = U + all_er(tri)*Area*((v(n1)*b1 + v(n2)*b2 + v(n3)*b3)^2 ...
        + (v(n1)*c1 + v(n2)*c2 + v(n3)*c3)^2);

end

U = U*eps0/2;
C = 2*U;            % Potential difference of 1 volt assumed

end
```

In the FEM analysis, the dielectric region is defined and each triangle's area is multiplied by the (relative) dielectric constant as needed. An array `all_er` is created for convenience in passing the dielectric region information to the capacitance calculated.

In `parse_msh_file2.m`, the −9999 boundary condition entry was created to specify a symmetry edge. In `struct15_6.geo` and `struct15_7.geo` this capability is used on the two left edge lines. Program `fem_2d.m` translates this boundary condition into the required "do nothing" for nodes along the symmetry line. In `struct15_7.geo`, the interface line between the dielectric and air regions shows up as a line requiring a boundary condition. Specifying −9999 for this line produces the desired result. Each time the program encounters these nodes in setting up the boundary conditions, it treats these nodes as if they were internal nodes; these nodes show up for every triangle using these nodes, on both sides of the dielectric interface, resulting in the correct coefficient matrix assembly.

The full boundary condition specification information is listed here:

Line N Number	Boundary Condition
1	0 (default)
2	0 (default)
3	−9999
4	1
5	−9999
6	0 (default)
7	0 (default)
8	−9999
9	1
10	1

The following table lists the results of this capacitance calculation as compared to published results for open microstripline:[7]

Relative Dielectric Constant	FEM Analysis	Published Results	Error (%)
1	30.6	27.4	12.0
2	47.6	44.5	7.0
10	182.8	177.0	3.3
20	351.0	342.1	2.6

The results are relatively poor for a relative dielectric constant of one (12%); the calculated capacitance is much too high, but improves as the dielectric constant increases. The extra capacitance can be attributed to the finite distance to the sidewall and topwall in the FEM calculation; the published results pertain to an open line (with no sidewall or topwall). As the dielectric constant increases (not changing the resolution, i.e. using the same `struct15_7.msh` file), the fraction of electric field energy stored in the dielectric increases and the presence of the sidewall and topwall become less and less significant.

This program, as written, is awkward. Two straightforward improvements would be to

1 Create a new file for the boundary condition specification: This would save a tedious and error-prone user input process.

2 Create a file containing the dielectric constant information or augment the triangle information file with this information. Operationally, both approaches are equivalent; however, the former approach can be written to maintain compatibility with earlier data files by having the program default the entire mesh to an air dielectric if no dielectric constant file is found.

PROBLEMS

15.1 Write a simple `gmsh` file for a circular coaxial cable cross section with inner radius 1 and outer radius 2. Start with the inner circle centered inside the outer circle (a concentric cable) and then move the inner circle in small increments in any direction toward an outer wall. Plot how the capacitance and the peak electric field magnitude vary with the position of the center conductor.

15.2 Referring to the structure of Problem 15.1, return the inner circle to the center and then modify the `.geo` file to allow subsections to be drawn as shown in Figure P15.2. The capacitance, voltage, and electric field distributions of the full line may then be found by considering the two (straight) radial lines, symmetry axes.

Calculate the capacitance and the peak electric field of the full structure for several values of the angle φ shown in Figure P15.2.

15.3 For values of φ approaching 90°, Figure P15.2 resembles the cross section of a parallel plate capacitor that is infinite in extent and is being modeled using symmetry axes. Can an approximate capacitance expression for the concentric coaxial cable be derived using only the ideal parallel plate capacitor equation?

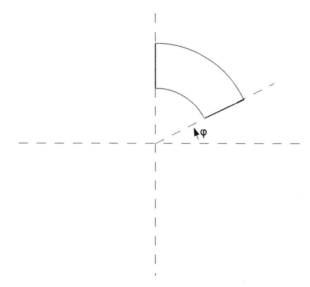

FIGURE P15.2   Sample subsection of concentric coaxial line.

15.4   A finite length of transmission line will have a series of resonance frequencies, that is, frequencies at which the line will support standing waves. This is directly analogous to the audiofrequencies produced by a violin string. (One of) these frequencies may be used, with supporting electronics, to create a frequency reference signal. Suppose that we wish to use a boxed microstrip line. Assume the line to be infinite in extent for a capacitance calculation, and ignore the fringing capacitances discussed in the MoM chapters. The resonance frequency of the line will vary, to a good approximation, with $1/\sqrt{C}$, where C is the center conductor to outer box capacitance. Unfortunately, the top of the outer box, which is a sheet of metal supported only at its edges, is easily flexed by outside forces (e.g., stresses due to mounting the microstrip package in a product of some sort), and even with identical microstip packages, the resulting resonance frequency is too unpredictable for practical use.

A   For the dimensions shown, calculate the relative variation in resonance frequency due to the box cover flexing. Approximate the flexing as a small change in overall box height.

B   Double the relative dielectric constant of the dielectric slab (from 10 to 20). Reduce the center conductor width to get the capacitance back to its original value. Now repeat part **a**. Have we made things better or worse? Here is the gmsh structure file for Problem 15.4:

```
// structP15_4.geo This is the same structure as struct51.geo,
// the dimensions have been changed

lc = .2;                      // background resolution

t = .025;                       // thickness of center conductor
w = .070;                     // 1/2 width of center conductor
h = 1.98;                      // height of box
d = 1.;                      // height of dielectric slab
b = 2.500;                       // 1/2 width of outer box
```

```
Point(1) = {0, 0, 0, lc};
Point(2) = {b, 0, 0, lc};
Point(3) = {b, h, 0, lc};
Point(4) = {0, h, 0, lc};
Point(5) = {0, d+t, 0, lc};
Point(6) = {w, d+t, 0, lc/5}; // higher resolution near the
conductor edges
Point(7) = {w, d, 0, lc/5};
Point(8) = {0, d, 0, lc};
Point(9) = {b, d, 0, lc};

Line(1) = {1, 2};
Line(2) = {2, 9};
Line(3) = {9, 7};
Line(4) = {7, 8};
Line(5) = {8, 1};

// Lower box
Line Loop(1) = {1, 2, 3, 4, 5};
Plane Surface(1) = {1} ;

Line(6) = {9, 3};
Line(7) = {3, 4};
Line(8) = {4, 5};
Line(9) = {5, 6};
Line(10) = {6, 7};

//upper box
Line Loop(2) = {-3, 6, 7, 8, 9, 10};
Plane Surface(2) = 2;
```

REFERENCES

1. Y. Kwon and H. Bang, *The Finite Element Method Using Matlab*, 2nd ed., CRC Press, Boca Raton, FL, 2000.

2. http://galileoandeinstein.physics.virginia.edu/142E/
   10_1425_web_ppt_pdfs/10_1425_web_Lec_17_CenterOfMassEtc.pdf.

3. http://www.heisingart.com/dvc/ch%2009%20center%20of%20mass.pdf.

4. http://faculty.trinityvalleyschool.org/hoseltom/handouts/Center%
   20of%20Mass3.pdf.

5. W. R. Smythe, *Static and Dynamic Electricity*, Section 4.14, McGraw Hill, New York, 1950.

6. M. V. K. Chari and S. J. Salon, *Numerical Methods in Electromagnetism*, Academic Press, San Diego, 2000.

7. http://www.cepd.com/calculators/microstrip.htm.

16

FEM in Three Dimensions

16.1   CREATING THREE-DIMENSIONAL MESHES

Our first task in developing three-dimensional (3d) FEM capability is to generate 3d meshes to describe our geometries. Fortunately, gmsh is well equipped to do this, with its two-dimensional (2d) meshing capabilities extending directly into three dimensions. As in the case of 2d meshes, there are many ways to generate a given geometry in three dimensions. In this chapter we will consider only one of these approaches. The gmsh documentation and many online sources should be consulted for a comprehensive discussion of this topic.

A problem that we will encounter using unstructured 3d meshes is that visualizing what we have done can be difficult; there is simply too much going on. In the examples in this chapter, many of the figures of meshed structures were generated using low-resolution meshing. This is necessary to make these figures at all useful. The .geo file accompanying these examples will show the resolution used to generate any results quoted from the FEM analysis. Also, the gmsh ability to rotate a figure in space was utilized in generating some of the figures. Since the rotation chosen for a figure was arbitrary (it looked good), details of the rotation are not given.

The gmsh script struct16_1.geo shows the construction of a cube:

```
// struct16_1.geo
// first experiments with 3D

lc = 1;

Point(1) = {-.5, -.5, -.5, lc};
Point(2) = {.5, -.5, -.5, lc};
```

Introduction to Numerical Electrostatics Using MATLAB, First Edition. Lawrence N. Dworsky.
© 2014 John Wiley & Sons, Inc. Published 2014 by John Wiley & Sons, Inc.

```
Point(3) = {.5, .5, -.5, lc};
Point(4) = {-.5, .5, -.5, lc};

Point(5) = {-.5, -.5, .5, lc};
Point(6) = {.5, -.5, .5, lc};
Point(7) = {.5, .5, .5, lc};
Point(8) = {-.5, .5, .5, lc};

Line(1) = {1, 2};
Line(2) = {2, 3};
Line(3) = {3, 4};
Line(4) = {4, 1};

Line(5) = {5, 6};
Line(6) = {6, 7};
Line(7) = {7, 8};
Line(8) = {8, 5};

Line(9) = {1, 5};
Line(10) = {2, 6};
Line(11) = {3, 7};
Line(12) = {4, 8};

Line Loop(1) = {1, 2, 3, 4}; // front
Line Loop(2) = {1, 10, -5, -9}; // bottom
Line Loop(3) = {-3, 11, 7, -12}; // top
Line Loop(4) = {5, 6, 7, 8}; // back
Line Loop(5) = {-9, -4, 12, 8}; // left
Line Loop(6) = {-10, 2, 11, -6}; // right

Plane Surface(1) = {1};
Plane Surface(2) = {2};
Plane Surface(3) = {3};
Plane Surface(4) = {4};
Plane Surface(5) = {5};
Plane Surface(6) = {6};

Surface Loop(1) = {1, 2, 3, 4, 5, 6};
Volume(1) = {1};

//Volume(1) = {10, 20, 30, 40, 50, 60};
```

Examining this file, first we see the eight points that define the vertices of the cube. Then we see the 12 lines that connect these points to form the edges of the cube, followed by the six line loops that make up the edges of the surfaces of the cube. Note the negative numbers in some of the line loop definitions. A line loop must "walk its way" around a surface, with all lines interconnected in a tail-to-head configuration (first defining number to second defining number). If a line is defined incorrectly, simply putting a minus sign in front of its number in the line loop definition has the effect of reversing the head and tail of the line.

Continuing, we see the six plane surface definitions for the six faces of the cube. We ultimately see two lines that we haven't seen before. The *surface loop* definition connects the six surfaces defined by line loops into one closed surface, just as the *line loop* definition connects lines into a closed loop. Finally, the *volume* definition identifies the volume enclosed by the surface loop, our cube.

Meshing the volume is accomplished with the Mesh – 3D command. The result of this is shown in Figure 16.1. Several points to note are:

1 Figure 16.1 is a rotated view of the figure that the gmsh screen first shows, as discussed above. This rotation is accomplished by grabbing the small coordinate axis in the lower right corner of the gmsh screen and dragging it around. The original orientation can be returned by clicking on the Z in the lower left corner of the gmsh screen. Clicking on Z means that we are looking down at a transparent X–Y plane. Similarly, the X or Y immediately to the left of the Z can be clicked on. In this simple example all three choices will yield the same picture.

2 As seen in the rotated figure, gmsh has zoned the cube into a collection of tetrahedrons. A *tetrahedron* is a four-sided figure in which each face is a triangle. Using tetrahedrons as our 3d elements is not a necessary choice, but it is the simplest way to move from triangles in two dimensions. In this simple example the grid appears to be structured. This is the result of the highly symmetric geometry of the cube, not an inherent property of gmsh.

3 The resolution chosen is very low.

4 This cube is an acceptable geometry for a bounding electrode box or for a cubical electrode inside some bounding region, but by itself there is no possible electrostatic FEM analysis. The surface of the cube will be a boundary condition, and we need (at least) two different boundary conditions (voltages) before there is a voltage distribution to find.

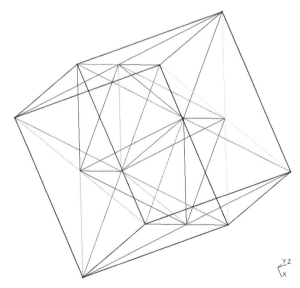

FIGURE 16.1 A meshed cube.

The gmsh script struct16_2.geo is a geometry description almost ready for FEM analysis:

```
// Struct16_2.geo
// Inner and outer box experiment

lc = 1;

// inner box

Point (1) = {-.5, -.5, -.5, lc};
Point (2) = {.5, -.5, -.5, lc};
Point (3) = {.5, .5, -.5, lc};
Point (4) = {-.5, .5, -.5, lc};

Point (5) = {-.5, -.5, .5, lc};
Point (6) = {.5, -.5, .5, lc};
Point (7) = {.5, .5, .5, lc};
Point (8) = {-.5, .5, .5, lc};

Line (1) = {1, 2};
Line (2) = {2, 3};
Line (3) = {3, 4};
Line (4) = {4, 1};

Line (5) = {5, 6};
Line (6) = {6, 7};
Line (7) = {7, 8};
Line (8) = {8, 5};

Line (9) = {1, 5};
Line (10) = {2, 6};
Line (11) = {3, 7};
Line (12) = {4, 8};

Line Loop (1) = {1, 2, 3, 4}; // front
Line Loop (2) = {1, 10, -5, -9}; // bottom
Line Loop (3) = {-3, 11, 7, -12}; // top
Line Loop (4) = {5, 6, 7, 8}; // back
Line Loop (5) = {-9, -4, 12, 8}; // left
Line Loop (6) = {-10, 2, 11, -6}; // right

Plane Surface (1) = {1};
Plane Surface (2) = {2};
Plane Surface (3) = {3};
Plane Surface (4) = {4};
Plane Surface (5) = {5};
Plane Surface (6) = {6};

Surface Loop (1) = {1, 2, 3, 4, 5, 6};
```

```
// outer box

Point(100) = {-1, -1, -1, lc};
Point(200) = {1, -1, -1, lc};
Point(300) = {1, 1, -1, lc};
Point(400) = {-1, 1, -1, lc};

Point(500) = {-1, -1, 1, lc};
Point(600) = {1, -1, 1, lc};
Point(700) = {1, 1, 1, lc};
Point(800) = {-1, 1, 1, lc};

Line(100) = {100, 200};
Line(200) = {200, 300};
Line(300) = {300, 400};
Line(400) = {400, 100};

Line(500) = {500, 600};
Line(600) = {600, 700};
Line(700) = {700, 800};
Line(800) = {800, 500};

Line(900) = {100, 500};
Line(1000) = {200, 600};
Line(1100) = {300, 700};
Line(1200) = {400, 800};

Line Loop(100) = {100, 200, 300, 400}; // front
Line Loop(200) = {100, 1000, -500, -900}; // bottom
Line Loop(300) = {-300, 1100, 700, -1200}; // top
Line Loop(400) = {500, 600, 700, 800}; // back
Line Loop(500) = {-900, -400, 1200, 800}; // left
Line Loop(600) = {-1000, 200, 1100, -600}; // right

Plane Surface(100) = {100};
Plane Surface(200) = {200};
Plane Surface(300) = {300};
Plane Surface(400) = {400};
Plane Surface(500) = {500};
Plane Surface(600) = {600};

Surface Loop(100) = {100, 200, 300, 400, 500, 600};

Volume(1) = {100, 1};
```

Script `struct16_2.geo` extends `struct16_1.geo` in that now there are two cubes, a smaller cube centered inside a larger cube, as seen in Figure 16.2.

The volume statement in `struct16_2.geo` specifies the surface loop of the larger, outer cube followed by the surface loop of the smaller, inner cube. The resultant meshing treats the inner cube as a *subtracted* region and creates tetrahedrons only in the region between (and on) the two cube surfaces. This is exactly the meshing we want for treating

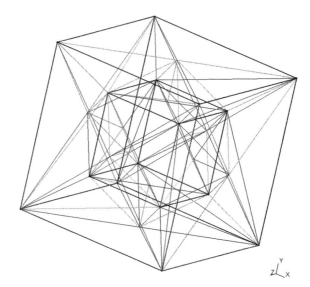

FIGURE 16.2   Two concentric cubes.

this as an electrostatics problem where the two cube surfaces are electrodes (and the outer cube surface is a bounding surface). Once again, in this example the resolution of the meshing is kept very low so that the figure is comprehensible to the viewer. A careful examination of Figure 16.2 will reveal that the resolution is so low that all of the nodes are located on the cube surfaces. In other words, there are no free variables (voltages) for an FEM analysis to find. The obvious answer here is to decrease the value of `lc` in `struct16_2.geo` sufficiently for the meshing to become meaningful. Fortunately, `gmsh` handles all the details.

Before addressing the details of 3d FEM analysis, consider one more example, `struct16_3.geo` — concentric spheres:

```
// struct16_3  Concentric spheres

// geometry parameters
cl = .1; // sphere resolution
r1 = 0.75; // inner sphere radius
r2 = 2.; // outer sphere radius

X0 = 0.; // Center of both spheres
Y0 = 0.;
Z0 = 0.;

// ---------- Inner sphere ----------------------

// Sphere points
Point(1) = { X0, Y0, Z0, cl*r1 }; // center
Point(2) = { X0, Y0, Z0 + r1, cl*r1 };
Point(3) = { X0 - r1, Y0, Z0, cl*r1 };
Point(4) = { X0, Y0, Z0 - r1, cl*r1 };
```

```
Point(5) = { X0 + r1, Y0, Z0, cl*r1 };
Point(6) = { X0, Y0 - r1, Z0, cl*r1 };
Point(7) = { X0, Y0 + r1, Z0, cl*r1 };

// 1/4 Circle segments for sphere
Circle(1) = { 2, 1, 3 };
Circle(2) = { 3, 1, 4 };
Circle(3) = { 4, 1, 5 };
Circle(4) = { 5, 1, 2 };
Circle(5) = { 3, 1, 6 };
Circle(6) = { 6, 1, 5 };
Circle(7) = { 5, 1, 7 };
Circle(8) = { 7, 1, 3 };
Circle(9) = { 2, 1, 6 };
Circle(10) = { 6, 1, 4 };
Circle(11) = { 4, 1, 7 };
Circle(12) = { 7, 1, 2 };

// 1/8 Sphere surface outlines
Line Loop(1) = { 1, 5, -9 };
Line Loop(2) = { 4, 9, 6 };
Line Loop(3) = { 2, -10, -5 };
Line Loop(4) = { 3, -6, 10 };
Line Loop(5) = { 1, -8, 12 };
Line Loop(6) = { 4, -12, -7 };
Line Loop(7) = { 2, 11, 8 };
Line Loop(8) = { 3, 7, -11 };

// 1/8 Sphere surface faces
Ruled Surface(1) = { 1 };
Ruled Surface(2) = { 2 };
Ruled Surface(3) = { 3 };
Ruled Surface(4) = { 4 };
Ruled Surface(5) = { 5 };
Ruled Surface(6) = { 6 };
Ruled Surface(7) = { 7 };
Ruled Surface(8) = { 8 };

// Sphere surface
Surface Loop(1) = { 1, 2, 3, 4, 5, 6, 7, 8 };

// Outer sphere -----------------------------------

// Sphere points
Point(100) = { X0, Y0, Z0, cl*r2 }; // center
Point(200) = { X0, Y0, Z0 + r2, cl*r2 };
Point(300) = { X0 - r2, Y0, Z0, cl*r2 };
Point(400) = { X0, Y0, Z0 - r2, cl*r2 };
Point(500) = { X0 + r2, Y0, Z0, cl*r2 };
Point(600) = { X0, Y0 - r2, Z0, cl*r2 };
Point(700) = { X0, Y0 + r2, Z0, cl*r2 };
```

```
// 1/4 Circle segments for sphere
Circle(100) = { 200, 100, 300 };
Circle(200) = { 300, 100, 400 };
Circle(300) = { 400, 100, 500 };
Circle(400) = { 500, 100, 200 };
Circle(500) = { 300, 100, 600 };
Circle(600) = { 600, 100, 500 };
Circle(700) = { 500, 100, 700 };
Circle(800) = { 700, 100, 300 };
Circle(900) = { 200, 100, 600 };
Circle(1000) = { 600, 100, 400 };
Circle(1100) = { 400, 100, 700 };
Circle(1200) = { 700, 100, 200 };

// 1/8 Sphere surface outlines
Line Loop(100) = { 100, 500, -900 };
Line Loop(200) = { 400, 900, 600 };
Line Loop(300) = { 200, -1000, -500 };
Line Loop(400) = { 300, -600, 1000 };
Line Loop(500) = { 100, -800, 1200 };
Line Loop(600) = { 400, -1200, -700 };
Line Loop(700) = { 200, 1100, 800 };
Line Loop(800) = { 300, 700, -1100 };

// 1/8 Sphere surface faces
Ruled Surface(100) = { 100 };
Ruled Surface(200) = { 200 };
Ruled Surface(300) = { 300 };
Ruled Surface(400) = { 400 };
Ruled Surface(500) = { 500 };
Ruled Surface(600) = { 600 };
Ruled Surface(700) = { 700 };
Ruled Surface(800) = { 800 };

// Sphere surface
Surface Loop(100) = { 100, 200, 300, 400, 500, 600, 700, 800 };

// Total structure --------------------------------

// Volume for meshing
Volume(1) = { 100, 1};
```

Program struct16_3.geo describes a pair of concentric spheres. There are several new gmsh commands to introduce in order to generate spheres (and later on, cylinders) and other shapes. Naturally the sphere descriptions could be mixed and matched with cube or brick descriptions. Concentric spheres is a good example to work with because the voltage profile and capacitance of this structure are easily found from basic considerations, so this example is very useful for verifying the calculations of an FEM analysis.

Looking at `gmsh35.geo`, the following items are relevant:

1   Although the centers of both spheres is the origin (0,0,0), the center location is specified as (X_0, Y_0, Z_0). This makes it convenient to move one (or both) of the spheres off center if so desired.

2   The characteristic lengths for each sphere is written as a coefficient (`cl`) multiplied by that sphere's radius. This automatically creates higher resolution for small spheres and, lower resolution for large spheres. This is convenient, for example, for setting up multiple outer boundary shells, as described in Chapter 15, for approximating an open boundary.

3   Each sphere is generated by creating eight $\frac{1}{8}$ sphere surface sections. There is nothing special about using eight surfaces (starting with four circle segments in each axis). Because `gmsh` requires circle arcs to be less than π (less than $\frac{1}{2}$ circle), at least three sections are needed per circle. Using four points in each plane, however, results in points are easy to generate (and verify) without any calculations.

4   The plane surface command used in the concentric cubes example (`struct16_2.geo`) has been replaced by the *ruled surface* command. This command is necessary for generating a well-behaved mesh on a curved (nonplanar) surface using a technique called *transfinite interpolation*.[1] For our purposes, we will say that "gmsh takes care of it" and move on.

Figure 16.3 shows the concentric spheres prior to meshing. Since the spherical surfaces are generated by meshing instructions (the ruled surface, surface loop, and volume commands), only the three orthogonal circles defining each sphere are seen.

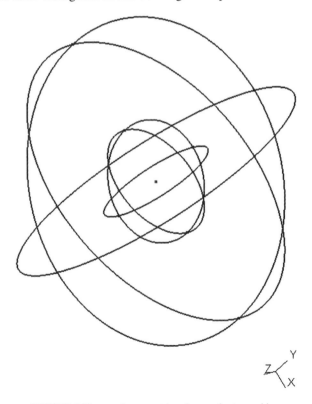

FIGURE 16.3   gmsh concentric spheres prior to meshing.

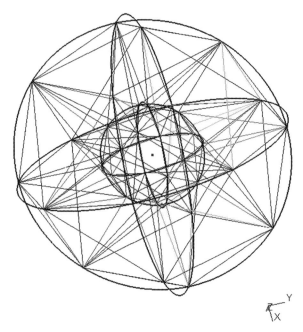

FIGURE 16.4 Concentric spheres after meshing.

Figure 16.4 shows the concentric spheres after meshing. A very low-resolution mesh is used for this figure ($cl = 1.0$). This number will be dropped to $cl = 0.1$ for actual FEM calculations. This figure shows that the desired geometry has, indeed, been generated; the region between two concentric spheres is meshed.

16.2 THE FEM COEFFICIENT MATRIX IN THREE DIMENSIONS

The derivation of the coefficient matrix terms in three dimensions using tetrahedron elements and a four-point function is a direct extension of the derivations of the 2d matrix terms in Chapter 13. We begin by defining the voltage anywhere in the tetrahedron with the function

$$V(x,y) = V_1 f_1(x,y,z) + V_1 f_2(x,y,z) + V_3 f_3(x,y,z) + V_4 f_4(x,y,z) \qquad (16.1)$$

where f_i are the basis functions and (x_1, y_1, z_1), and so on, are the vertices of the tetrahedron:

$$\begin{aligned}
f_1 &= a_1 + b_1 x + c_1 y + d_1 z \\
f_2 &= a_2 + b_2 x + c_2 y + d_2 z \\
f_3 &= a_3 + b_3 x + c_3 y + d_3 z \\
f_4 &= a_4 + b_4 x + c_4 y + d_4 z
\end{aligned} \qquad (16.2)$$

Then a_i, b_i c_i, and d_i are determined by the following four sets of conditions:

$$
\begin{bmatrix}
1 & x_1 & y_1 & z_1 \\
1 & x_2 & y_2 & z_2 \\
1 & x_3 & y_3 & z_3 \\
1 & x_4 & y_4 & z_4
\end{bmatrix}
\begin{bmatrix}
a_1 \\ b_1 \\ c_1 \\ d_1
\end{bmatrix}
=
\begin{bmatrix}
1 \\ 0 \\ 0 \\ 0
\end{bmatrix}
\tag{16.3}
$$

$$
\begin{bmatrix}
1 & x_1 & y_1 & z_1 \\
1 & x_2 & y_2 & z_2 \\
1 & x_3 & y_3 & z_3 \\
1 & x_4 & y_4 & z_4
\end{bmatrix}
\begin{bmatrix}
a_1 \\ b_1 \\ c_1 \\ d_1
\end{bmatrix}
=
\begin{bmatrix}
0 \\ 1 \\ 0 \\ 0
\end{bmatrix}
\tag{16.4}
$$

... and so on.

The coefficient matrix in the equations above is related to the volume of the tetrahedron by[2]

$$
\text{vol} = \frac{1}{6}
\begin{vmatrix}
1 & x_1 & y_1 & z_1 \\
1 & x_2 & y_2 & z_2 \\
1 & x_3 & y_3 & z_3 \\
1 & x_4 & y_4 & z_4
\end{vmatrix}
\tag{16.5}
$$

The electric field components are given by

$$
-E_x = \frac{\partial V}{\partial x} = V_1 \frac{\partial f_1}{\partial x} + V_2 \frac{\partial f_2}{\partial x} + V_3 \frac{\partial f_3}{\partial x} + V_4 \frac{\partial f_4}{\partial x} = V_1 b_1 + V_2 b_2 + V_3 b_3 + V_4 b_4
$$

$$
-E_y = V_1 c_1 + V_2 c_2 + V_3 c_3 + V_4 c_4
$$

$$
-E_z = V_1 d_1 + V_2 d_2 + V_3 d_3 + V_4 d_4
\tag{16.6}
$$

and then

$$
E^2 = E_x^2 + E_y^2 + E_z^2 = \left(V_1 b_1 + V_2 b_2 + V_3 b_3 + V_4 b_4\right)^2 + \left(V_1 c_1 + V_2 c_2 + V_3 c_3 + V_4 c_4\right)^2
$$
$$
+ \left(V_1 d_1 + V_2 d_2 + V_3 d_3 + V_4 d_4\right)^2
\tag{16.7}
$$

The energy stored in the tetrahedron is

$$
U = \frac{\varepsilon}{2} \iiint_T E^2 \, dx \, dy \, dz = E^2 v_T
\tag{16.8}
$$

where the integral is over the tetrahedron T and v_T is the volume of the tetrahedron given by equation (16.5).

For tetrahedron n with nodes n_1, n_2, n_3, and n_4, the terms that must be added to the coefficient matrix, found by taking the partial derivatives with respect to V_1, V_2, V_3, and V_4, are

To row n_1, column n_1:

$$
v_t\left(b_1^2 + c_1^2 + d_1^2\right)
\tag{16.9}
$$

To (n_1, n_2):

$$v_t(b_1 b_2 + c_1 c_2 + d_1 d_2) \tag{16.10}$$

To (n_1, n_3):

$$v_t(b_1 b_3 + c_1 c_3 + d_1 d_3) \tag{16.11}$$

To (n_1, n_4):

$$v_t(b_1 b_4 + c_1 c_4 + d_1 d_4) \tag{16.12}$$

To (n_2, n_1):

$$v_t(b_2 b_1 + c_2 c_1 + d_2 d_1) \tag{16.13}$$

To (n_2, n_2):

$$v_t\left(b_2^2 + c_2^2 + d_2^2\right) \tag{16.14}$$

and so on.

16.3 PARSING THE gmsh FILES AND SETTING BOUNDARY CONDITIONS

The file parsing script has been rewritten to handle the 3d mesh; here is the MATLAB script parse_msh_file_3d.m—a 3d gmsh parsing script:

```
% parse_msh_file_3d.m
% Read the mesh file generated by gmsh and parse it
% This version is for 3d tetrahedron (4 node) structures

%   Element types handled are 1 = line, 2 = 3 node triangle, 15 = point
%       4 = tetrahedron

  filename = uigetfile('*.msh');              % read the file
  fid = fopen(filename, 'r');

  tline = fgetl(fid);                         % read the first file line
  if (strcmp(tline, '$MeshFormat')) == 0      % Verify file type
    'Problem with this file'
    nodes_3d = []; tets_3d = []; bcs_3d = [];
    fclose (fid);
    return
  end

  tline = fgetl(fid);                         % look for version number
  [header] = sscanf(tline, '%f %d %d');
  if (header(1) ~ = 2.2), 'Possible gmsh version incompatibility', end;
```

```
tline = fgetl (fid) ;                        % Look for $EndMeshFormat
if (strcmp (tline, '$EndMeshFormat')) == 0
  'No $EndMeshFormat Line found'
  nodes_3d = [] ; tets_3d = [] ; bcs_3d = [] ;
  fclose (fid) ;
  return
end

tline = fgetl (fid) ;                        % Look for $Nodes
if (strcmp (tline, '$Nodes')) == 0
  'No $Nodes Line found'
  nodes_3d = [] ; tets_3d = [] ; bcs_3d = [] ;
  fclose (fid) ;
  return
end

nr_nodes = fscanf (fid, '%d', 1) ;           % get number of nodes
nodes_3d = zeros (nr_nodes, 3) ;             % allocate nodes array
for i = 1 : nr_nodes                         % get nodes information
  temp = fscanf (fid, '%g', 4) ;
  n = temp (1) ;
  nodes_3d (n, :) = temp (2:4) ;
end
fgetl (fid) ;                                % Clear the line

tline = fgetl (fid) ;                        % Look for $EndNodes
if (strcmp (tline, '$EndNodes')) == 0
  'No $EndNodes Line found'
  nodes = [] ; tet = [] ; bcs_3d = {} ;
  fclose (fid) ;
  return
end

tline = fgetl (fid) ;                        % Look for $Elements
if (strcmp (tline, '$Elements')) == 0
  'No $Elements Line found'
  nodes_3d = [] ; tets_3d = [] ; bcs_3d = [] ;
  fclose (fid) ;
  return
end

tets_3d = [] ; bcs_3d = [] ;
nr_elements = fscanf (fid, '%d', 1) ;        % get number of elements
for i = 1: nr_elements                       % process elements information
  temp = fscanf (fid, '%d %d', 2) ;          % get element type
  if temp (2) == 15 | temp (2) == 1          % defined point or line
    fgetl (fid) ;                            % clear the (file) line
  elseif temp (2) == 2                       % defined triangle
    temp2 = fscanf (fid, '%d %d %d %d %d %d', 6) ;
    bcs_3d = [bcs_3d; [temp2 (3:6)', 88888.8]] ;
```

```
     fgetl(fid);
   elseif temp(2) == 4                    % created mesh tetrahedron
     temp3 = fscanf(fid, '%d %d %d %d %d %d %d', 7);
     tets_3d = [tets_3d; temp3(4:7)'];
     fgetl(fid);
   else
     'Unknown type in mesh file'
     nodes_3d = []; tets_3d = []; bcs_3d = [];
     fclose(fid);
     return
   end
end
fclose(fid);


nr_bc_lines = length(bcs_3d);
'bcs file: '
bcs_3d
disp('Set bcs for all surfaces: ');
for i = 1 : nr_bc_lines
  if bcs_3d(i,5) == 88888.8              % check for dummy index
    itri = bcs_3d(i,1);
    fprintf('Surface nr %d : \n', itri)
    j = input('Enter bc: ')              % decide on the bc
    for k = 1 : nr_bc_lines              % set the bc
      if bcs_3d(k,1) == itri, bcs_3d(k,5) = j; end;
    end
    'bcs file: '
    bcs_3d
  end
end

filename = input('Name for output files, NO OVERWRITE CHECKING: ', 's');
nodes_file_name = strcat('nodes_3d_', filename, '.txt');
tets_file_name = strcat('tets_3d_', filename, '.txt');
bcs_file_name = strcat('bcs_3d_', filename, '.txt');
fid = fopen(nodes_file_name, 'w');
fprintf(fid, '%g\t%g\t%g\r\n', nodes_3d');
fclose(fid);
fid = fopen(tets_file_name, 'w');
fprintf(fid, '%d\t%d\t%d\t%d\r\n', tets_3d');
fclose(fid);
fid = fopen(bcs_file_name, 'w');
fprintf(fid, '%d\t%d\t%d\t%d\t%f\r\n', bcs_3d');
fclose(fid);
```

The logic of this script is the same as in previous versions except that now the bc setting is prompted by defined surfaces rather than by defined lines. This is why the face groupings for the surfaces (in this example, the two spheres) are numbered to clarify the linkage between the faces and their respective volume surfaces. The FEM model will work with any potential difference between the electrodes (spheres in this example), but the capacitance calculation to be described assumes a 1 V potential difference between these electrodes, so it is prudent to use this value if the capacitance calculation is of interest.

The FEM analysis is, as the analysis indicated it would be, a direct extensions of the 2d analysis programs: There is almost nothing to comment on about these programs.

- MATLAB script fed_3d.m — 3d FEM analysis program:

```
% 3d FEM with triangles
% fem_3d.m

close all

filename = input ('Generic name for input files: ', 's');
nodes_file_name = strcat ('nodes_3d_', filename, '.txt');
tets_file_name = strcat ('tets_3d_', filename, '.txt');
bcs_file_name = strcat ('bcs_3d_', filename, '.txt');

% get node data from file
dataArray = load (nodes_file_name);
nds.x = dataArray (:,1); nds.y = dataArray (:,2); nds.z =
dataArray (:,3);
nr_nodes = length (nds.x)

% get tetrahedrons data from file
tets = (load (tets_file_name)) ';
[tets_width, nr_tets] = size (tets);

% get bcs data from file
%[bcs_nodes, bcs_volts] = get_bcs_3d (bcs_file_name);
dataArray = load (bcs_file_name);
bcs_nodes = dataArray (:,2:4); bcs_volts = dataArray (:,5);
nr_bcs = length (bcs_volts)

% allocate the matrices
a = sparse (nr_nodes, nr_nodes);
f = zeros (nr_nodes,1);
b = zeros (4,1); c = zeros (4,1); d = zeros (4,1);
dd = ones (4,4);

% generate the matrix terms
for tet = 1: nr_tets              % Walk through all the triangles
  n1 = tets (1,tet); n2 = tets (2,tet);
  n3 = tets (3,tet); n4 = tets (4,tet);
  dd (1,2) = nds.x (n1); dd (1,3) = nds.y (n1); dd (1,4) = nds.z (n1);
  dd (2,2) = nds.x (n2); dd (2,3) = nds.y (n2); dd (2,4) = nds.z (n2);
  dd (3,2) = nds.x (n3); dd (3,3) = nds.y (n3); dd (3,4) = nds.z (n3);
  dd (4,2) = nds.x (n4); dd (4,3) = nds.y (n4); dd (4,4) = nds.z (n4);
  vol = det (dd) /6;              % volume of the tetrahedron
```

```
temp = dd\ [1;0;0;0]; b(1) = temp(2); c(1) = temp(3); d(1) = temp(4);
temp = dd\ [0;1;0;0]; b(2) = temp(2); c(2) = temp(3); d(2) = temp(4);
temp = dd\ [0;0;1;0]; b(3) = temp(2); c(3) = temp(3); d(3) = temp(4);
temp = dd\ [0;0;0;1]; b(4) = temp(2); c(4) = temp(3); d(4) = temp(4);

a(n1,n1) = a(n1,n1) + vol*(b(1)^2 + c(1)^2 + d(1)^2);
a(n1,n2) = a(n1,n2) + vol*(b(1)*b(2) + c(1)*c(2) + d(1)*d(2));
a(n1,n3) = a(n1,n3) + vol*(b(1)*b(3) + c(1)*c(3) + d(1)*d(3));
a(n1,n4) = a(n1,n4) + vol*(b(1)*b(4) + c(1)*c(4) + d(1)*d(4));

a(n2,n1) = a(n2,n1) + vol*(b(2)*b(1) + c(2)*c(1) + d(2)*d(1));
a(n2,n2) = a(n2,n2) + vol*(b(2)^2 + c(2)^2 + d(2)^2);
a(n2,n3) = a(n2,n3) + vol*(b(2)*b(3) + c(2)*c(3) + d(2)*d(3));
a(n2,n4) = a(n2,n4) + vol*(b(2)*b(4) + c(2)*c(4) + d(2)*d(4));

a(n3,n1) = a(n3,n1) + vol*(b(3)*b(1) + c(3)*c(1) + d(3)*d(1));
a(n3,n2) = a(n3,n2) + vol*(b(3)*b(2) + c(3)*c(2) + d(3)*d(2));
a(n3,n3) = a(n3,n3) + vol*(b(3)^2 + c(3)^2 + d(3)^2);
a(n3,n4) = a(n3,n4) + vol*(b(3)*b(4) + c(3)*c(4) + d(3)*d(4));

a(n4,n1) = a(n4,n1) + vol*(b(4)*b(1) + c(4)*c(1) + d(4)*d(1));
a(n4,n2) = a(n4,n2) + vol*(b(4)*b(2) + c(4)*c(2) + d(4)*d(2));
a(n4,n3) = a(n4,n3) + vol*(b(4)*b(3) + c(4)*c(3) + d(4)*d(3));
a(n4,n4) = a(n4,n4) + vol*(b(4)^2 + c(4)^2 + d(4)^2);
end


%bcs
for i = 1: nr_bcs
  if bcs_volts(i) ~ = -9999          % Line 1 of 2 for problem 16.3
    for j = 1 : 3
      n = bcs_nodes(i,j);
      a(n, :) = 0; a(n,n) = 1; f(n) = bcs_volts(i);
    end
  end                  % Line 2 of 2 for problem 16.3
end


%    There might be a few nodes such as a sphere center that are not
%       connected to anything. We could eliminate them and renumber
%       or, simpler, just stop them from making trouble:
for i = 1 : nr_nodes
    if a(i,i) == 0
      a(i, :) = 0; a(i,i) = 1;
    end
end


v = a\f;


% calculate capacitance
C = get_tet3d_cap(v, nr_nodes, nr_tets, nds, tets)
```

- MATLAB function get_tet3d_cap.m:

```
function C = get_tet3d_cap(v, nr_nodes, nr_tets, nds, tets)
% Calculate the capacitance of a 3d tetrahedron structure
% Simple linear approx. function is assumed

  eps0 = 8.854;
  b = zeros(4,1); c = zeros(4,1); d = zeros(4,1); dd = ones(4,4);
  U = 0;

  for tet = 1: nr_tets      % Walk through all the triangles
    n1 = tets(1,tet); n2 = tets(2,tet);
    n3 = tets(3,tet); n4 = tets(4,tet);

    dd(1,2) = nds.x(n1); dd(1,3) = nds.y(n1); dd(1,4) = nds.z(n1);
    dd(2,2) = nds.x(n2); dd(2,3) = nds.y(n2); dd(2,4) = nds.z(n2);
    dd(3,2) = nds.x(n3); dd(3,3) = nds.y(n3); dd(3,4) = nds.z(n3);
    dd(4,2) = nds.x(n4); dd(4,3) = nds.y(n4); dd(4,4) = nds.z(n4);
    vol = det(dd)/6;

    temp = dd\[1;0;0;0]; b(1) = temp(2); c(1) = temp(3); d(1) =
temp(4);
    temp = dd\[0;1;0;0]; b(2) = temp(2); c(2) = temp(3); d(2) =
temp(4);
    temp = dd\[0;0;1;0]; b(3) = temp(2); c(3) = temp(3); d(3) =
temp(4);
    temp = dd\[0;0;0;1]; b(4) = temp(2); c(4) = temp(3); d(4) =
temp(4);

    U = U + vol*( (v(n1)*b(1) + v(n2)*b(2) + v(n3)*b(3) + v(n4)*b(4))^2 ...
        + (v(n1)*c(1) + v(n2)*c(2) + v(n3)*c(3) + v(n4)*c(4))^2 ...
        + (v(n1)*d(1) + v(n2)*d(2) + v(n3)*d(3) + v(n4)*d(4))^2 );
  end

  U = U*eps0/2;
  C = 2*U;       % Potential difference of 1 volt assumed

end
```

One small addition to the 3d FEM program is the code in lines 73–80 of fed_3d.m. Because of the nature of gmsh, the nodes that define the centers of the circle arcs that ultimately define the spheres show up as defined nodes in the .msh file. These nodes are added into the node count and appear as empty rows in the coefficient matrix (the nodes don't connect to anything, they're just reference points). The program could be written to sniff out these nodes, eliminate them and reduce the node count, but this could complicate the program debugging process because node numbers have been changed from the original gmsh nonde numbers. Instead, the simple task of just making these nodes dummy boundary conditions is performed. Since there are very few of these nodes in a structure description, the overall efficiency of the solution is not diminished meaningfully, and the node numbering is preserved. Also, since these nodes are not part of any tetrahedron, the dummy boundary condition does not contribute to the total energy (capacitance) calculation.

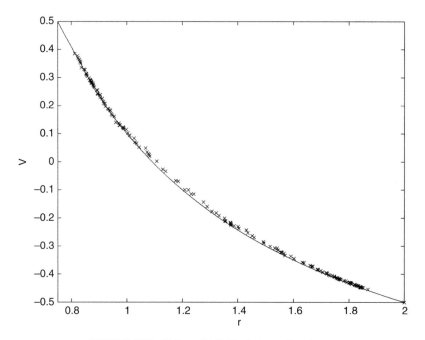

FIGURE 16.5 Voltage distribution between the spheres.

With a gmsh resolution of cl = 0.1, fem_3d.m predicts a capacitance of 137.2 pF. The exact capacitance may be calculated from basic principles, namely

$$V_1 = \frac{Q}{4\pi\varepsilon_0 r_1}$$
$$V_2 = \frac{Q}{4\pi\varepsilon_0 r_2} \qquad (16.15)$$

so that

$$V_1 - V_2 = \frac{Q}{4\pi\varepsilon_0}\left[\frac{1}{r_1} - \frac{1}{r_2}\right] \qquad (16.16)$$

$$C = \frac{Q}{|V_1 - V_2|} = \frac{4\pi\varepsilon_0}{(1/r_1) - (1/r_2)} = 133.5\,pF \qquad (16.17)$$

Figure 16.5 shows the FEM predicted voltage distribution between the spheres (tickmarks ×) overlaid on the exact solution (continuous line).

16.4 OPEN BOUNDARIES AND CYLINDERS IN SPACE

```
// struct16_4    2 Cylinders

// geometry
cl = .5;
rc = 1.; // cylinder radius
```

```
rs = 7.; // sphere radius
xs = 3; // cylinder center-center separation
zs = 4; // cylinder height

// First cylinder ----------------------------------

Point (1) = {-xs/2, 0, -zs/2, cl}; // center of first circle
Point (2) = {-xs/2 + rc, 0, -zs/2, cl};
Point (3) = {-xs/2, rc, -zs/2, cl};
Point (4) = {-xs/2-rc, 0, -zs/2, cl};
Point (5) = {-xs/2, -rc, -zs/2, cl};

Circle (1) = {2, 1, 3}; // 1/4 circle segments
Circle (2) = {3, 1, 4};
Circle (3) = {4, 1, 5};
Circle (4) = {5, 1, 2};

Line Loop (1) = {1, 2, 3, 4}; // bottom of cylinder
Plane Surface (1) = {1};

Point (6) = {-xs/2, 0, zs/2, cl}; // center of second circle
Point (7) = {-xs/2 + rc, 0, zs/2, cl};
Point (8) = {-xs/2, rc, zs/2, cl};
Point (9) = {-xs/2-rc, 0, zs/2, cl};
Point (10) = {-xs/2, -rc, zs/2, cl};

Circle (5) = {7, 6, 8}; // 1/4 circle segments
Circle (6) = {8, 6, 9};
Circle (7) = {9, 6, 10};
Circle (8) = {10, 6, 7};

Line Loop (2) = {5, 6, 7, 8};
Plane Surface (2) = {2};

Line (10) = {2, 7}; // connecting lines for cylinder walls
Line (11) = {3, 8};
Line (12) = {4, 9};
Line (13) = {5, 10};

Line Loop (3) = {5, -11, -1, 10};
Line Loop (4) = {6, -12, -2, 11};
Line Loop (5) = {7, -13, -3, 12};
Line Loop (6) = {8, -10, -4, 13};

Ruled Surface (3) = {3};
Ruled Surface (4) = {4};
Ruled Surface (5) = {5};
Ruled Surface (6) = {6};

// Cylinder Surface
Surface Loop (1) = {1, 2, 3, 4, 5, 6};
```

```
// Second Cylinder -------------------------------
Point(21) = {xs/2, 0, -zs/2, cl}; // center of second circle
Point(22) = {xs/2+rc, 0, -zs/2, cl};
Point(23) = {xs/2, rc, -zs/2, cl};
Point(24) = {xs/2-rc, 0, -zs/2, cl};
Point(25) = {xs/2, -rc, -zs/2, cl};

Circle(21) = {22, 21, 23}; // 1/4 circle segments
Circle(22) = {23, 21, 24};
Circle(23) = {24, 21, 25};
Circle(24) = {25, 21, 22};

Line Loop(21) = {21, 22, 23, 24}; // bottom of cylinder
Plane Surface(21) = {21};

Point(26) = {xs/2, 0, zs/2, cl}; // center of second circle
Point(27) = {xs/2+rc, 0, zs/2, cl};
Point(28) = {xs/2, rc, zs/2, cl};
Point(29) = {xs/2-rc, 0, zs/2, cl};
Point(30) = {xs/2, -rc, zs/2, cl};

Circle(25) = {27, 26, 28}; // 1/4 circle segments
Circle(26) = {28, 26, 29};
Circle(27) = {29, 26, 30};
Circle(28) = {30, 26, 27};

Line Loop(22) = {25, 26, 27, 28};
Plane Surface(22) = {22};

Line(30) = {22, 27}; // connecting lines for cylinder walls
Line(31) = {23, 28};
Line(32) = {24, 29};
Line(33) = {25, 30};

Line Loop(23) = {25, -31, -21, 30};
Line Loop(24) = {26, -32, -22, 31};
Line Loop(25) = {27, -33, -23, 32};
Line Loop(26) = {28, -30, -24, 33};

Ruled Surface(23) = {23};
Ruled Surface(24) = {24};
Ruled Surface(25) = {25};
Ruled Surface(26) = {26};

// Cylinder Surface
Surface Loop(21) = {21, 22, 23, 24, 25, 26};

// Outer sphere -----------------------------------------

// points
Point(41) = {0, 0, 0, cl}; // center
```

```
Point(42) = {0, 0, rs, cl};
Point(43) = {-rs, 0, 0, cl};
Point(44) = {0, 0, -rs, cl};
Point(45) = {rs, 0, 0, cl};
Point(46) = {0, -rs, 0, cl};
Point(47) = {0, rs, 0, cl};

// 1/4 circle segments
Circle(41) = {42, 41, 43};
Circle(42) = {43, 41, 44};
Circle(43) = {44, 41, 45};
Circle(44) = {45, 41, 42};
Circle(45) = {43, 41, 46};
Circle(46) = {46, 41, 45};
Circle(47) = {45, 41, 47};
Circle(48) = {47, 41, 43};
Circle(49) = {42, 41, 46};
Circle(50) = {46, 41, 44};
Circle(51) = {44, 41, 47};
Circle(52) = {47, 41, 42};

// 1/8 sphere surface outlines

Line Loop(41) = {41, 45, -49};
Line Loop(42) = {44, 49, 46};
Line Loop(43) = {42, -50, -45};
Line Loop(44) = {43, -46, 50};
Line Loop(45) = {41, -48, 52};
Line Loop(46) = {44,-52, -47};
Line Loop(47) = {42, 51, 48};
Line Loop(48) = {43, 47, -51};

// 1/8 sphere ruled surface

Ruled Surface(41) = {41};
Ruled Surface(42) = {42};
Ruled Surface(43) = {43};
Ruled Surface(44) = {44};
Ruled Surface(45) = {45};
Ruled Surface(46) = {46};
Ruled Surface(47) = {47};
Ruled Surface(48) = {48};

// Sphere surface
Surface Loop(41) = {41, 42, 43, 44, 45, 46, 47, 48};

// build the volume

// Volume(1) = {1};
// Volume(21) = {21};
Volume(41) = {41, 1, 21};
```

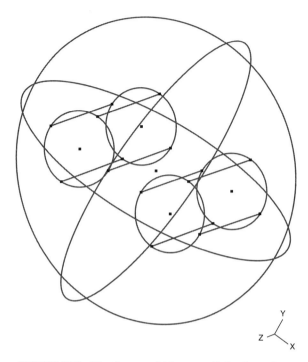

FIGURE 16.6 Wire frame model for two cylinders in a sphere.

The gmsh script struct16_4.geo shows the construction of two cylinders and a surrounding sphere. The cylinders are constructed in the same manner as the spheres; ruled surfaces of a $\frac{1}{4}$ cylinder are constructed, circles for the top and bottom of the cylinder are constructed, and then they are all joined together as a surface loop. The final volume command establishes the region to be meshed as the insider of the sphere absent the two cylinders.

The wire frame model for all of this, when the outer sphere radius is 4.0, is shown in Figure 16.6.

The capacitance between the two cylinders, in an unbounded region, is approximated by setting the cylinder potentials to +0.5 and −0.5, setting the outer-sphere potential to 0, and increasing the sphere radius until the sphere no longer contributes to the capacitance. Increasing the sphere radius by adding lower and lower resolution shells, as described in Chapter 15, is an approach that takes more time to set up but ultimately makes better use of computer resources.

PROBLEMS

16.1 Create a gmsh file for Figure P16.1. Run the FEM analysis and calculate the capacitance of the structure.

16.2 We can model a small screw as a metallic right cylinder. Modify the structure of structp16_1.geo by adding such a cylinder, of radius 0.25 cm, to the center

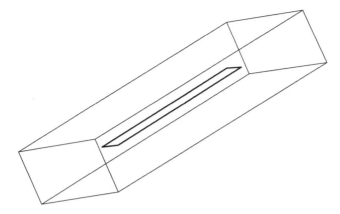

Dimensions (cm):
Outer box = 10L x 5W x 2H
Inner conductor = 6L x 1W x .025H
Inner conductor is centered in L & W, bottom is 1cm from
    bottom of box

FIGURE P16.1   Structure and dimensions for Problem 16.1.

of the top of the box. Calculate the capacitance as a function of the depth that this
screw reaches into the box.

16.3 `structp16_1.geo` has symmetry planes $X = 0$ and $Y = 0$. $Z = 0$ looks a bit like a
symmetry plane, but it is not a symmetry plane because the strip conductor is not
centered in Z in the box. Write a new version of `structp16_1.geo` showing only
one-fourth of the structure to utilize these symmetries. Rewrite whichever MATLAB
programs are necessary to perform the calculations on this data file.

16.4 The goals of this problem and Problem 16.5 are to add dielectric material capability to
the 3d FEM analysis. The first task is to modify `fem_3d.m` to properly handle individual tetrahedron dielectric constants. Rather than "patching" the dielectric region
definition into the FEM program as was done with the 2d analysis, modify the
file structure so that the dielectric information is part of the tetrahedron data
file. A separate small program will then be necessary to supply the correct dielectric
information to the file.

16.5 Improve the dielectric calculation accuracy of Problem 16.4.

REFERENCES

1. P. Knupp and S. Steinberg, *Fundamentals of Grid Generation*. CRC Press, Boca Raton, FL, 1994.

2. `http://mathworld.wolfram.com/Tetrahedron.html`.

17

Electrostatic Forces

17.1 INTRODUCTION

The basis of electrical phenomena is the *Coulomb force* — the force exerted by a charged particle on other charged particles. The electric field due to a collection of charged particles is defined as the force on a small "test" particle, such as an electron. The charge of the test particle is assumed to be so small as to allow virtually no perturbation of the electric field due to the test particle's presence. In practice, this is almost always an excellent approximation for use with low-density electron charge distributions (*low density* here means that the motion of the electrons is not influenced by that of the other electrons). Typically, electronic devices are built in high-vacuum containers (CRT displays, scanning electron microscopes), and electron motion in a static electric field is an exercise in solving Newton's equations. We solve the following equation, usually numerically, where $\vec{X}$ is the vector of the position coordinates of the charged particle:

$$\vec{F} = m\,\vec{a} = m\frac{d^2\,\vec{X}}{dt^2} = q\,\vec{E} \tag{17.1}$$

When the electric field forces are strong enough to actually move one of the electrodes, then we cannot use equation (17.1), which assumes that the electric field is constant in time. Instead, we must use the principle of *virtual work*,[1] which states that for a movement (of an electrode or some other conductor or dielectric that contributes to the stored energy calculation), in the X direction, for example, we obtain the following equation, where U_E is the stored energy in the electric field.

Introduction to Numerical Electrostatics Using MATLAB, First Edition. Lawrence N. Dworsky.
© 2014 John Wiley & Sons, Inc. Published 2014 by John Wiley & Sons, Inc.

$$F_x = \frac{\partial U_E}{\partial x} \qquad (17.2)$$

When U_E increases with a movement in the $+X$ direction, equation (17.1) predicts a positive force; therefore, F_x is the force pushing the electrode in the $+X$ direction, and the work done by this force increases the energy stored in the electric field.

In the strictest definition, since the electric field is changing with time, we are no longer dealing with an electrostatic system. In practical-scale systems, however, the speed of mechanical motion of masses is so low in comparison to the speed of light that we are basically still dealing with an electrostatic model with parameters changing very slowly. A numeric value that is often quoted to justify this assertion is the approximate ratio of the speed of light to the speed of sound (elastic wave propagation), which, in most solids, is equal to 10^5. Sometimes the term *quasistatic* is used to summarize a situation where the electric field is changing as a result of movement(s) in the structure over the time period of interest, but these movements are so slower than the speed of light that electrostatic analysis is still valid.

17.2 ELECTRON BEAM ACCELERATION AND CONTROL

Analyzing an electron (or an electron beam) trajectory in a vacuum is accomplished by first finding the electric field distribution in the structure and then solving equation (17.1) numerically. Equation (17.1) is a *vector equation*, meaning that there are two equations for a two-dimensional (2d) problem, three equations for a three-dimensional (3d) problem, and so on. Since these equations do not couple, their solutions are handled by solving them separately; consequently we are solving ordinary differential equations (ODEs). MATLAB has many tools available for this task,[2] and online tutorials are available.[3–5]

To keep things simple and not get diverted into a study of numerical ODE solutions, we'll use the simplest numerical integration technique available, namely, equal forward time steps.

Since

$$v_x = \frac{dx}{dt} \qquad (17.3)$$

(the x component of) equation (17.1) may be rewritten as

$$\frac{dv_x}{dt} = \frac{q}{m} E_x \qquad (17.4)$$

For small time steps Δt, we obtain

$$v_x(\text{new}) = v_x(\text{old}) + \frac{q}{m} E_x \, \Delta t \qquad (17.5)$$

and then

$$x_{\text{new}} = x_{\text{old}} + v_x \, \Delta t \qquad (17.6)$$

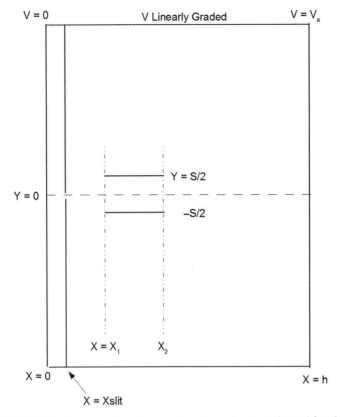

FIGURE 17.1   Simple structure demonstrating electron beam steering and focusing.

Consider the 2d structure shown in Figure 17.1. The outer box consists of an $X = 0$ wall set at 0 V, an $X = h$ wall set at V_a volts, and two Y walls (the exact dimensions don't matter) on which the voltage is linearly graded between 0 and V_a volts.

Electrons will be inserted at (0,0) (the cathode) at essentially 0 velocity. They will be accelerated to $X = h$ (the anode) by the anode potential V_a (see Figure 17.1). When they strike the anode, each electron will have acquired $-eV_a$ energy, where e is the electron charge.

Setting the energy put into an electron (which started at rest) is equal to its kinetic energy

$$\frac{mv_x^2}{2} = -eV_a \qquad (17.7)$$

we can find the velocity of the electron:

$$v_x = \sqrt{\frac{-2eV_a}{m}} \qquad (17.8)$$

If the electron has velocity components in y and/or z as well as in x, then the total velocity must be used to calculate the kinetic energy.

As an aside, kinetic energy in meter–kilogram–second (mks). units is expressed in joules. In practice, an electron's energy is often expressed in electronvolts (eV). One electronvolt is the amount of energy gained by an electron when it is accelerated through a one-volt potential difference. Numerically, it is one joule multiplied by the charge on one electron: $1 \text{ eV} = 1.6 \times 10^{-19}$ joules (J).

Returning to Figure 17.1, an electron staring at rest at (0,0) will accelerate toward the anode, passing through a small slit in an electrode whose voltage is determined by its placement so as to cause minimal deviation in the electric field:

$$V_{\text{slit}} = \frac{x_{\text{slit}}}{h} V_a \qquad (17.9)$$

The purpose of the slit electrode will be discussed below.

Next, the electron passes between two parallel plates (control electrodes) whose voltages are set using a combination of two parameters, V_f and V_s, where

$$\begin{aligned} V_{\text{upper}} &= \frac{V_f + V_s}{2} \\ V_{\text{lower}} &= \frac{V_f - V_s}{2} \end{aligned} \qquad (17.10)$$

The MATLAB program `electrons_1.m` solves the electrostatic problem for the structure described above and then calculates and displays the electron's trajectory:

- [MATLAB script `electrons_1.m`]—electron trajectory calculation program:

```
% electrons_1.m   2d electron steering and focusing

clear; close all;

q = -1.602e-19;     m = 9.109e-31;          % electron properties

global x_max;  global y_max;
y_max = 100;     % height of the box, (nr of rows = 1st indec)
x_max = 500;     % with of the box, (nr of cols = 2nd index)
Va = 1000;       % anode voltage

n_max = x_max*y_max;
a = spalloc(n_max,n_max,5*n_max);    % coefficient array
b = zeros(n_max,1);                   % forcing function array

% set up internal node coefficient array

for i = 2 : y_max - 1
  for j = 2 : x_max - 1
    n = get_n(i,j);
    a(n,n) = -4;
    a(n,n-1) = 1;
    a(n,n + 1) = 1;
    a(n,n + x_max) = 1;
    a(n,n-x_max) = 1;
```

```
    end
end

% set up outer box bcs
for i = 1 : y_max
  n = get_n(i,x_max);        % anode end
  a(n,:) = 0; a(n,n) = 1;   b(n) = Va;
  n = get_n(i,1);            % cathode end
  a(n,:) = 0; a(n,n) = 1; b(n) = 0;
end

% Graded walls
for j = 1 : x_max
  n_low = get_n(1,j);
  n_high = get_n(y_max,j);
  Vj = Va*(j-1)/(x_max-1);
  a(n_low,:) = 0; a(n_low,n_low) = 1; b(n_low) = Vj;
  a(n_high,:) = 0; a(n_high,n_high) = 1; b(n_high) = Vj;
end

% emission slit
x_slit = 20;
y_slit_1 = y_max/2-2; y_slit_2 = y_max/2 + 2;
V_slit = Va*(x_slit-1)/(x_max-1);
for i = 2 : y_slit_1
  n = get_n(i,x_slit);
  a(n,:) = 0;
  a(n,n) = 1;
  b(n) = V_slit;
end
for i = y_slit_2 : y_max-1
  n = get_n(i,x_slit);
  a(n,:) = 0;
  a(n,n) = 1;
  b(n) = V_slit;
end

% control electrodes
x1 = x_slit + 90; x2 = x1 + 50;
y_up = y_slit_2;
y_down = y_slit_1;
V_bal = 220; V_diff = 0.3;
V_up = V_bal + V_diff/2; V_down = V_bal - V_diff/2.;

for j = x1 : x2
  n = get_n(y_up,j);
  a(n,:) = 0; a(n,n) = 1; b(n) = V_up;
  n = get_n(y_down,j);
  a(n,:) = 0; a(n,n) = 1; b(n) = V_down;
end

% solve it
V = a\b;
```

```
% An xy profile will be very useful

Vxy = zeros (x_max, y_max);
for i = 1 : y_max
  for j = 1 : x_max
    n = get_n(i,j);
    Vxy(i,j) = V(n);
  end
end


% ---------- electron trajectory section

dt = 1.e-8;

% Set up a picture of the structure then add the trajectories

figure(1)
axis ([0, x_max, y_max/2-25,y_max/2+25])    % outer box
hold on
line ([x_slit,x_slit], [y_max/2-25, y_slit_1], 'Color', [0 0 0])
line ([x_slit,x_slit], [y_slit_2, y_max/2+25], 'Color', [0 0 0])
line ([x1,x2], [y_up,y_up], 'Color', [0 0 0]);
line ([x1,x2], [y_down,y_down], 'Color', [0 0 0]);
xlabel ('X')
ylabel ('Y')

angs = [-.3, -.2, -.1, .1, .2, .3];
for angle = angs      % degrees
  angle

  x = [1];   y = [y_max/2];   vx = [0];   vy = [0];
  t = 0;
  set_angle_flag = 1;
  while x(end) < x_max & y(end) < y_max & y(end) > 0
    t = [t, t(end) + dt];
    [Ex,Ey] = get_E(x(end),y(end),V);
    vx = [vx, vx(end) + q*Ex*dt/m];
    x = [x, x(end) + vx(end)*dt];
    vy = [vy, vy(end) + q*Ey*dt/m];
    y = [y, y(end) + vy(end)*dt];

    if x(end) >= x_slit & set_angle_flag == 1
      set_angle_flag = 0;
      v = sqrt (vx(end)^2 + vy(end)^2)
      vx(end) = v*cos(angle*pi/180);
      vy(end) = v*sin(angle*pi/180);
    end
  end

  plot(x(1:100:end),y(1:100:end),'k')
  y(end)
end
```

- [MATLAB function get_n.m]:

```
function n = get_n(i,j)

  global x_max; global y_max;

  n = j + (i-1)*x_max;

end
```

- [MATLAB function get_ij.m]:

```
function [ i_out,j_out ] = get_ij(n)

  global x_max; global y_max;

  i = ceil(n/x_max - 1);
  j = n - 1 - i*x_max;
  i_out = i + 1;
  j_out = j + 1;

end
```

- [MATLAB function get_E.m]:

```
function [ Ex, Ey ] = get_E(x,y,V)
% Calculate the electric field based upon position

  global x_max; global y_max;

  j = floor(x); i = floor(y);
  n1 = get_n(i,j); V1 = V(n1);
  n2 = get_n(i,j + 1); V2 = V(n2);
  n3 = get_n(i + 1,j + 1); V3 = V(n3);
  n4 = get_n(i + 1,j); V4 = V(n4);

  x_l = x - j; y_l = y - i;
  Ex = (V1-V2)*(1-y_l) + (V4-V3)*y_l;
  Ey = (V1-V4)*(1-x_l) + (V2-V3)*x_l;

end
```

The structure parameters, as defined in this program, are

Outer box: x_max = 500, y_max = 100
Slit: x_slit = 20, slit_width = 4
Control electrodes: x1 = x_slit + 110, x2 = x1 + 140, separation = 4

Script electrons_1.m uses a simple finite difference (FD) program and voltage/field interpolation scheme based on the analyses presented in Chapter 10.

For the parameters as listed and $\Delta t = 1 \times 10^{-8}$, `electrons_1.m` predicts an electron velocity at the anode of $1.8755 - 10^7$ (m/s), which is [see equation (17.2)] correct to five decimal places. Note that the dimensions chosen for this example are somewhat unrealistic — a 500-m-long vacuum chamber is not a common item. These numbers were chosen for convenience in providing an example. In practice, the length of the vacuum chamber could be approximately 0.25 m; everything else (including Δt) would be scaled accordingly.

A real electron source does not generate all electrons at exactly zero energy and at exactly the same place. The purpose of the slit is to provide for an accelerating region (to the left of the slit) where the ideal electron source's electrons can travel in the $+X$ direction, unperturbed by anything happening at the control electrodes. At the slit, we perturb the Y velocity of the electrons by a chosen amount while maintaining its total velocity. Examining the trajectories of electrons beginning at different small angles, distributed about zero degrees, allows us to gain some insight as to the real issues and techniques involved in an electron beam structure such as a cathode ray tube (CRT).

When V_s, the *scanning* voltage, is zero, both control electrodes are at the same *focus* voltage V_f.

In all three diagrams (Figure 17.2a–c), electrons were started with initial angles of $\pm 0.3°$, $\pm 0.2°$, and $\pm 0.1°$. In Figure 17.2a, $V_f = 220$ V and has very little effect on the electron trajectories. In Figure 17.2b, $V_f = 140$ V; the outer electron trajectories are being bend toward the anode, resulting in a tighter bunching of trajectories at the anode. In Figure 17.2c, $V_f = 120$ V; the outer trajectory electrons are being bent too far — their trajectories cross each other on the way to the anode, resulting in a poorly bunched distribution of electrons.

This electron behavior is in many ways analogous to optical focusing, and is called *electron optics.*[6] The control electrodes used here are a crude form of an electron lens that will have a defined focal length. In practice, an electron lens will be cylindrically symmetric and may be composed of several *elements*.

Once the electron beam has been focused, it must be aimed at a desired location on the anode. This is accomplished here by the voltage V_s, which is applied differentially to the control electrodes. Figure 17.3 is a repeat of Figure 17.1 but with $V_f = 0.3$ V. The differential scanning voltage has relocated the electron beam location at the anode in the $+Y$ direction.

An actual electron lens system is much more complicated than the simple example portrayed here. While the focusing structure is cylindrically symmetric, there must be both X and Y deflection electrodes so that any desired position on the anode may be addressed by the electron beam.

A problem with electrostatic beam focusing and steering arises when high anode voltages (25 KV is typical) are used in structures such as CRTs. Since the electrons accelerate to much higher velocities at higher anode voltages, they spend less time in the region of any given electrode structure, and hence these electrode structures lose effectiveness.

A practical answer to this problem is to abandon electrostatic focusing and use magnetic field focusing. In a magnetostatic field

$$\overrightarrow{F} = q \, \overrightarrow{v} \, X \, \overrightarrow{B} \tag{17.11}$$

where B is the magnetic field. Since the force on the electron is proportional to its velocity (and the field strength), this system is effective at high anode voltages. Inspection of the back of a CRT television, for example, will reveal a doughnut-shape coil structure called a *yoke* wrapped around the neck of the CRT. The yoke provides the magnetic field necessary for controlling the electron flow accurately and effectively.

(a)

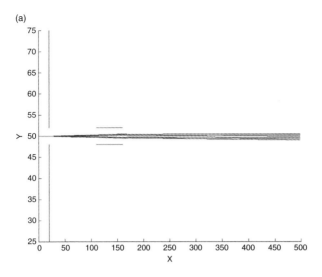

(b)

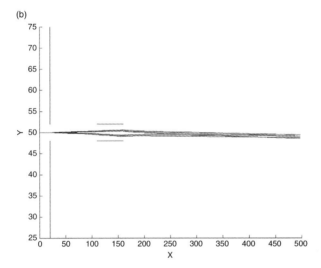

(c)

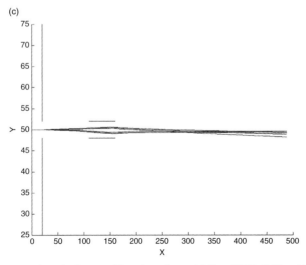

FIGURE 17.2 Electron trajectories for several focusing voltages: (a) $V_f = 220$ V; (b) $V_f = 140$ V; (c) $V_f = 120$ V.

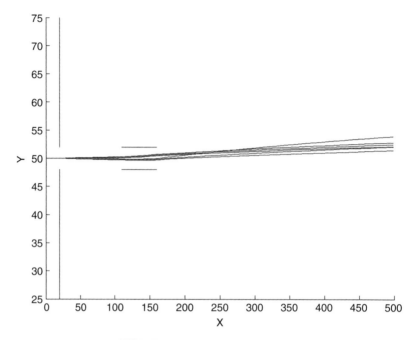

FIGURE 17.3   Deflected electron beam.

17.3   THE ELECTROSTATIC RELAY (SWITCH)

In the context here, a *relay* is an electrically controlled electrical switch. A typical commercial relay is shown (schematically) in Figure 17.4.

As shown in the figure, two electrical contact points are positioned facing each other. One of the contacts is mounted on a rigid arm; the other, on an arm that can pivot (move). The contacts are separated by a spring that is connected to the latter (pivoting) support. When the electromagnet is energized by applying a voltage to the control wires, the pivoting arm is drawn toward the electromagnet until the contacts touch and the switch is closed.

A common variation on this structure is for the pivot mount to be replaced by a fixed mount and the spring eliminated. The upper arm is made of a springy material that at rest keeps the contacts apart. The magnetic force causes the upper arm to bend until the contacts touch.

Historically, electrical relays have always used magnetic force to move one of the contacts. When arm sizes can be measured in millimeters, electrostatic forces at practical voltages are simply too small to be of use.

The advent of microelectromechanical machining (MEMM) has made the electrostatic relay a practical device.[7,8]

Figure 17.5 shows, schematically, the essential components of a MEMM relay. Two conducting plates of surface area A face each other. One of the plates is rigidly mounted; the other is suspended by a spring with a spring constant K. At equilibrium, there is zero force compressing or extending the spring, and the two plates are separated by a distance g (gravitational force is ignored). When a voltage V is a applied between the plates, the Coulomb force causes the upper plate to move toward the lower plate. The instantaneous position of the upper plate is $u(y)$:

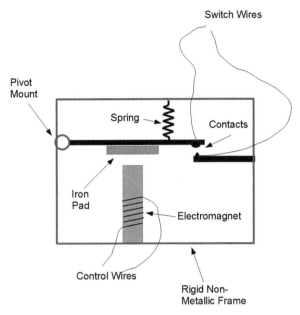

FIGURE 17.4   Typical electromagnetic relay.

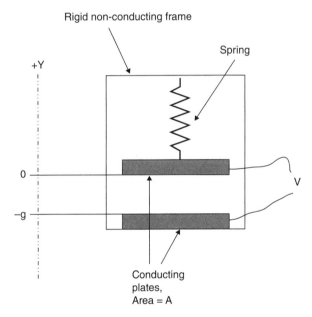

FIGURE 17.5   Essential structure of simple MEMM relay.

$$-g \leq u \leq 0 \qquad (17.12)$$

Assuming that the plate separation is small compared to A, we can approximate the capacitance between the plates using the ideal parallel plate capacitor relationship:

$$C = \frac{A\varepsilon_0}{u(y) + g}$$

(17.13)

The energy stored in the electric field is

$$U = \frac{1}{2}CV^2 = \frac{1}{2}\frac{A\varepsilon_0 V^2}{u + g}$$

(17.14)

The force on the upper plate is a combination of the spring force and the Coulomb force:

$$F = -Ku + \frac{\partial U}{\partial u} = -Ku - \frac{A\varepsilon_0 V^2}{2(u + g)^2}$$

(17.15)

The equilibrium position s is the position where the spring force and the Coulomb force balance:

$$0 = 2Ku(u + g)^2 + A\varepsilon_0 V^2$$

(17.16)

It is convenient at this point to normalize this equation. Let

$$s \equiv \frac{u}{g} \quad -1 \le s \le 0$$

(17.17)

and

$$\beta^2 \equiv \frac{A\varepsilon_0}{2Kg^3}$$

(17.18)

Substituting equations (17.17) and (17.18) into (17.16), we obtain

$$s(s + 1)^2 + \beta^2 V^2 = 0$$

(17.19)

Equation (17.19), describing the equilibrium position of the upper plate(s) as a function of voltage (βV), is plotted in Figure 17.6.

Starting at (0,0), the displacement s is zero when the voltage βV is zero. As the voltage increases (note that the sign of the voltage does not matter), the displacement increases until it reaches a peak of $\frac{-1}{3}$ at $|\beta V| = \sqrt{\frac{4}{27}} \approx 0.39$. Further increases of displacement until $s = -1$ (the electrodes are touching) require less voltage, until again zero voltage is required for the electrodes to touch.

This unusual behavior is better understood by reference to Figure 17.7.

Figure 17.7 shows the net force on the movable electrode plotted against the electrode's position for several voltages. Consider first the upper curve, $(\beta V)^2 = 0.25$. This curve crosses the net force = 0 line twice, at $s \approx -0.08$ and $s \approx -0.70$; thus, these values of s are equilibrium points for this voltage.

At $s \approx -0.08$, the slope of the force curve as it passes through equilibrium is such that the moving electrode is pushed toward this point. If s is not large enough, the force is negative so

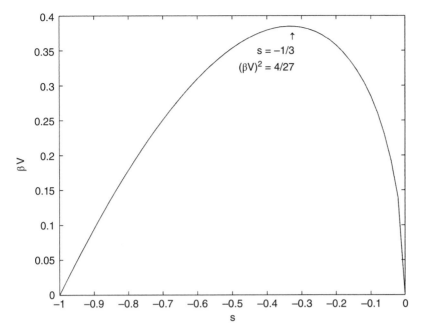

FIGURE 17.6 Plot of equation (17.19).

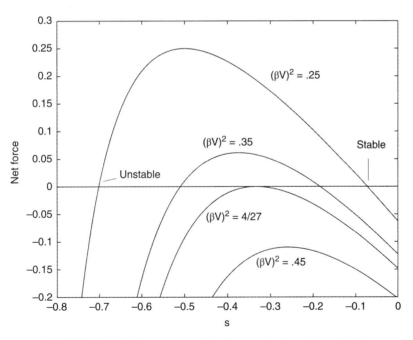

FIGURE 17.7 Net force versus position for several voltages (normalized).

as to push the electrode down, decreasing s (remember, s is negative). If s is too large (too negative), the force is positive, pushing s toward zero. Hence $s \approx -0.08$ is a *stable equilibrium point*. This is analogous to a ball rolling into a depression in the floor — small movements result in a force returning the ball to the depression.

At $s \approx -0.70$, the forces are directed so as to move the electrode away from the equilibrium point. If the electrode is moved a small amount toward $s = 0$, the positive force will return the electrode to the stable equilibrium point. If the electrode is moved a small amount toward $s = -1$, the negative force will send the electrode toward contacting the fixed electrode. This is called an *unstable equilibrium condition*, analogous to balancing a ball on the tip of a pencil.

As βV increases, the stable equilibrium point moves toward more negative values of s, as shown in both Figures 17.7 and 17.10. At $(\beta V)^2 \geq \frac{4}{27}$, however, Figure 17.7 shows that there are no longer any equilibrium points; the force is always negative, driving the moving electrode down toward the fixed electrode.

If the two electrodes touch, the voltage across them is, of course, short-circuited and goes to zero, causing the electrodes to spring apart. The mass of the moving electrode in conjunction with the spring have a resonance frequency that determines how quickly the electrodes move apart and then start coming together again (as soon as they're apart, the voltage is reestablished). The magnetic version of this is known as a "door buzzer."

If a thin dielectric layer is added to either of the electrodes, the voltage is never short-circuited and at a high enough voltage, the electrodes will "latch" at $s = -1$, squeezing on the dielectric layer. This device is, of course, no longer a switch — the electrodes never make contact.

Figure 17.8 shows one way of designing an electrostatic relay that circumvents the abovementioned problem. The structure in Figure 17.8 has separate control and switch wire connections, as did the magnetic relay shown in Figure 17.4. This is an important property for a useful relay — it is almost always desirable to have the control and switched circuits isolated from each other.

In this device, the moving electrode and spring have been combined into a single cantilevered arm. The springiness of the arm keeps the contacts apart when no voltages are present. When a sufficiently high control voltage is applied, the arm bends downward until the switch contacts touch, preventing further motion. A limitation of this structure is that the voltage being switched must not be so high as to cause significant electrostatic force. Since the force is proportional to the cross-sectional area of the electrodes, making the switch

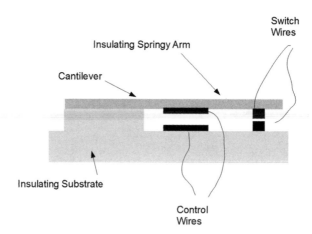

FIGURE 17.8   MEM relay with separate control electrodes.

electrodes' surface area much smaller than the control electrodes' surface area (assuming both voltages to be approximately equal) is usually adequate to achieve this goal.

The structure in Figure 17.8 may also be used as a variable-voltage capacitor. Adjusting the electrode gaps so that the control voltage can be conveniently set to levels below $(\beta V)^2 = \frac{4}{27}$ allows for varying s between 0 and $\frac{-1}{3}$ with the capacitance between the *switch* electrodes varying accordingly.

The basic electrostatic switch structure can be modified to allow for several other applications. Since the moving mass–spring system has a natural mechanical resonance, an electric oscillator circuit can be coupled to the device to create an electromechanical resonator. Such a resonator has the interesting property that the resonance can be varied, or "tuned" by adjusting the control voltage since the electrostatic force due to the control voltage interacts with the spring constant and creates an "effective" spring constant.[7]

If a voltage is applied to the basic electrostatic switch structure to bring the movable electrode to, say, $s = \frac{-1}{6}$, then a stable equilibrium is created. If the voltage source used to establish this bias condition has very high internal impedance, then it behaves like a current source for times which are short as compared to the RC time constant of the source–capacitor system (see Chapter 2 for a discussion of RC networks and time constants).

Now, assume that the movable electrode is a thin membrane or an electrode mounted on a thin membrane. Vibrations in the air near this membrane will cause the membrane to vibrate, which, in turn, will cause the voltage across the capacitor to oscillate. This oscillating voltage can be sensed, and we now have a device known as a *DC condenser microphone*. [*Notes*: (1) the abbreviation DC (direct current) is commonly applied for any electrical system with fixed voltages present, (2) *condenser* is another (old) name for capacitor].

Figure 17.9 shows (schematically) another approach to building an electrostatic capacitor/position sensor.

The structure in Figure 17.9 is an interlaced "comb" of moving and fixed electrodes. Since the capacitance (for motion near the position shown) is due primarily to the sidewall to sidewall proximity of the electrodes, the capacitance is, to first order, a linear function of position.

For Figure 17.9, let X_0 be the position of the fingers when the spring is at rest. For a system of n fingers of thickness w and a finger–finger gap of g, we obtain

$$C \approx \frac{-\varepsilon_0 n w x}{g} \tag{17.20}$$

The electric field stored energy is

$$U_E = -\frac{1}{2}\frac{\varepsilon_0 n w x}{g}V^2 \tag{17.21}$$

and the Coulomb force is

$$F = -\frac{1}{2}\frac{\varepsilon_0 n w}{g}V^2 \tag{17.22}$$

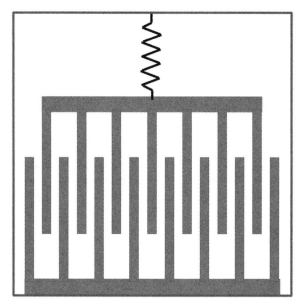

FIGURE 17.9 Electrostatic comb structure capacitor.

The force on the spring is

$$F = -k(x + x_0) \tag{17.23}$$

Setting the sum of these forces equal to zero, we obtain

$$x = -x_0 - \frac{1}{2} \frac{\varepsilon_0 nw}{gk} V^2 \tag{17.24}$$

The response is, of course, always downward (an attractive force result), but in this case x is linear in V^2 with no unstable region. If a small bias voltage V_0 is applied and then a small AC. signal is superimposed on this bias voltage, the response (x) will be approximately linear. Conversely, if a small motion is applied (accelerometer or microphone) when there is a bias voltage present, an approximately linear response voltage will be developed.

17.4 ELECTRETS AND PIEZOELECTRICITY: AN OVERVIEW

When a conventional dielectric material is polarized (e.g., used as the dielectric layer in a capacitor that has a voltage across it), the dielectric material is polarized—that is, a dipole moment is induced in the dielectric that serves to reduce the electric field in the dielectric. This dipole moment, in turn, produces a surface charge which, in turn, increases the capacitance of the capacitor from its airdielectric value. When the voltage is removed, the polarization and the surface charge vanish.

In some cases, if the dielectric is heated sufficiently and then cooled back to room temperature with the charging voltage applied, the polarization "freezes" into the material. Even after the charging voltage and electrodes are removed, the polarization, charge, and resulting external electric field remain. This material, in this state, is called an *electret*. An electret's properties are the electrical analog of the magnetic field and magnetic moment that are "frozen into" a permanent magnet.

If the airgap in a DC condenser microphone is partially replaced with electret material (leaving the diaphragm electrode free to vibrate), then no bias, or polarizing, voltage is necessary. Note that this is only one of several possible configurations for an electret microphone.[9]

Certain crystalline materials contain aligned dipole moments along anisotropic axes. Quartz is an example of such a material. When a voltage is applied across the material (in the proper axis), the dipoles try to align with the applied field and the material stresses. This stress could be microscopic or quite macroscopic, depending on the material. If the same material is mechanically stressed (again, choice of axes is important), the dipoles rotate and a charge is developed at some surfaces. If these surfaces are electroded, a voltage is developed between them.

These materials are called *piezoelectric*. The fact that the piezoelectric effect is bidirectional (with voltage producing stress and stress producing voltage) is derivable from basic thermodynamic considerations.

There is a second category of materials, made up of some *ceramics*, which can be polarized into a piezoelectric state in the same manner as electrets are polarized. Barium titanate is an example of these materials. These materials are typically referred to as *poled* piezoelectric materials.

The piezoelectric effect is used in loudspeakers and micropositioners (voltage to stress) and igniters for flame ignition (stress to voltage). A carefully dimensioned plate or tuning fork of piezoelectric material can be excited to vibrate at one of its natural mechanical resonance frequencies and, coupled to an electrical oscillator circuit, will oscillate and act as a frequency reference. Every quartz clock and watch made contains such a quartz crystal resonator.

The equations governing piezoelectricity are coupled elastic wave equation and electrostatic equations. Since the ratio of the speed of light to the speed of sound in most materials is approximately 10^5, the time derivatives of mechanical variables (stress, strain) can be significant, while the time derivatives of electrical variables can be totally insignificant, thereby making this a true electrostatic-realm situation.

Elastic wave analysis involves tensor mathematics. These are not necessarily difficult analyses, but the background discussion requires several chapters in a book and are outside the scope of this text. The interested reader is directed to the references.[10–12]

17.5 POINTS ON A SPHERE

A practical problem that arises often in many disciplines is the need to uniformly disperse some number of points ($n > 1$) on a sphere. At the outset we must realize that thus far this is an inadequately defined problem. The first issue is that a given distribution of points can be moved ("slid") around on the surface in two axes. We resolve this issue by first fixing one point arbitrarily, say, at the north pole: $(\ominus, \varphi) = (0,0)$. This eliminates the $\ominus$ sliding problem. Then we constrain the second point in φ; say, to $\phi = 0$.

We can now start writing solutions. Some of the solutions can be written easily:

For $n = 2$: $(\Theta, \varphi) = (0,0)$, $(\pi, 0)$
For $n = 3$: $(\Theta, \varphi) = (0,0)$, $(\pi/3, 0)$, $(\pi/3, \pi)$
For $n = 4$: $(\Theta, \varphi) = (0,0)$, $(\pi, 0)$, $(\pi/2, 0)$, $(\pi/2, \pi)$
For $n = 5$: $(\Theta, \varphi) = (0,0)$, $(\pi/2, 0)$, $(\pi/2, 2\pi/3)$, $(\pi/2, 4\pi/3)$, $(\pi, 0)$
For $n = 6$: $(\Theta, \varphi) = (0,0)$, $(\pi/2, 0)$, $(\pi/2, \pi/2)$, $(\pi/2, \pi)$, $(\pi/2, 3\pi/2)$, $(\pi, 0)$

At $n = 7$ we find several problems preventing us from writing a solution set. Simple symmetry arguments are no longer of any use. We need a definition of *uniformly distributed*. There is the possibility that, for a given definition, an exact solution might not exist. If an exact solution does exist, it might not be unique. If an exact solution does not exist, can we define a *best approximation*, which also might not be unique. Finally, once we have our definition in order, can we create an algorithm to produce the array of points?

One way to approach this problem is to convert it to an electrostatic problem. Assume that each point on the surface of the sphere has a charge of +1 and is free to move about the surface (obeying the two constraints given above). The points will arrange themselves so as to minimize the energy of the system. Since the potential between two points is $1/r$, where r is the geodesic distance (the straight line cutting through the sphere), the penalty for being too close to some point outweighs the benefit of being too far from some other point, and the points should settle themselves in a compromise distribution, which we can define as *uniformly* distributed.

For $r_{i,j}$ the distance between points i and j, in an arbitrary set of units, we obtain

$$V_{i,j} = \frac{1}{r_{i,j}} \tag{17.25}$$

The potential of point i is

$$V_i = \sum_{i \neq j} \frac{1}{r_{i,j}} \tag{17.26}$$

and the total potential energy of the system, again in arbitrary units, is

$$V_{\text{tot}} = \sum_i V_i \tag{17.27}$$

The — or possibly a — distribution that minimizes V_{tot} is our desired distribution.

Numerical search procedures that find an extreme of a *figure of merit* such as V_{tot} in the expression above form a sophisticated field by themselves.[13]

To consider the distributions generated using this definition without being sidetracked by the details of the search algorithm, the program `points_on_sphere.m` was written using a two-phase random search algorithm:

1 Try 10,000 random combinations of (θ, ϕ), evaluate V_{tot}, and save the best results (the distribution yielding the lowest value of V_{tot}.)

2 Iterate (an additional) 40,000 times using a normal distribution of random numbers with a mean of the latest best solution and a sigma (σ) of 1% of this same mean.

Here is the file `points_on_sphere.m` — a program to generate a uniform point distribution on a sphere:

```
% playing with locating points on a sphere by minimizing electrostatic
% energy

close
global radius
radius = 1.0;

n = 19;      % total number of points

V_max = 1.e20;
nr_tries = 50000

for i_try = 1 : nr_tries

% point one is fixed by default to (theta,phi) = (0,0)
% point two is partially constrained to (theta,0)
  if i_try < 10000
    t = acos(2*rand(n,1) - 1); t(1) = 0;
    p = 2*pi*rand(n,1); p(1) = 0; p(2) = 0;
  else
    t = t_max + .01*t_max.*randn(n,1);
    if t > pi/2, t = pi - t; end;
    p = p_max + .01*p_max.*randn(n,1);
    if p > 2*pi, p = 4*pi - p; end;
  end
  [thetas,phis] = meshgrid(t,p);

  rij = dist(thetas, phis, thetas', phis');
  rij(1:n+1 : end) = Inf;    % set diagonal terms to infinity
  Vij = sum(1./rij);
  V_tot = sum(Vij);

if V_tot < V_max
   V_max = V_tot;
   t_max = t;
   p_max = p;
  end

end
V_max

x_plot = radius.*cos(p_max).*sin(t_max);
y_plot = radius.*sin(p_max).*sin(t_max);
z_plot = radius.*cos(t_max);

Tri = fliplr(convhulln([x_plot,y_plot,z_plot]));
```

```
trimesh(Tri,x_plot,y_plot,z_plot)
axis square
C = [.5,.5,.5; .5,.5,.5; .5,.5,.5];
colormap(C)

quality = std(Vij)/mean(Vij)/sqrt(length(Vij))
```

After settling on a distribution, the program creates a triangular meshing of the points found and generates graphics to show the results.

Figure 17.10 shows the results produced by `points_on_sphere.m` for $n = 79$ (an arbitrarily chosen number).

As n increases, this program produces poorer and poorer results because of its very crude search algorithm. Semechko's MATLAB program handles the search procedure much more elegantly than the simple system shown here, with correspondingly better (and much more rapidly generated) results for large values of n.[14] Inspection of his well-documented code shows that while implementing the algorithm takes only a few lines of code, implementing the search procedure is a nontrivial task.

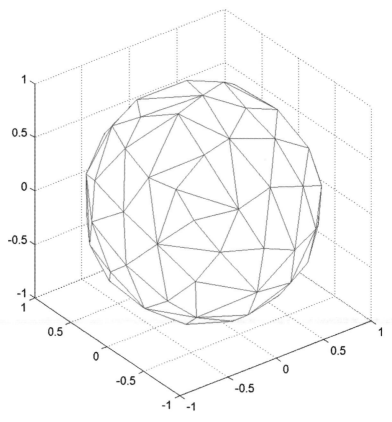

FIGURE 17.10   Diagram showing 79 points on sphere distribution example.

PROBLEMS

17.1 Figure P17.1 is a repeat of Figure 15.14.

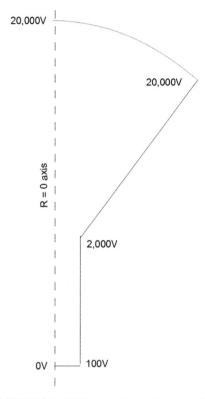

FIGURE P17.1   CRT shell with bc voltages specified.

Figure P17.1 shows the voltages at the endpoints of all line and curve segments. The voltages along the lines connecting these points should be linearly graded as boundary conditions. Assume that there are several electron emitters along the cathode end of the structure $(z = 0)$. These emitters are at $r = 0.01$, 0.05, and 0.1. Assume that these emitters place electrons into the vacuum at zero velocity. Write a MATLAB program to calculate the trajectories of electrons starting from each of these emitters.

17.2 To the structure of Problem 17.1, add a metallic hollow cylinder with the following dimensions: inner radius $r = 0.20$, outer radius $r = 0.25$, lower height $z = 0.15$, and upper height $z = 0.25$. Set various fixed voltages on this cylinder and examine the resulting electron trajectories.

17.3 Figure P17.4a shows a cross section of an axisymmetric structure consisting of: (1) an outer metallic cylinder of radius 5 and height 8; (2) an inner hollow metallic cylinder of inner radius 1.0, outer radius 1.5, and height 2.5 sitting at the base of the

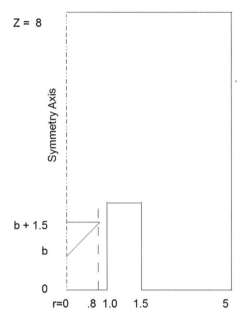

FIGURE P17.4a   Structure for Problem 17.3.

outer cylinder (and connected to this base); and (3) a circular wedge of radius 0.8 and height 1.5 located with the base of the wedge at height b above the base.

(a)   Construct a gmsh.geo file to mesh this structure. Calculate the capacitance as a function of b for $0 < b < 3$.

(b)   Assume that there is a spring of constant $k = 1$ attached between the top of the wedge and the top of the outer cylinder, and that the wedge can move up and down (it is held laterally by unshown restraints). The spring has 0 tension when $b = 3.0$. Calculate the position of the wedge as a function of the voltage V between the wedge and the cylinders.

17.4   Write a program to distribute a number of electrons in a circle of radius 1 by minimizing the electrostatic energy of the system.

REFERENCES

1. http://en.wikipedia.org/wiki/Virtual_work.

2. http://www.mathworks.com/support/solutions/en/data/1-JDKMLK/index.html.

3. http://www.jhu.edu/motn/relevantnotes/usingmatlab_ode.pdf.

4. http://www.math.pitt.edu/~sussmanm/2071Spring08/lab01b/index.html.

5. http://laser.ceb.cam.ac.uk/wiki/images/e/e5/NumMeth_Handout_7.pdf.

6. P. Hawkes and E. Casper, *Electron Optics*, Academic Press, 1989.

7. C. Liu, *Foundations of MEMs*, 2nd ed. Pearson Education, 2011.

8. L. Dworsky and M. Chason, *Electrostatically Switched Integrated Relay and Capacitor*, US Patent **5**,051,643, 1991.

9. http://en.wikipedia.org/wiki/Electret_microphone.

10. APC International Limited, *Piezoelectric Ceramics: Principles and Applications*, 2011.

11. W. Nelson, ed., *Piezoelectric Materials: Structure, Properties and Applications*, Nova Science Publishers, 2010.

12. V. Bottom, *Introduction to Quartz Crystal Unit Design*, Van Nostrand Reinhold, 1982.

13. S. Teukolsky, W. Vetterling, and B. Flannery, *Numerical Recipes*, 3rd ed., The Art of Scientific Computing (series), Cambridge Univ. Press, 2007. (earlier versions of this book are available online).

14. http://www.mathworks.com/matlabcentral/fileexchange/37004-uniform-sampling-of-a-sphere.

A

Interfacing with Other Languages

(*Note*: The MATLAB coding described in this appendix has been tested on a Windows 7 computer. The equivalent system calls should work on a Unix or Mac computer, but have not been tested.)

Sometimes it is desirable to perform some calculations in a language other than MATLAB. Possible reasons for this are that (1) a complicated, working, calculation program already exists; or (2) it is difficult or even impossible to vectorize a MATLAB program but desirable to utilize the calculation speed of some fully compiled optimized code in a language such as C or FORTRAN.

MATLAB has built-in capabilities for doing this.[1,2] The MATLAB capabilities are significant and merit investigation if this type of interface is desired. There are limitations, however, as to what compilers and languages are supported. The technique described in this appendix is a "quick and dirty" patch that works well provided you can write or modify and compile code with a compiler and language of your choice into an executable program.

The procedure is very straightforward and utilizes MATLAB's capability of executing a system command and then waiting for its completion:

1   Prepare data in MATLAB using whatever script or functions are necessary.
2   Save the necessary data to a file.
3   Execute the external program.
4   The external program reads the data file, does its work, and writes its results to a second data file. The external program may open and read and/or write other files if necessary.

Introduction to Numerical Electrostatics Using MATLAB, First Edition. Lawrence N. Dworsky.
© 2014 John Wiley & Sons, Inc. Published 2014 by John Wiley & Sons, Inc.

5 MATLAB reads the results from the appropriate data file(s) and proceeds with calculations, graphics, and so on.

Following is the MATLAB script `run_ppt4.m`, with a calling program for different versions of `ppt4.m`:

```
% run_ppt4.m  a convenience script for running and plotting ppt4 results
% with a Fortran interface

rad = 0.25; step = .1;
y1 = 15 + rad + step; y2 = 40;
y = [y1];
while y(end) < y2
    y = [y, y(end) + step];
    step = 1.25*step;
end
y = y(1 : end-1)

tic
%  Comment out the undesired choice below

%  This section calls the fortran program --------------

x = 0
xy_fortran = [x, y]'
save ppt4_data.dat xy_fortran -ascii
dos 'ppt4_fortran.exe'
fid = fopen('ppt4_ftn_out.dat','rt'); % open the file
volts = (fscanf(fid, '%g'))'; % read the file

%  This section runs the program in Matlab ----------------

%  volts = ppt4(x, y)

%  ----------------------------------------------
toc

y = [15 + rad, y, 40]; volts = [1.0, volts, 0];

coefs = polyfit(y, volts, 3)
curve = polyval(coefs,y);

figure(1);
plot(y, volts, 'xk', y, curve, 'k')

figure(2);
E_curve = - (3*coefs(1)*y.^2 + 2*coefs(2)*y + coefs(3));
plot (y, E_curve, 'k')
```

MATLAB script `run_ppt4.m` is an example of this procedure, using the probabilistic potential theory program `ppt4.m`, from Chapter 11, as an example. The run_ppt4.m program is written to call either the MATLAB or FORTRAN version of the program (simply comment out the undesired choice).

The FORTRAN source code ppt4_ftn.f90, the FORTRAN version of ppt4.m, is as follows:

```fortran
! fortran version of ppt4.m

implicit integer (t)
implicit real*8 (a-h, o-s, u-z)
real*8 y_pt(1000)   ! y scan data points
real*8 walls(100,5)   ! wall data
real*8, allocatable :: volts(:), dists(:,:), results(:)
real*8 stuff(3) ! return from boundary checkers: 1 = distance, 2,3 =
closest point
integer hit_boundary

pi = 4*atan(1.0)
call srand(12345)

! get the data
call gather_data(x_pt, nr_pts, y_pt, nr_walls, walls)
allocate (volts(nr_pts), dists(nr_walls,4), results(nr_pts))

print *, 'nr_pts = ', nr_pts
print *, 'x_pt = ', x_pt
do i = 1, nr_pts
  print *, i, y_pt(i)
end do

do i = 1, nr_walls
  print '(5f10.3)', (walls(i,j), j = 1,5)
end do

tic = time8()
do i_nr = 1, nr_pts                ! starting point counter
  nr_iters = 50000
  nr_ones = 0

  do iter = 1, nr_iters         ! iteration counter
    hit_boundary = 0
    x = x_start
    y = y_pt(i_nr)

    do while (hit_boundary < = 0)     ! wandering point loop
      do i_wall = 1, nr_walls         ! examine all the boundary conditions
        if (walls(i_wall,1) .eq. 1) then
          call hor_line(i_wall, walls, x, y, stuff )
        else if (walls(i_wall,1) .eq. 2) then
          call vert_line(i_wall, walls, x, y, stuff )
        else
          call circle_dist(i_wall, walls, x, y, stuff )
        end if
```

```
  do j = 1, 3
    dists(i_wall,j) = stuff(j)
  end do
  dists(i_wall,4) = walls(i_wall,5)        ! keep the voltage with us
end do

d = 1.e20
index = 0
do i = 1, nr_walls
  if (dists(i,1) < d) then
    d = dists(i,1)
    index = i
  end if
end do

rad_step = dists(index,1)
p1x = dists(index,2)
p1y = dists(index,3)
v = dists(index,4)
phi = atan2(p1y-y , p1x-x)
p2x = x + rad_step*cos(phi + pi/2)
p2y = y + rad_step*sin(phi + pi/2)
p3x = x + rad_step*cos(phi + pi)
p3y = y + rad_step*sin(phi + pi)
p4x = x + rad_step*cos(phi + 3*pi/2)
p4y = y + rad_step*sin(phi + 3*pi/2)
volts = dists(index,4) ;

prob = rand(0)
if (prob < 1./4.) then
  x = p1x
  y = p1y
else if (prob < 1./2.) then
  x = p2x
  y = p2y
else if (prob < 3./4.) then
  x = p3x
  y = p3y
else
  x = p4x
  y = p4y
end if

!     print *, prob, x, y

! if prob < 1/4 we hve an extraction

  if (prob < 1./4.) then
    if (v .eq. 0) then ! check for voltage
      hit_boundary = 1
    else
      hit_boundary = 2
```

```fortran
        end if
      end if
    end do

  if (hit_boundary .eq. 2) nr_ones = nr_ones + 1
!    print *, 'hit_boundary = ', hit_boundary
!    pause

  end do

  v_at_point = (1.*nr_ones)/nr_iters
  print *, 'y, v = ', y_pt(i_nr), v_at_point

  results(i_nr) = v_at_point
end do
toc = time8() - tic
print *, 'execution time = ', toc

open (3, file = 'ppt4_ftn_out.dat')
do i = 1, nr_pts
  print '(2f12.5)', y_pt(i), results(i)
  write (3,'(f12.5)') results(i)
end do
close (3)

stop
end

! ---------------------------------------------

subroutine circle_dist(i_wall, walls, pt_x, pt_y, stuff)

implicit real*8 (a-h, o-z)
real*8 walls(100,5), stuff(3)

rad = walls(i_wall,2)
x0 = walls(i_wall,3)
y0 = walls(i_wall,4)

stuff(1) = sqrt((pt_x - x0)**2 + (pt_y - y0)**2) - rad
theta = atan2(pt_y - y0, pt_x - x0)
stuff(2) = x0 + rad*cos(theta)
stuff(3) = y0 + rad*sin(theta)

return
end

! ---------------------------------------------

subroutine hor_line(i_wall, walls, pt_x, pt_y, stuff)

implicit real*8 (a-h, o-z)
real*8 walls(100,5), stuff(3)
```

```fortran
y = walls(i_wall,2)
xl = walls(i_wall,3)
xr = walls(i_wall,4)

if ((pt_x > = xl) .and. (pt_x < = xr)) then
  stuff(1) = abs(y - pt_y)
  stuff(2) = pt_x
else if (pt_x < xl) then
  stuff(1) = sqrt((xl - pt_x)**2 + (y - pt_y)**2)
  stuff(2) = xl
else
  stuff(1) = sqrt((xr - pt_x)**2 + (y - pt_y)**2)
  stuff(2) = xr
end if
stuff(3) = y

return
end

! --------------------------------------------

subroutine vert_line(i_wall, walls, pt_x, pt_y, stuff )

implicit real*8 (a-h, o-z)
real*8 walls(100,5), stuff(3)

x = walls(i_wall,2)
yd = walls(i_wall,3)
yu = walls(i_wall,4)

if ((pt_y > = yd) .and. (pt_y < = yu)) then
  stuff(1) = abs(x - pt_x)
  stuff(3) = pt_y
else if (pt_y < yd) then
  stuff(1) = sqrt((x - pt_x)**2 + (yd - pt_y)**2)
  stuff(3) = yd
else
  stuff(1) = sqrt((x - pt_x)**2 + (yu - pt_y)**2)
  stuff(3) = yu
end if
stuff(2) = x

return
end

! --------------------------------------------

subroutine gather_data(x_pt, nr_pts, y_pt, nr_walls, walls)

implicit real*8 (a-h, o-z)
real*8 y_pt(1000) ! y scan data points
real*8 walls(100,5) ! wall data
```

```
open (3, file = 'ppt4_data.dat')
read (3,*), x_pt
nr_pts = 0
do
  nr_pts = nr_pts + 1
  read (3,*, end = 100) y_pt(nr_pts)
end do
100 close (3)
nr_pts = nr_pts - 1

open (3, file = 'data1.txt')
nr_walls = 0
do
  nr_walls = nr_walls + 1
  read (3,*, end = 200) (walls(nr_walls,j), j = 1,5)
end do
200 close (3)
nr_walls = nr_walls - 1

return
end
```

The FORTRAN program `ppt4_ftn.f90` was written to mimic the programming of the MATLAB code.

Additional notes are as follows:

1 The FORTRAN version runs almost 100 times faster than the MATLAB version. This doesn't mean that neither version has room for improvement; this is simply an example of what can be done quickly and easily.

2 Separate tests showed that the disk read–write times consume only ~0.1 second for relatively small files.

3 This technique could be used for large data files and/or multiple read–write operations by creating a RAM disk rather than by using the actual disk drive. In the case of SSD drives, individual judgment calls are necessary.

REFERENCES

1. http://www.mathworks.com/support/compilers/interface.html.

2. http://www.karenkopecky.net/Teaching/Cclass/MatlabCallsC.pdf.

Index

Abraham, H., 157
AC, 41, 46, 49
Adams, A.T., 122
Alternating current, *see* AC
Amplifier, power, *see* Power amplifier
Ang, W.T., 162
Anode, 45–48
Assembling coefficient matrix,
   see Coefficient matrix
Axisymmetric
  structure–FD formulation, 196–206
  structure–FEM formulation, 337–348

Bang, H., 288, 373
Barycenter, 124–127
Basis function, 270–271, 279–280
Battery, 15
Bell curve, *see* Normal distribution
Bias, electrostatic force devices, 49
Bias point, 45–46
Binomial probability
  as error predictor for random walks, 240–242
  mean, 241
  standard deviation, 241
Bird, J., 50
Bottom, V., 421

Boundary condition, 23, 103, 166–167
  in FEM analysis, 318–319
  periodic, 225
  symmetric, *see* Symmetry
  treatment of irregular in FD, 180–181
Boundary element method, 160
Boxed microstrip, *see* Microstrip, boxed
Brownian motion, 237

Cambrell, G.K., 182, 225
Capacitance
  calculation using energy in field, 178–180
  calculation using Gauss' law, 175–177
  concentric cylinders, 19
  concentric spheres, 17–18, 149–152
  cylinder, between planes, 116–121
  defined, 15
  ideal parallel plate, 16
  lower bound approximation, 207–214
  multielectrode structures, 227–228, 324–326
  mutual, 74–78
  parallel plate capacitor, 70–78
  parasitic, 78, 228
  self, 74–78
  series connected capacitors, 24
  upper bound approximation, 180

Introduction to Numerical Electrostatics Using MATLAB, First Edition. Lawrence N. Dworsky.
© 2014 John Wiley & Sons, Inc. Published 2014 by John Wiley & Sons, Inc.